For over 35 years, biological scientists have come to rely on the research protocols and methodologies in the critically acclaimed *Methods in Molecular Biology* series. The series was the first to introduce the step-by-step protocols approach that has become the standard in all biomedical protocol publishing. Each protocol is provided in readily-reproducible step-by-step fashion, opening with an introductory overview, a list of the materials and reagents needed to complete the experiment, and followed by a detailed procedure that is supported with a helpful notes section offering tips and tricks of the trade as well as troubleshooting advice. These hallmark features were introduced by series editor Dr. John Walker and constitute the key ingredient in each and every volume of the *Methods in Molecular Biology* series. Tested and trusted, comprehensive and reliable, all protocols from the series are indexed in PubMed.

Methods in Molecular Biology

Series Editor
John M. Walker
School of Life and Medical Sciences
University of Hertfordshire
Hatfield, Hertfordshire, UK

For further volumes:
http://www.springer.com/series/7651

Bunyaviruses

Methods and Protocols

Edited by

Hani Boshra

Faculty of Veterinary Medicine, Department of Pathology, Fundamental and Applied Research for Animals and Health (FARAH), University of Liège, Liège, Belgium

Humana Press

Editor
Hani Boshra
Faculty of Veterinary Medicine
Department of Pathology
Fundamental and Applied Research
for Animals and Health (FARAH)
University of Liège
Liège, Belgium

ISSN 1064-3745 ISSN 1940-6029 (electronic)
Methods in Molecular Biology
ISBN 978-1-0716-4337-2 ISBN 978-1-0716-4338-9 (eBook)
https://doi.org/10.1007/978-1-0716-4338-9

This Humana imprint is published by the registered company Springer Science+Business Media, LLC, part of Springer Nature.
The registered company address is: 1 New York Plaza, New York, NY 10004, U.S.A.

Preface

From their discovery nearly 100 years ago, bunyaviruses have been implicated in a variety of public health outbreaks, ranging from the highly pathogenic Crimean-Congo hemorrhagic fever virus (CCHFV) to veterinary pathogens such as Schmallenberg virus (SBV). Bunyaviruses can also act as zoonotic pathogens (e.g., Rift Valley fever virus), which can be fatal in both animal and human hosts. Furthermore, bunyaviruses such as Tospoviruses can also exclusively infect plant varieties and are a concern to the agricultural industry. Therefore, it is not surprising that interest in the study of these viruses has greatly increased over the past few decades, covering a large spectrum of experimental techniques. In this edition of the Methods in Molecular Biology series, we will try to provide detailed protocols for the study of bunyaviruses, which cover a wide variety of topics, ranging from virus manipulation and purification, to the expression of recombinant proteins, to the experimental study of bunyavirus infection in both small and large animals for virological and immunological analysis. A chapter dedicated to plant bunyaviruses has also been included, as this rapidly developing field has often been overlooked by virologists studying the class *Bunyaviricetes.* The authors of each chapter have been individually chosen based on their previous expertise with each technique, and every attempt has been made to ensure that each protocol can be replicated by scientists with a basic level of molecular biology.

Liege, Belgium *Hani Boshra*

Contents

Contributors

M. Altinli • *Bernhard Nocht Institute for Tropical Medicine (BNITM), Hamburg, Germany*

Stephanie E. Ander • *Department of Immunology and Microbiology, University of Colorado School of Medicine, Aurora, CO, USA*

Akeel Baig • *Hamilton, ON, Canada*

Christina Barry • *Public Health Agency of Canada (PHAC), Winnipeg, MB, Canada*

Calixte Bayrou • *Faculty of Veterinary Medicine, Clinical Department of Production Animals, FARAH Research Center, University of Liège, Liège, Belgium*

Alexander Bello • *Public Health Agency of Canada (PHAC), Winnipeg, MB, Canada*

Éric Bergeron • *Viral Special Pathogens Branch, Division of High-Consequence Pathogens & Pathology, Centers for Disease Control & Prevention, Atlanta, GA, USA*

Lotty Birnberg • *Animal Health Research Center (CReSA-IRTA), Universitat Autònoma de Barcelona, Cerdanyola del Vallès, Barcelona, Spain*

P. Böhmer • *Bernhard Nocht Institute for Tropical Medicine (BNITM), Hamburg, Germany*

Belén Borrego • *Animal Health Research Centre (CISA), Centro Nacional Instituto de Investigación y Tecnología Agraria y Alimentaria (INIA, CSIC), Madrid, Spain*

Hani Boshra • *Faculty of Veterinary Medicine, Department of Pathology, Fundamental and Applied Research for Animals and Health (FARAH), University of Liège, Liège, Belgium*

Alejandro Brun • *Animal Health Research Centre (CISA), Centro Nacional Instituto de Investigación y Tecnología Agraria y Alimentaria (INIA, CSIC), Madrid, Spain*

Núria Busquets • *Animal Health Research Center (CReSA-IRTA), Universitat Autònoma de Barcelona, Cerdanyola del Vallès, Barcelona, Spain*

Catherine Cêtre-Sossah • *CIRAD, UMR ASTRE, Montpellier, France; ASTRE, University of Montpellier, CIRAD, INRAe, Montpellier, France*

Bennett J. Davenport • *Department of Immunology and Microbiology, University of Colorado School of Medicine, Aurora, CO, USA*

S. Matthijn De Boer • *Wageningen Bioveterinary Research (WBVR), RALelystad, The Netherlands*

Daniel Desmecht • *Faculty of Veterinary Medicine, Department of Morphology and Pathology, FARAH Research Center, University of Liège, Liège, Belgium*

Stuart Dowall • *United Kingdom Health Security Agency (UKHSA), Salisbury, Wiltshire, UK*

Alyssa B. Evans • *Department of Microbiology and Cell Biology, Montana State University, Bozeman, MT, USA*

Aura R. Garrison • *United States Army Medical Research Institute of Infectious Diseases, Fort Detrick, MD, USA*

Joseph W. Golden • *United States Army Medical Research Institute of Infectious Diseases, Fort Detrick, MD, USA*

Danyil Grybchuk • *Life Science Research Centre, Faculty of Science, University of Ostrava, Ostrava, Czechia; Central European Institute of Technology, Masaryk University, Brno, Czechia*

ROGER HEWSON • *United Kingdom Health Security Agency (UKHSA), Salisbury, Wiltshire, UK*

TETSURO IKEGAMI • *Department of Pathology, The Sealy Institute for Vaccine Sciences, The Center for Biodefense and Emerging Infectious Diseases, The University of Texas Medical Branch at Galveston, Galveston, TX, USA*

EDUARDO JURADO-COBENA • *Department of Microbiology and Immunology, The University of Texas Medical Branch at Galveston, Galveston, TX, USA*

ELIF KARAASLAN • *Viral Special Pathogens Branch, Division of High-Consequence Pathogens & Pathology, Centers for Disease Control & Prevention, Atlanta, GA, USA*

EMMA KENNEDY • *United Kingdom Health Security Agency (UKHSA), Salisbury, Wiltshire, UK*

ALEXEI YU KOSTYGOV • *Life Science Research Centre, Faculty of Science, University of Ostrava, Ostrava, Czechia; Zoological Institute, Russian Academy of Sciences, St. Petersburg, Russia*

PEYTON LABONTE • *Department of Entomology and Plant Pathology, North Carolina State University, Raleigh, NC, USA*

LUDOVIC MARTINELLE • *Faculty of Veterinary Medicine, CARE-FEPEX experimental Station, Fundamental and Applied Research for Animal and Health (FARAH) Center, University of Liege, Liege, Belgium*

LUIS MARTINEZ-SOBRIDO • *Texas Biomedical Research Institute, San Antonio, TX, USA*

ANITA K. MCELROY • *Pediatric Infectious Disease and Center for Vaccine Research, University of Pittsburgh School of Medicine, Pittsburgh, PA, USA*

HOLLY MCINNES • *Department of Entomology and Plant Pathology, North Carolina State University, Raleigh, NC, USA*

THOMAS E. MORRISON • *Department of Immunology and Microbiology, University of Colorado School of Medicine, Aurora, CO, USA*

NICOLE MUELLER • *Public Health Agency of Canada (PHAC), Winnipeg, MB, Canada*

AITOR NOGALES • *Animal Health Research Centre (CISA), Centro Nacional Instituto de Investigación y Tecnología Agraria y Alimentaria (INIA, CSIC), Madrid, Spain*

RYAN PACKER • *Department of Entomology and Plant Pathology, North Carolina State University, Raleigh, NC, USA*

AURÉLIE PÉDARRIEU • *CIRAD, UMR ASTRE, Montpellier, France; ASTRE, University of Montpellier, CIRAD, INRAe, Montpellier, France*

M. REUTER • *Bernhard Nocht Institute for Tropical Medicine (BNITM), Hamburg, Germany*

ALYSSIA ROBINSON • *Public Health Agency of Canada (PHAC), Winnipeg, MB, Canada*

DORITH ROTENBERG • *Department of Entomology and Plant Pathology, North Carolina State University, Raleigh, NC, USA*

CLAUDE SAEGERMAN • *Faculty of Veterinary Medicine, Research Unit of Epidemiology and Risk analysis applied to Veterinary sciences (UREAR-ULiège), Fundamental and Applied Research for Animal and Health (FARAH) Center, University of Liege, Liege, Belgium*

E. SCHNETTLER • *Bernhard Nocht Institute for Tropical Medicine (BNITM), Hamburg, Germany; Faculty of Mathematics, Informatics and Natural Sciences, University Hamburg, Hamburg, Germany*

FLORINE SCHOLTE • *Viral Special Pathogens Branch, Division of High-Consequence Pathogens & Pathology, Centers for Disease Control & Prevention, Atlanta, GA, USA*

MICHAEL B. SHERMAN • *Department of Biochemistry and Molecular Biology, University of Texas Medical Branch at Galveston, Galveston, TX, USA; Sealy Center for Structural Biology and Molecular Biophysics, University of Texas Medical Branch, Galveston, TX, USA*

THOMAS J. SMITH • *Department of Biochemistry and Molecular Biology, University of Texas Medical Branch at Galveston, Galveston, TX, USA; Sealy Center for Structural Biology and Molecular Biophysics, University of Texas Medical Branch, Galveston, TX, USA*

ALYSSA STULBERG • *Public Health Agency of Canada (PHAC), Winnipeg, MB, Canada*

ANNE-SOPHIE VAN LAERE • *Faculty of Veterinary Medicine, Department of morphology and Pathology, FARAH Research Center, University of Liège, Liège, Belgium*

ANNA E. WHITFIELD • *Department of Entomology and Plant Pathology, North Carolina State University, Raleigh, NC, USA*

HEIDI WOOD • *Public Health Agency of Canada (PHAC), Winnipeg, MB, Canada*

JIAO XU • *State Key Laboratory of Virology, School of Public Health, Wuhan University, Wuhan City, Hubei Province, China*

XUE-JIE YU • *State Key Laboratory of Virology, School of Public Health, Wuhan University, Wuhan City, Hubei Province, China*

VYACHESLAV YURCHENKO • *Life Science Research Centre, Faculty of Science, University of Ostrava, Ostrava, Czechia*

XIANKUN ZENG • *United States Army Medical Research Institute of Infectious Diseases, Fort Detrick, MD, USA*

Chapter 1

An Introduction to the Bunyaviruses

Hani Boshra

Abstract

The bunyaviruses are an ever-expanding group of RNA viruses that have been linked to a variety of different diseases around the world. First characterized nearly a century ago, over 500 different types of bunyaviruses have been characterized thus far, with hosts ranging from mammals to plants to single-celled organisms. As many of the currently described bunyaviruses have been found to be vector-borne, with transmission being mediated by either insects or rodents, these viruses have incorporated immune-evasive molecules into their relatively small genome. As these viruses have been implicated in a number of public health threats, there is an increased interest in performing experiments that could improve our understanding of these infectious agents. Therefore, the objectives of this book are outlined in this chapter, with a variety of techniques being described for the study of a variety of different bunyaviruses.

Key words Bunyaviruses, Bunyaviricetes, RNA viruses, Emerging diseases

1 The Bunyaviruses (Class: Bunyaviricetes)

Bunyaviruses represent one of the largest groups of RNA viruses, representing over 500 known species [1]. First characterized in Uganda in the 1940s, bunyaviruses have since been found to be associated with a variety of febrile and hemorrhagic illnesses. Primarily transmitted via mosquito or rodent vectors, bunyaviruses have become an increasingly popular research focus, due to their association with zoonotic outbreaks. Unsurprisingly, this increased research interest has led to the discovery (and characterization) of novel bunyaviruses, which have now been shown to infect a wide variety of hosts, ranging from animals to plants (i.e. Tospoviruses), to even single-celled organisms (i.e. Leishbunyaviruses) [1].

In fact, the rapidly increasing number of novel bunyaviruses being discovered has led to their phylogenetic classification being changed from family (formally *Bunyaviridae*) to order (formally *Bunyavirales*), to the class *Bunyaviricetes* (as of this publication) [2]. This new class comprises two orders, the *Elliovirales* and the *Hareavirales*. As with previous taxonomical classifications, these

Hani Boshra (ed.), *Bunyaviruses: Methods and Protocols*, Methods in Molecular Biology, vol. 2893,
https://doi.org/10.1007/978-1-0716-4338-9_1,

orders represent a divergence of clades based on the sequence of the RNA-based RNA polymerase [1]. While the order *Elliovirales* is comprised of eight families (*Cruliviridae*, *Fimoviridae*, *Hantaviridae*, *Konkoviridae*, *Peribunyaviridae*, *Phasmaviridae*, *Tospoviridae*, and *Tulasviridae*), the order Hareavirales consists of seven families (*Discoviridae*, *Leishbuviridae*, *Mypoviridae*, *Nairoviridae*, *Phenuiviridae*, and *Wupedeviridae*). As there have been multiple major taxonomical reorganizations over the past decade, it is likely that more will follow, as more (and diverse) genomes are discovered.

Although the genomic organization of bunyaviruses varies in genome size, morphologies, and translated protein products, a typical bunyavirus genome is composed of the following [3, 4]:

1. An RNA-dependent RNA polymerase (RdRp);
2. At least two glycoproteins, expressed at the exterior of the virion
3. A nucleoprotein
4. A nonstructural protein [1] which is usually expressed on either/both the M- and S-segments.

All of these components play a role in the infection and replication of the virus. In the case of the RdRp, this protein plays a key role both in the replication of the negative-sense genome, as well as the synthesis of the positive-sense mRNA [4]. The glycoproteins expressed at the exterior of the genome not only play a structural role but are also used in the uptake of the virus by the host cell [5]. Depending on the type of bunyavirus studied, it has been shown that phagocytic uptake may be mediated by pH-dependent endocytosis, or receptor-related phagocytic uptake [6]. The nucleoprotein plays a structural role in the packaging of the viral genome through the formation of ribonucleoprotein complexes in the interior of the virion [7], while the nonstructural proteins have been previously shown to play roles in the inhibition of the host antiviral response. A schematic of the replication process of a typical bunyavirus is shown in Fig. 1.

Although bunyaviruses continue to be explored using a variety of molecular, immunological, and pathological perspectives, the ever-expanding knowledge base of these viruses has created a need for a standardized level of virological techniques, which may be used (with some alterations—depending on the type of bunyavirus being studied) in order to deduce mechanisms associated with tropism, immunology, and molecular characterization of individual viral components. In essence, the study of bunyaviruses cannot be simply limited to understanding the properties of the currently known members of this class of viruses; rather, certain techniques should be mastered (or at the very least, understood) to permit the discovery of novel avenues of investigation into this ever-expanding field of research.

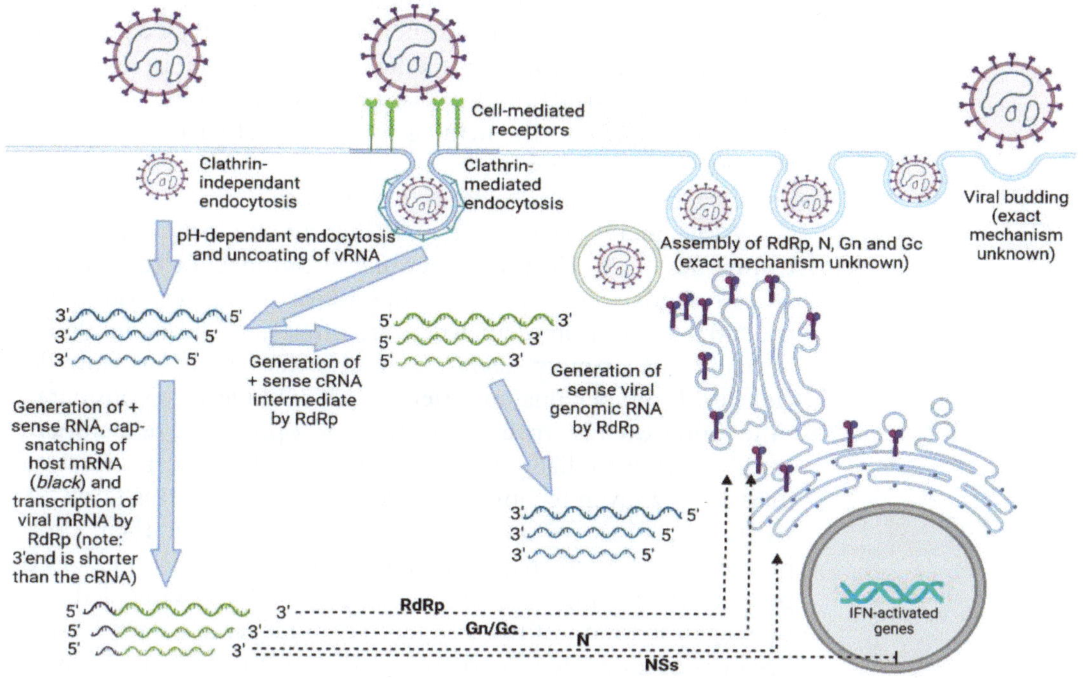

Fig. 1 A schematic of a typical bunyavirus genome as well as an overview of the infection/replication cycle of bunyaviruses. Schematic of our current understanding of the replication cycle of a typical bunyavirus. Bunyaviruses may be able to infect a cell either through receptor-mediated endocytosis (which may involve clathrin) or through clathrin-independent endocytosis. Upon viral entry, uncoating of the virion results in the presentation of multi-segmented (−) sense RNA in the host cytoplasm. An RNA-dependent RNA polymerase (RdRp) from the infectious particle enables the replication of the viral genome into a positive sense RNA intermediate. A "cap-snatching" mechanism enables the attachment of 5′ host mRNA, resulting in translatable viral mRNA. Translation of the viral mRNA results in the synthesis of all the components of an infectious virus (i.e., glycoprotein N, glycoprotein C, nucleoprotein, and RdRp) as well as up to two nonstructural proteins, NSs and NSm. NSs has been shown to be capable of inhibiting the transcription of interferon-induced genes. The (+) sense intermediate RNA can also serve as a template for genomic replication by the RdRp into (−) sense genomic RNA. The figure was generated using icons and templates from Biorender.com. This figure was reproduced from Boshra et al. [1] under the Creative Commons Attribution license, from a review written in the journal *Viruses* with permission from the author (H. Boshra)

2 The Purpose of This Book

When I started my journey into the world of bunyaviruses more than 15 years ago, one of the first (of many) obstacles I encountered involved the mastery of the laboratory techniques used to study these kinds of viruses. At the time, molecular techniques varied considerably between laboratories and research groups. Indeed, since their discovery nearly 80 years ago, advances in the discovery, detection, pathology, experimental modeling, proteomic/genomic analysis, as well as the immunological characterization of bunyaviruses have changed considerably; and fortunately for those newly entering the field, the experimental protocols used to study these viruses (as well as the diseases they may cause) have become more standardized. Therefore, a book dedicated to providing a wide array of protocols for studying this class of viruses would be of particular help to those new to the field, as well as expert virologists who are interested in studying novel aspects of these viruses that would normally be outside their realm of expertise.

While many scientists try to incorporate new experimental techniques by replicating experiments outlined in the Materials and Methods section of a scientific journal, the steps outlined in the "M&M" section of an article are often unclear, incomplete, or in many cases, simply reference other articles with the same deficiencies. From personal experience, this could lead to considerable time unnecessarily troubleshooting or optimizing experimental steps, which were likely already known to other researchers with a mastery of a given technique. Here, the chapters in this book were written by those who have first-hand expertise in their chosen protocol, with each chapter designed to properly describe: (1) the purpose of carrying out the selected protocol; (2) a detailed, step-by-step description of the selected technique, and; (3) an extensive "Notes" section, dedicated to pointing out potential pitfalls and possible points of concern for each step.

While this book describes experimental protocols for a variety of viruses of the class *Bunyaviricetes*, a protocol described for use on one particular species of bunyavirus may be applicable to another species, as long as consideration is given to the respective tropism of each virus (on a cellular and species level) as well as taking the appropriate safety measures applicable for each virus (i.e. biosafety level, laboratory containment, and personal protective equipment). It is our hope that the reader will be able to use the protocols described in this book to further advance our knowledge of these remarkably interesting and important viruses, with the hopes that future generations will continue to make novel discoveries and advancements in this field (. . .and hopefully, contribute new protocols in the future!).

3 Topics Covered in This Book

3.1 Use of Rift Valley Fever Virus Expressing NanoLuc Luciferase for the Assessment of Neutralizing Antibodies and Antivirals

Determining the presence of neutralizing antibodies against bunyaviruses is an important technique in molecular diagnostics. Whether it is determining whether an individual has been previously exposed to infection, or simply determining the efficacy of a potential vaccine candidate, being able to perform virus neutralization assays is an important technique in determining host exposure to bunyaviruses. Here, Borrego et al. describe a detailed protocol using a recombinantly expressed (and attenuated) Rift Valley fever virus. This recombinant virus possesses a luciferase reporter gene that can be used to detect the presence of either neutralizing antibodies or antiviral compound activity in as little as 48 h. Furthermore, this technique can be modified as a method to screen novel antiviral compounds.

3.2 A Serological Assay for the Detection of Antibodies Generated Against California Serogroup Virus Infection

The use of monoclonal antibodies in the detection of pathogenic bunyaviruses is a readily used technique in the immunosurveillance of animals which may be in the acute phase of infection (i.e. when IgM-specific antibodies are being generated in response to an initial viral infection). One method for such detection of the California serogroup of orthobunyaviruses (CSG) involves the use of enzyme-linked immunosorbent assays (ELISA). A protocol describing such a technique is described by Stulberg, Barry et al., and includes both the preparatory steps (i.e. viral antigen purification), as well as the steps involved in carrying out an IgM antibody capture enzyme-linked immunosorbent assay (MAC-ELISA) for CSGs.

3.3 Molecular Diagnosis of Phlebovirus Riftense Infection Using an L-Segment-Based Real-Time RT-PCR Method

As molecular diagnostics are considered among the most common techniques used in the field of bunyavirus biology, Pédarrieu and Cêtre-Sossah describe a protocol for the detection of Rift Valley fever virus (also known as *Phlebovirus riftense*) using real-time RT-PCR. Not only is this technique widely used in reference laboratories around the world to detect this pathogen but is also of significant importance for the detection of other bunyaviruses, as the L-segment is the most conserved of all three segments of the genome; and thus, serves as a point of reference for both viral detection, as well as the phylogenetic classification.

3.4 Ultracentrifugation Purification of Cache Valley Virus Using Iodixanol

Virus purification is of particular importance in many experimental applications. Whether a pure virus preparation is needed for structural biological studies, or the generation/validation of reagents for immunological analysis (i.e. validation of diagnostic kits, the study of immunological interactions in vivo, or the study of virus tropism), the ability to purify nearly homogeneous preparations of bunyaviruses may be necessary. In this book, Baig describes a protocol for the purification of Cache Valley virus using ultracentrifugation in a density-gradient medium, resulting in a native, structurally intact preparation of the virus.

3.5 Cryogenic Electron Microscopy of Rift Valley Fever Virus

As electron microscopy continues to emerge as an essential technique in the field of structural biology, Sherman and Smith provide several detailed protocols associated with the cryo-electron microscopy of Rift Valley fever virus (RVFV). These protocols cover the preparation of virus suspensions for cryo-EM, as well as detailed steps in image acquisition and processing for the generation of a structural model of the virus. Furthermore, as RVFV is a Biosafety Level-3 (BSL3) pathogen, this chapter also provides an overview of the general etiquette of working in BSL3 laboratories.

3.6 Detection and Characterization of Antiviral Proteins Against Bunyaviruses in Mosquito-Derived Cells

As bunyaviruses have been characterized as primarily vector-borne, understanding factors that may inhibit/enhance viral replication while in a transmission vector may provide significant insights into understanding the mechanisms required for both replication in the host, as well as the regulatory proteins used by the vector to regulate the infectivity of the bunyavirus in question. In order to address this issue, Böhmer et al. describe methods to evaluate these factors through either quantitative PCR, viral titration, or quantification using a reporter gene system (i.e. luciferase).

3.7 Vector Competence Assays for RVFV in Mosquitoes

As many bunyaviruses are transmitted via insect vectors such as mosquitoes, a protocol describing a method for evaluating the ability of an insect vector in transmitting these types of viruses is provided. Here, Birnberg and Busquets describe protocols for both the rearing of mosquitos and assessing their ability to act as a vector for Rift Valley fever virus through blood feeding, virus propagation in insect cell lines, experimental infections, mosquito saliva extraction, as well as quantification of viral loads through reverse-transcription quantitative PCR (RT-qPCR).

3.8 Experimental Infection of Ticks with SFTSV

Following the previous chapter on the quantification of antiviral factors in mosquitos, Xu and Yu present a protocol for the experimental infection of ticks using two different methods; this first method describes a protocol for the microinjection of ticks with severe fever with thrombocytopenia virus (SFTSV), while the second protocol serves to naturally infect ticks through feeding on an animal (in this case, a rabbit) previously infected with SFTSV. These techniques are important, as standardizing experimental infection of bunyaviruses is of particular importance when studying the host immune response, or when evaluating novel prophylactic measures (i.e. antivirals or vaccines) against pathogenic viruses.

3.9 Methods for Thrips Transmission and Maintenance of Tomato Spotted Wilt Virus (TSWV)

While animal bunyaviruses can be transmitted by insect vectors (i.e. arboviruses), plant bunyaviruses may also be transmitted via insects. Therefore, as plant hosts differ considerably with regards to their experimental manipulation, LaBonte et al. have presented a detailed experimental protocol for the dissemination of tospoviruses through mechanical transmission, as well as through the manipulation of thrips, their natural insect transmission vector.

3.10 In Vitro Evaluation of Bunyavirus T-Cell Immunity

When studying the immune response (either through natural infection or evaluating individual immune-stimulating components such as vaccines), measuring T-cell responses may play an important role in understanding the host immune response to infection or vaccination. These days, studying T-cell responses in small animal models serves as a commonly used experimental approach, with purified T-cells analyzed following experimental infection or vaccination. To this end, McElroy provides experimental protocols for studying

CD4 and CD8 T-cell activation/stimulation via flow cytometry, as well as through the use of an enzyme-linked immunospot (ELISPOT) assay. In this chapter, this technique is used to study the ex vivo response of T-cells to components of Rift Valley fever virus.

3.11 Analysis of Leishbuviridae from Trypanosomatids

This book also includes one of the most novel techniques in the study of bunyavirus; that is, the discovery of novel bunyaviruses from a nonanimal host species. In this chapter, Grybchuk et al. provide a detailed protocol for the discovery of a novel bunyavirus species, which is specific to a unicellular eukaryote. This discovery utilizes Next Generation Sequencing (NGS) techniques, which, combined with bioinformatic analysis, can be used to identify and assemble whole bunyavirus genomes starting from the extracted RNA of infected host cells.

3.12 Recombinant Protein Production and Purification of Rift Valley Fever Nucleoprotein from Escherichia Coli Expression Systems

In this chapter, Desmecht and Boshra present a protocol for the generation of purification of recombinant Rift Valley fever virus nucleoprotein. This technique involves the expression of the cDNA of the nucleoprotein in a bacterial expression vector. As the nucleoprotein of bunyaviruses tends to be the most antigenic component of the virus, most diagnostic test kits tend to use these proteins as the substrate or target for the detection of antibodies following exposure to a bunyavirus of interest. This technique may also be of interest to both structural biologists (who may wish to study the structure of the ribonucleoprotein complex found inside the virion of nucleoproteins), as well as immunologists who would like to study the cell-mediated immunity induced by bunyavirus nucleoproteins.

3.13 Bunyaviruses: Expression of Recombinant Glycoproteins from Insect Cells

This chapter, written by de Boer, presents a protocol for the recombinant expression of glycoproteins Gn and Gc of Rift Valley fever virus, generating virus-like particles that retain some of the functions of the natural, mammalian-derived virion. The products generated from this protocol could be used for immunological studies, as well as a potential vaccine candidate.

3.14 Murine Models to Study Crimean-Congo Hemorrhagic Fever Hepatic Injury

As many bunyaviruses are pathogenic to humans and large animals, studying the effects of these viruses on their natural host may prove to be difficult due to practical limitations. Therefore, being able to reproduce viral pathogenesis in small animal models is a useful alternative when studying severe disease. Here, Golden et al. present a protocol for the study of hepatic injury due to Crimean-Congo hemorrhagic fever virus, using a mouse model. This mouse model, when treated with IFN-I-specific monoclonal antibodies, can reproduce hepatic injury similar to that observed following human infection. Techniques describing viral challenge, liver harvest and fixation, RNA in situ hybridization, and the quantification of viral load are provided in detail. The ability to replicate bunyavirus pathology is indispensable when studying the immune response following a natural infection of a virus, or when evaluating the efficacy of novel prophylactic measures (i.e. vaccines, antivirals) against a virus of interest.

3.15 A Review on Reliable and Standardized Animal Models to Study the Pathogenesis of Schmallenberg Virus in Ruminant Natural Host Species

While small animal models are an important step in evaluating the virulence and pathogenesis of bunyaviruses, larger animals such as sheep and goats also play an important role when evaluating experimental infections of bunyaviruses that are known to infect livestock. In this chapter, Martinelle and Saegerman provide an in-depth review of the key factors associated with the experimental infection of Schmallenberg virus (SBV) in sheep/goats. Topics such as the route of inoculation preparation, route of infection, as well as evaluating the ability of SBV in crossing the placental barrier are discussed.

3.16 Schmallenberg Virus, from the Necropsy Room to the Laboratory

Continuing with the theme of Schmallenberg virus (SBV), Cayrou et al. describe a detailed method for detecting the presence of SBV in naturally infected ruminants; and include characterizing SBV infection using veterinary pathological techniques. These techniques include performing a necroscopy on animals suspected of being infected with SBV, as well as protocols related to histopathology, serological detection, and RT-PCR—all of which are described in detail.

3.17 Reverse Genetics System for Crimean-Congo Hemorrhagic Fever Virus

While virus-like particles have the capacity to mimic the structural proteins located at the exterior of the bunyavirus virion, researchers may wish to work with a more complete virion, in order to study other aspects of virus replication. Furthermore, depending on the virulence of the bunyavirus in question, a high-containment facility (such as a BSL3 or BSL4 laboratory) may not be readily available. To address these issues, Scholte et al. present a protocol describing the generation of virus-like replicon particles, using a reverse-genetics system. This technique describes using such a system based on the genome of Crimean-Congo hemorrhagic fever virus (normally a BSL4 pathogen). However, the generation of virus-like replicon particles enables the generation of infectious particles, which are not replicative or pathogenic, enabling their use in standard laboratory BSL2 conditions. It should also be noted that the protocol for the generation of virus particles through reverse genetics can also be applied to other bunyaviruses, should the use of recombinant viruses be necessary (i.e. mutagenesis studies or the evaluation of chimeric proteins).

3.18 Recombinant Vaccine Production: Production of a Recombinant CCHF MVA Vaccine

Another popular approach in the development of vaccine candidates against pathogenic bunyaviruses is the use of an attenuated strain of vaccinia for the recombinant expression of bunyavirus antigens. In this case, Kennedy et al. describe the generation of a recombinant modified vaccinia Ankara (MVA) expressing both the nucleoprotein and glycoprotein components of Crimean-Congo haemorrhagic fever virus (CCHFV). Both of these components have been previously shown to induce an immune response in animal models, with the nucleoprotein inducing a cell-mediated response, while the glycoproteins induce the production of neutralizing antibodies. Therefore, the protocol described in this chapter could be applied to other bunyaviruses, taking into account the antigen of interest.

3.19 Efficient Expression of Oropouche Virus Nonstructural Proteins NSs and NSm

In bunyaviruses, the roles of nonstructural proteins remain a point of particular controversy, as studies using mutagenesis and reverse genetics have yielded varying (and sometimes contradictory) findings. However, it has been shown that the nonstructural protein NSs can inhibit the host type I IFN immune response in some bunyaviruses [1]. Here, Jurado-Cobena and Ikegami present a protocol for the recombinant expression of the NSs and NSm proteins of Oropouche virus, a bunyavirus of increasing concern to public health in South America.

3.20 Systemic Clodronate Liposome Treatment for Depletion of Blood-Contacting Phagocytes and Investigation of Their Role in the Control of Bunyavirus Viremia

An important aspect in the study of bunyaviruses involves elucidating the role of phagocytes in the host immune response. One such method involves the comparison of the immune response of a healthy animal model compared with one where host phagocytes are either inactivated or depleted. In this chapter, Ander et al. describe the use of clodronate-containing liposomes to temporarily deplete phagocytes from mice. The technique will also describe the inoculation and detection of La Crosse virus in phagocyte-depleted mice, in comparison to control mice (i.e. those with intact, functional phagocytes). This technique can also be modified for use in other members of the Bunyaviricetes, as mouse models are available for a wide variety of pathogenic bunyaviruses.

References

1. Boshra H (2022) An overview of the infectious cycle of bunyaviruses. Viruses 14(10):2139. https://doi.org/10.3390/v14102139
2. Walker PJ, Siddell SG, Lefkowitz EJ et al (2022) Recent changes to virus taxonomy ratified by the international committee on taxonomy of viruses (2022). Arch Virol. https://doi.org/10.1007/s00705-022-05516-5
3. Reguera J, Malet H, Weber F et al (2013) Structural basis for encapsidation of genomic RNA by La Crosse Orthobunyavirus nucleoprotein. Proc Natl Acad Sci USA 110(18):7246–7251. https://doi.org/10.1073/pnas.1302298110
4. Gerlach P, Malet H, Cusack S et al (2015) Structural insights into bunyavirus replication and its regulation by the vRNA promoter. Cell

161(6):1267–1279. https://doi.org/10.1016/j.cell.2015.05.006

5. Elliott RM (1990) Molecular biology of the bunyaviridae. J Gen Virol 71(Pt 3):501–522
6. Hofmann H, Li X, Zhang X et al (2013) Severe fever with thrombocytopenia virus glycoproteins are targeted by neutralizing antibodies and can use DC-SIGN as a receptor for pH-dependent entry into human and animal cell lines. J Virol 87:4384–4394
7. Ter Horst S, Conceicao-Neto N, Neyts J et al (2019) Structural and functional similarities in bunyaviruses: perspectives for pan-bunya antivirals. Rev Med Virol 29:e2039

Chapter 2

Use of Rift Valley Fever Virus Expressing NanoLuc Luciferase for the Assessment of Neutralizing Antibodies and Antivirals

Belén Borrego, Luis Martinez-Sobrido, Alejandro Brun, and Aitor Nogales

Abstract

Rift Valley fever (RVF) is an arboviral zoonotic disease affecting many African countries with the potential to spread to other geographical areas. In this chapter we describe the use of a replication-competent recombinant (r)RVFV expressing NanoLuc Luciferase (Nluc) for in vitro studies. The determination of parameters such as neutralizing antibodies in serum samples, or the antiviral activity of drugs is usually carried out using standard assays based on the assessment of cytopathic effect on cell cultures. The use of a virus encoding a traceable reporter protein allows to correlate the presence or absence of infection with the detection of the product in the infected cultures, thus tracking the level of RVFV infection in an objective, quantitative manner. In addition to this quantitative measurement of results, our protocol offers two other advantages, such as a shorter time to read, given that 48 h post-infection the production of the reporter protein is enough to give an accurate result, and the use of an attenuated virus, which reduces the risk of exposure.

Key words RVFV, Reporter genes, Luciferase, NanoLuc, In vitro, Neutralizing antibodies, Antivirals, Ribavirin

1 Introduction

Rift Valley fever virus (RVFV), belonging to the genus Phlebovirus, family *Phenuiviridae* and class *Bunyaviricetes*, is a lipid bilayer enveloped virus and its genome consists of three single-stranded, negative-sense or ambisense RNA segments differing in size: Small (S), Medium (M) and Large (L) [1, 2]. The RNA segment L encodes the viral RNA-dependent RNA polymerase (RdRp), which is necessary for viral replication and transcription [3, 4]. The M segment encodes a single open reading frame (ORF), which produces the envelope glycoproteins Gn and Gc, and nonstructural (NS) proteins: NSm and the 78 kDa Gn/NSm fusion protein [1, 5–7]. The S segment encodes two proteins in an

Hani Boshra (ed.), *Bunyaviruses: Methods and Protocols*, Methods in Molecular Biology, vol. 2893,
https://doi.org/10.1007/978-1-0716-4338-9_2,

ambisense strategy: the nucleoprotein (N), which protects the viral RNA [8], and the multifunctional NSs protein, which is considered the main virulence factor of RVFV [9, 10].

RVFV is a mosquito-borne virus that infects humans as well as many domestic and wild ruminant species, causing high rates of abortion and neonatal mortality, which poses a major economic impact in low-income African countries. Most infected humans usually develop mild symptoms although severe complications can develop, leading to encephalitis or to a, usually fatal, hemorrhagic disease. Importantly, RVFV has been classified by the World Health Organization (WHO) as one of the eight emerging pathogens that "pose the greatest public health risk due to their epidemic potential and/or where there are no or insufficient countermeasures" (https://www.who.int/activities/prioritizing-diseases-for-research-and-development-in-emergency-contexts). Therefore, considering the concerns for human and animal health and the economic impact due to RVFV infections, the development of new vaccines and anti-RVFV reagents for human or animal use is of great relevance [11–13]. Moreover, all these features make RVF a paradigm of disease for the One Health concept.

The implementation of reverse genetics methods for bunyaviruses and RVFV has been key for the generation of recombinant (r)RVFV to study multiple aspects of its biology, including viral replication, the function of viral proteins, the identification of host factors, and the determinants of viral pathogenesis [14–18]. Importantly, the development of reverse genetics for RVFV has also allowed the manipulation of the genome for the incorporation of heterologous sequences, such as reporter genes, including fluorescent or bioluminescent proteins [14, 19]. Notably, reporter-expressing rRVFV represents an excellent choice to detect the virus in infected cells and tissues, and to quantify the viral replication using the expression of the reporter genes as a readout. In this chapter, we describe the propagation of an Nluc-expressing rRVFV, where, as a difference from previously reported systems, the foreign gene was inserted into the viral RNA segment S without deleting the NSs gene (Fig. 1). Furthermore, this chapter describes a detailed in vitro protocol for using this rRVFV-Nluc in the identification and characterization of antiviral compounds that could be also adapted for high-throughput screening (HTS) assay format for antiviral drug discovery. In addition, we describe the use of rRVFV-Nluc to assess the levels of neutralizing antibodies from serum samples, which can be used with serum samples from multiple species (human, cattle, goats, sheep, rodents, etc.). Therefore, the provided protocols could be a very useful resource for the screening of novel compounds with antiviral activity, or to evaluate the presence of serum neutralizing antibodies induced by RVF vaccine candidates.

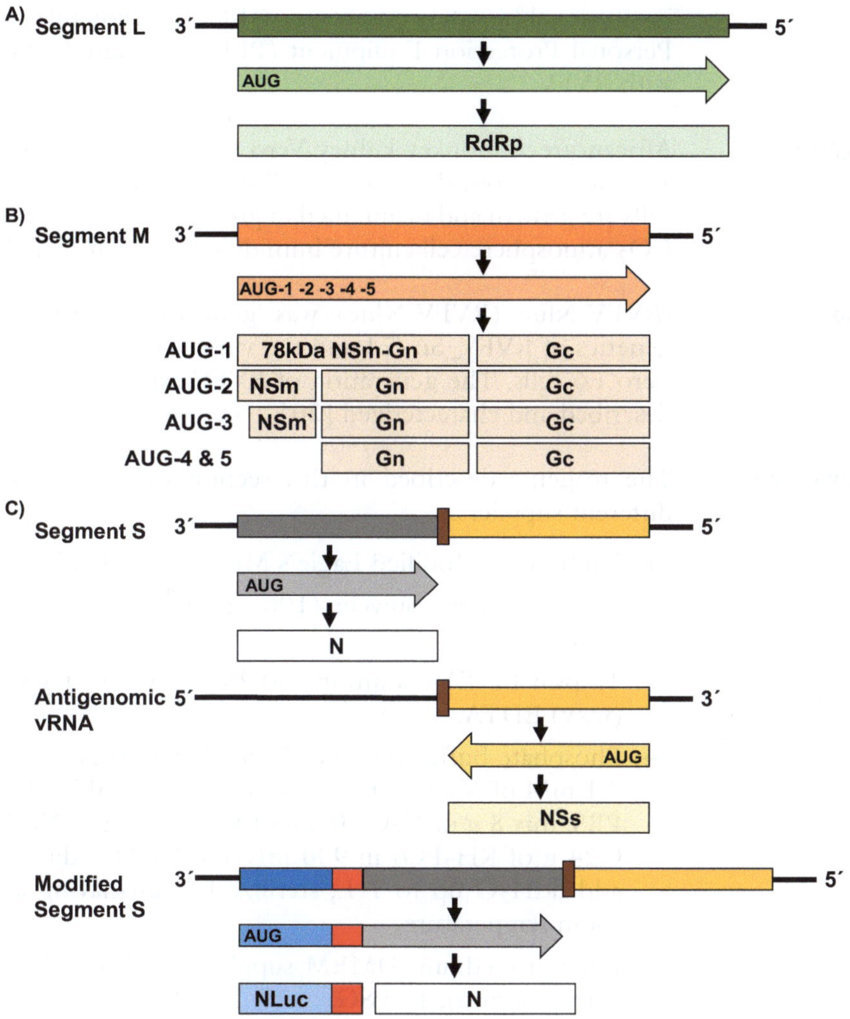

Fig. 1 Genomic organization of RVFV and modified segment S. Genomic sequence (top), transcribed mRNAs (middle) with the start codons (AUG), and the resulting proteins (bottom) are indicated for viral segments L (**a**), M (**b**), and S or modified S (**c**). Untranslated regions (UTR) flanking each segment are indicated with black lines. Proteins encoded by the M segment are produced by posttranslational cleavage of translated polyproteins. The nucleoprotein is generated by an mRNA transcribed from the vRNA. NSs is produced from an mRNA transcribed from the antigenomic copy vRNA generated during viral replication. The brown box represents the RVFV S-segment intergenic transcription termination site. Nluc and porcine teschovirus (PTV) 2A sequences are indicated with blue and red boxes, respectively

2 Materials

2.1 Biosafety

This chapter describes potential applications of RVFV expressing Nluc reporter gene. These studies must be performed under biosafety level 3 (BSL-3) laboratory conditions (*see* **Note 1**). All materials used with infectious RVFV-Nluc should be fully decontaminated before disposal, following the appropriate institutional biosafety committee (IBC) recommendations. Proper Personal Protection Equipment (PPE) is required when working with RVFV.

2.2 Cell Lines

African green monkey kidney Vero E6 cells (ATCC/CRL/1586) are available from the American Type Culture Collection. Vero E6 cells are growth and maintained in growth medium at 37 °C in a 5% CO_2 atmosphere cell culture humidified incubator (*see* **Note 2**).

2.3 Viruses

rRVFV-Nluc (RVFV-Nluc) was generated in the background genetics of RVFV_56/74 and the virus is routinely propagated in Vero E6 cells. The generation of RVFV expressing Nluc has been described and characterized [20].

2.4 Reagents and Labware

The reagents described in this section can be purchased from different suppliers.

1. Dulbecco's Modified Eagle's Medium (DMEM).
2. Penicillin/Streptomycin (100 μg/mL), 2 mM L-Glutamine (PSG 100×).
3. Trypsin-EDTA solution: 0.25% (w/v) trypsin, 0.02% (w/v) EDTA.
4. Phosphate buffered saline (PBS): 138 mM NaCl, 3 mM KCl, 8.1 mM of Na_2HPO_4, 1.5 mM KH_2PO_4, pH 7.4. To prepare PBS, mix 8 g of NaCl, 0.2 g of KCl, 1.44 g of Na_2HPO_4, and 0.24 g of KH_2PO_4 in 900 mL of ddH_2O. Adjust pH to 7.4, add ddH_2O up to 1 L, sterilize by autoclaving and store at room temperature.
5. Growth medium: DMEM supplemented with 5% FBS (Fetal Bovine serum), 1× PSG.
6. Overlay medium: 1% carboxy-methyl-cellulose in DMEM supplemented with 10% FBS.
7. Fixation solution: 10% Formalin (*see* **Note 3**).
8. Crystal violet 0.3% (w/v) solution: Mix 3 g of crystal violet, 50 mL of formaldehyde, 100 ml ethanol, and add ddH_2O up to 1 L. Store at room temperature.
9. Nano-Glo luciferase assay system (Promega).
10. Ribavirin.
11. Cell culture plasticware: T-75 cell culture flask, 6-well plates, 96-well plates, serological pipettes (5, 10 and 25 mL), universal filtered pipette tips (20-, 200-, and 1000-μL), polypropylene conical Falcon tubes (15-mL and 50-mL), and white flat-bottom 96-well microplates.

2.5 Instrumentation

1. Hemocytometer or automatic cell counter.
2. Benchtop refrigerated microcentrifuge.
3. Laboratory 5% CO_2 cell culture humidified incubators at 37 °C for maintenance of cells.
4. ClassII biosafety cabinet.
5. Light inverted microscope.
6. −80 °C freezer.
7. Luminescence microplate reader.

2.6 Software

1. GraphPad Prism (GraphPad Software Inc., Ver. 8.0).
2. Excel (Microsoft).

3 Methods

3.1 Propagation of RVFV-Nluc

Use this protocol to prepare virus stocks for all the in vitro assays described in this book chapter.

1. A day before infection, seed Vero E6 cells in a T75 flask so that on the next day the cells will be at 90% confluence, approximately 8×10^6 cells/T-75.
2. Place the cells overnight in a 37 °C, 5% CO_2 humidified incubator.
3. Next day (before infection), check confluence of the cell monolayer using an inverted microscope (*see* **Note 4**).
4. Wash cells with 10 mL of pre-warmed, sterile 1× PBS.
5. Next, inoculate Vero E6 cells (in the T75 flask) with RVFV-Nluc at a multiplicity of infection (MOI) of 0.01 diluted in 8 mL of growth medium.
6. Incubate for 1 h in a 37 °C, 5% CO_2 humidified incubator.
7. Afterwards, remove the inoculum and add 12 mL of growth medium and monitor daily for CPE. CPE should start appearing at 48 h post-infection (hpi). Incubation can be prolonged beyond 72 hpi if deemed appropriate to maximize the amount of virus recovered from the culture (*see* **Note 5**).
8. Collect the cell culture supernatant into a 15 mL conical tube and clear by centrifugation at 880×*g* for 10 min at 4 °C.
9. Transfer the cleared supernatant into a new 15 mL conical tube, homogenize by pipetting up and down, and prepare aliquots of the desired volume.
10. Store aliquots at −80 °C (*see* **Note 6**).
11. Determine viral titers by plaque assay or 50%Tissue Culture Infectious Dose ($TCID_{50}$).

3.2 Plaque Assay to Determine Viral Titers of RVFV-Nluc

1. Seed 6-well plates with Vero E6 cells (5×10^5 cells/well) in 2 mL growth medium. Place the plates overnight in a 37 °C, 5% CO_2 humidified incubator. Prepare the appropriate number of plates and wells to evaluate viral titers by triplicate (*see* **Note 7**).
2. Next day (before infection), check confluence of the cell monolayer using an inverted microscope (*see* **Notes 4** and **8**).
3. To determine viral titers, thaw one aliquot on ice and prepare dilutions in infection medium. Serially dilute 1:10 the samples in growth medium (from 10^{-1} to 10^{-8} dilutions).
4. Remove medium from wells and add 1 mL of each virus dilution. Place 1 h in a 37 °C, 5% CO_2 incubator.
5. In the meantime, prepare the overlay medium.
6. Remove virus inoculum and overlay cells in each well with 2 mL of overlay medium.
7. Incubate 2–3 days at 37 °C in a 5% CO_2 incubator.
8. Add 2 mL of fixation solution to each well and let stand for 15–30 min at room temperature. Then, remove medium, wash to eliminate traces of overlay medium and add 1 mL of crystal violet solution to each well. Incubate 20 min at room temperature.
9. Remove crystal violet solution from the wells.
10. Wash the plates with distilled water and let wells to dry (Fig. 2a).

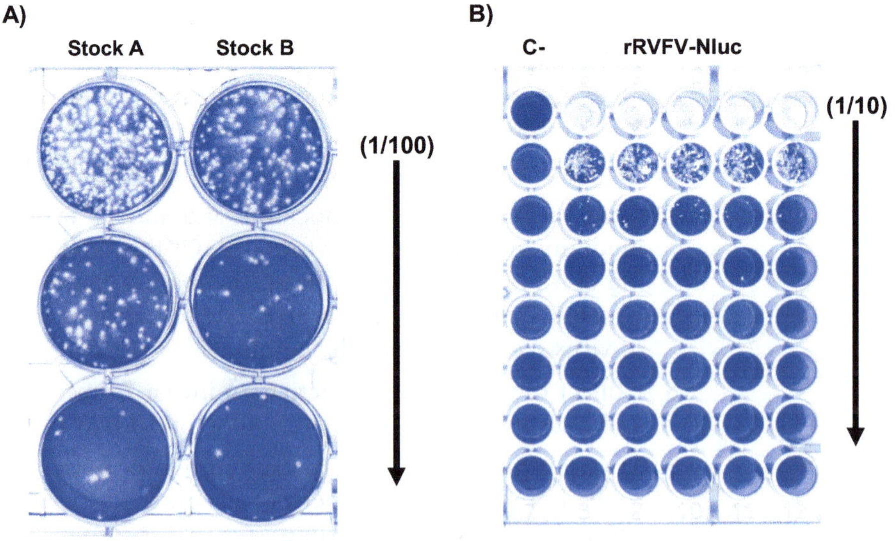

Fig. 2 Examples of two assays for the assessment of RVFV-Nluc viral titers. (**a**) Determination of titers in PFU/mL by plaque assay and crystal violet staining using 10 fold serial dilutions of viral stocks. (**b**) Viral titers determined by $TCID_{50}$. Tenfold viral stock dilutions were made in quintuplicates. Mock-infected cells (C–) used as negative control

11. Count virus plaques and calculate the viral titer in plaque-forming units per milliliter (PFU/mL) using the following formula: PFU/mL = number of plaques × virus dilution/volume of inoculum (mL). The mean value and standard deviation can be calculated on Excel or Prism software.

3.3 Titration of RVFV-NLuc Tissue Culture Infectious Dose 50 ($TCID_{50}$)

1. Seed 2×10^4 of Vero E6 cells/well in 96-well plates the day before the infection to reach confluence the next day.
2. Confirm cell monolayers before infection (*see* **Note 4**).
3. Make tenfold serial dilutions of the samples (from 10^{-1} to 10^{-8} dilutions) in growth medium. Starting with the most diluted sample, add 50 μL/well of the virus dilution to each of the wells in the 96-well plate (*see* **Note 7**).
4. After 1 h of viral absorption, remove the virus inoculum and add 100 μL of growth medium.
5. Place the plates at 37 °C in the 5% CO_2 humidified incubator for 3–4 days (*see* **Note 9**).
6. Remove the tissue culture medium from the 96-well plates, wash with PBS, and add 100 μL/well of crystal violet solution (*see* **Note 10**).
7. Incubate for 1 h at room temperature, remove the crystal violet solution, and wash the wells three times with distilled water. Dry the plates at room temperature (Fig. 2b).
8. Determine viral titers ($TCID_{50}$/mL) using the Reed and Muench approach [21].

3.4 Nluc-Based Assay to Evaluate Antivirals

We describe here the antiviral effect of Ribavirin as a proof of concept (Fig. 3). However, other compounds may be assessed using the same protocol.

1. A day before infection, seed Vero E6 cells in 96-well microplates so that on the next day, the cells will be at 90% confluence, approximately 2×10^4 cells/well. Each condition should be conducted in triplicates.
2. To evaluate antivirals, infect Vero E6 cells with 200 PFUs of RVFV-Nluc per well in a total volume of 50 μL, except for the last column (column 12). Dilute the viral stock in growth medium to obtain the proper viral concentration (200 PFUs/well/50 μL).
3. Allow viral absorption to occur at RT for 1 h.
4. In the meantime, serially dilute (1:2) ribavirin (1000 μM starting concentration) using growth medium in a separate 96-well microplate. Leave columns 11 and 12 with only growth medium for a negative control.

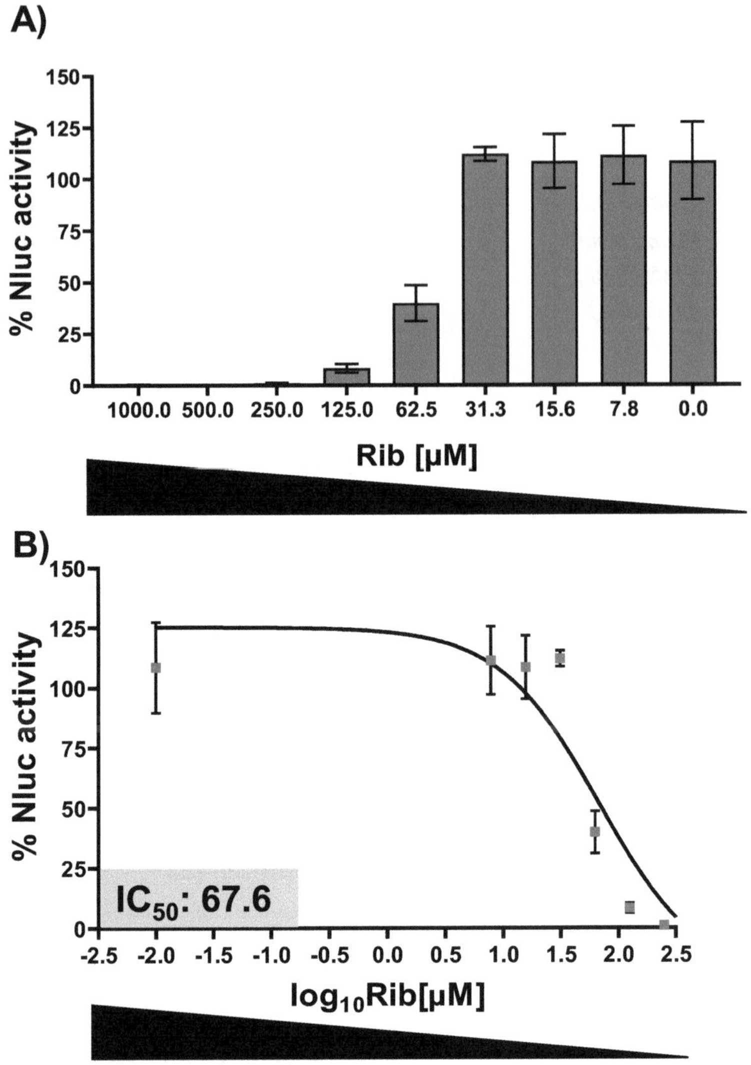

Fig. 3 A Nluc-based microneutralization assay for the identification of RVFV antivirals. Vero E6 cells (96-well microplate, 2×10^4 cells/well, and triplicates) were infected with 200 PFU of rRVFV-Nluc virus for 1 h, before being incubated with twofold serial dilutions of ribavirin (1000 μM starting concentration). After 48 h post-infection, percent of viral infection was determined by evaluating Nluc activity in the tissue culture supernatant using a microplate reader (**a**), and the percent inhibition 50% (IC_{50}) was calculated using sigmoidal dose–response curves (**b**). Both mock-infected and rRVFV-Nluc-infected cells in the absence of ribavirin were included as controls and for normalization of percent infection

5. After viral infection, carefully remove the viral inoculum, and transfer 150 μL of the serially diluted antiviral 96-well microplate to the 96-well microplate containing RVFV-Nluc-infected cells.

6. Incubate the microplates in a 5% CO_2 incubator at 37 °C for 48 h.
7. Evaluate the antiviral effect by determining the presence of Nluc in cell culture supernatants from infected cells (*see* **Note 11**).
8. Prepare the luciferase assay solution by diluting the Nluc substrate 1:50 with the provided dilution buffer. In a white flat-bottom 96-well microplate for luciferase assays, mix 25–50 μL of luciferase assay solution with 25–50 μL of the collected cell culture supernatant samples. Incubate the mix for 2–3 min at room temperature before measuring the luminescence using a luminometer.
9. Determine the 50% inhibitory concentration (IC_{50}) of ribavirin using sigmoidal dose-response curves using Prism software. Use untreated, infected cells (column 11 in the 96-well microplate) and untreated, uninfected cells (column 12 in the 96 well microplate) as the negative controls and for the normalization of percent infection based on Nluc signal.

3.5 Nluc-Based Assay to Evaluate Neutralizing Antibodies

We describe here, the assessment of neutralizing antibodies present in sheep serum obtained after animal was vaccinated with an experimental vaccine developed in the laboratory [22, 23] (Fig. 4). However, other sera samples from vaccinated or infected animals may be evaluated using the same protocol.

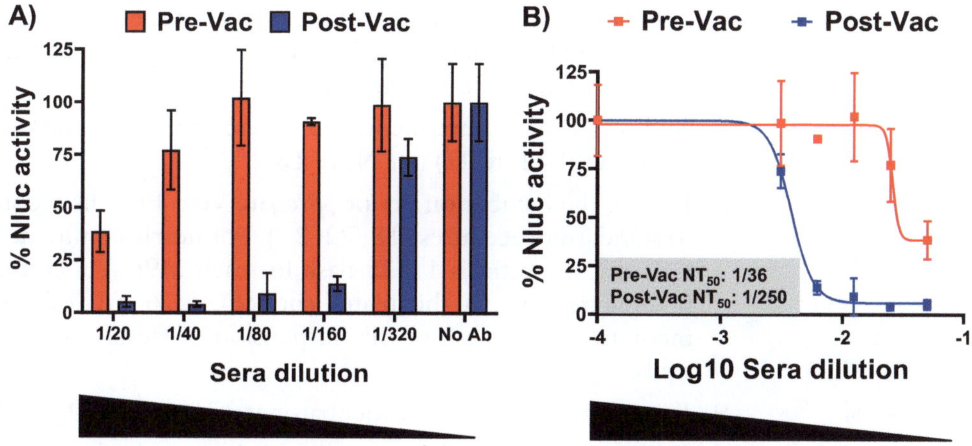

Fig. 4 A Nluc-based microneutralization assay for the identification of RVFV neutralizing antibodies. 100 $TCID_{50}$ of rRVFV-Nluc were incubated with twofold serial dilutions of serum (1/20–1/320) for 1 h. Then freshly trypsinized Vero E6 cells were added to virus–serum mixtures as described in the protocol. After 48 h post-infection, percent of viral infection was determined by evaluating Nluc activity in the tissue culture supernatant using a microplate reader (**a**), and the percent inhibition 50% (IC_{50}) was calculated using sigmoidal dose-response curves (**b**). Mock-infected and rRVFV-Nluc-infected cells in the absence of serum were included as controls and used for normalization of percent infection

1. Design the distribution of the samples to be tested in the 96-well microplates, in rows or columns, considering: (i) the number of samples; (ii) number of replicates: duplicate or quadruplicate for each sample; and (iii) sample dilutions (first and last dilution; x-fold dilution). Based on all these points, distribute your samples and prepare the required number of 96-well microplates. In this protocol we describe the testing of samples in 8 twofold dilutions starting from 1/10 (row A) to 1/1280 (row H) (*see* **Note 12**).
2. To dilute the samples add 50 μL of culture medium to all wells in the plate except row A, and 90 μL to all wells in row A, where the first dilution (1/10) of samples is made.
3. Add 10 μL of sample (*see* **Note 13**) into the wells of row A according to the determined distribution and mix well by pipetting. Then, take 50 μL (twofold dilutions) from wells in row A to wells in row B, mix well by pipetting, and proceed in the same way for the following rows (*see* **Note 12**).
4. Prepare a dilution of the virus stock in culture medium, so that 50 μL corresponds to 100 $TCID_{50}$ as calculated previously; and add 50 μL of this virus dilution per well. Incubate for 30–60 min at 37 °C (*see* **Note 14**).
5. In addition to the sample(s) plate(s), prepare a plate (or at least four columns-quadruplicates) for control wells, and add 50 μL of culture medium to all wells.
6. Prepare serial tenfold dilutions at least dilutions from (1/10) to (1/1000) from the prepared virus dilution (estimated at 100 $TCID_{50}$ per well/per 50 μL): by adding 50 μL to 450 of culture medium, and add 50 μL of the corresponding tenfold dilution of virus (rows A to D).
7. Add 50 μL of culture medium alone to the rest of the control (plate) wells (E to H) (*see* **Note 15**).
8. During the incubation period, prepare Vero E6 cells according to standard procedures [15, 22–24]. For nearly confluent Vero E6 cells, the ratio is 1 T25 flask for each M96 (subculture 1: 0.75 according to the plates surface), in 6 mL of growth medium. Add 50 μL of this suspension of freshly cells to the wells.
9. Incubate the plates in the incubator at 37 °C/5% CO_2 for 48 h.
10. Evaluate the neutralizing activity of the samples by determining the presence of Nluc in cell culture supernatants from infected cells. Collect 25 μL of the supernatant from the wells and follow the procedure as described in point **step 7**, Subheading 3.4. Infected plates can be left in the incubator and checked for CPE and stained with crystal violet at later times pi, as previously described.

4 Notes

1. Depending on each country's institutional regulations and permits, RVFV handling and storage is performed under a specific biosafety risk level. For instance, in some institutions, RVFV must be handled in BSL-4 facilities.
2. It is important to keep track of the cell passage number, since it can affect virus rescue efficiency and/or virus propagation. The use of late passage cells (above passage 30–35) is not recommended. In addition to Vero E6 cells other susceptible cell lines such as Vero (ATCC-CCL81), or BHK-21 (ATCC-CCL10) cells can be used.
3. Prepare the fixation solution in a fume hood to prevent formaldehyde exposure.
4. We recommend visualizing the cells under a light microscope to confirm the presence of a healthy cell monolayer, the cell confluence and the absence of contaminations.
5. CPE induced by RVFV infection in Vero E6 cells is characterized by rounding of the cells and disruption of the cell monolayer. We recommend collecting the supernatant 2–3 days post-infection when the CPE is 80–90%.
6. Virus stock should be aliquoted into small volumes to prevent multiple freeze–thaw cycles, which can reduce RVFV titers.
7. We recommend performing viral titrations in triplicate and calculating viral titers using the triplicate average.
8. A 90% confluent monolayer of Vero E6 cells by the time of infection is recommended. Confluent monolayers give rise to smaller-size virus plaques.
9. Titration of RVFVs by $TCID_{50}$ assay can be also done in an easier way by distributing the tenfold serial dilutions of the virus stock in 96-well plates as described in point 3, and then adding 50 μL of a suspension of freshly trypsinized cells to the wells. For nearly confluent Vero E6 cells, the ratio is 1 T25 flask for each M96 (subculture 1:0.75 according to the plates surface), in 6 mL of growth medium. Next steps of the protocol continue the same.
10. Likewise, you can add 50 μL/well of crystal violet solution (point 6) without removing the tissue culture medium from the 96-well plates.
11. Refer to the manufacturer's recommendations and optimize the assay if required.
12. In a basic distribution of 96-well microplates, four samples can be tested, each in triplicate (columns 1 to 3; 4 to 6 and so on), in 8 twofold dilutions starting from 1/10 to 1/1280 (rows A

to H). Depending on the expected results this pattern can be modified, i.e. testing only four dilutions per sample if close to negative results are expected; testing tenfold dilutions if samples are strongly positive; or starting from a higher dilution if positive results are expected (biological samples such as sera can have toxic effects).

13. Heat inactivation of serum (56 °C for 30 min) before testing is recommended when handling samples that can contain infectious RVFV.
14. According to international standards, 100 $TCID_{50}$/well (50% tissue culture infective dose in 50 μL under the conditions of the test) is the appropriate amount of virus to add, within an accepted range (e.g. 32–320 $TCID_{50}$). Values below 30 $TCID_{50}$ may result in poor infection levels and false neutralization scores. Conversely, higher amounts of virus may cause samples with low neutralization titres to be missed.
15. Control wells must be included to ensure that the virus dilution was correct and to provide the reading corresponding to 100% infection.

Acknowledgements

Research on RVFV was funded by grant PID2021-122567OB-I00 MCIN/AEI /10.13039/501100011033/ and by ERDF "A way of making Europe." Research was also partially funded by ISIDORe JRA Programme. ISIDORe has received funding from the European Union's Horizon Europe research and innovation programme under grant agreement number 101046133.

References

1. Boshra H (2022) An overview of the infectious cycle of bunyaviruses. Viruses 14(10). https://doi.org/10.3390/v14102139
2. Tercero B, Narayanan K, Terasaki K, Makino S (2021) Characterization of the molecular interactions that govern the packaging of viral RNA segments into Rift Valley fever Phlebovirus particles. J Virol 95(14):e0042921. https://doi.org/10.1128/jvi.00429-21
3. Wang X, Hu C, Ye W, Wang J, Dong X, Xu J et al (2021) Structure of Rift Valley fever virus RNA-dependent RNA polymerase. J Virol 96: e01713–e01721. https://doi.org/10.1128/JVI.01713-21
4. Ferron F, Weber F, de la Torre JC, Reguera J (2017) Transcription and replication mechanisms of Bunyaviridae and Arenaviridae L proteins. Virus Res 234:118–134. https://doi.org/10.1016/j.virusres.2017.01.018
5. Bian T, Hao M, Zhao X, Zhao C, Luo G, Zhang Z et al (2023) A Rift Valley fever mRNA vaccine elicits strong immune responses in mice and rhesus macaques. NPJ Vaccines 8(1):164. https://doi.org/10.1038/s41541-023-00763-2
6. Chapman NS, Zhao H, Kose N, Westover JB, Kalveram B, Bombardi R et al (2021) Potent neutralization of Rift Valley fever virus by human monoclonal antibodies through fusion inhibition. Proc Natl Acad Sci USA 118(14). https://doi.org/10.1073/pnas.2025642118
7. Terasaki K, Kalveram B, Johnson KN, Juelich T, Smith JK, Zhang L et al (2021) Rift Valley fever virus 78kDa envelope protein attenuates virus replication in macrophage-

derived cell lines and viral virulence in mice. PLoS Negl Trop Dis 15(9):e0009785. https://doi.org/10.1371/journal.pntd.0009785
8. Mottram TJ, Li P, Dietrich I, Shi X, Brennan B, Varjak M et al (2017) Mutational analysis of Rift Valley fever phlebovirus nucleocapsid protein indicates novel conserved, functional amino acids. PLoS Negl Trop Dis 11(12): e0006155. https://doi.org/10.1371/journal.pntd.0006155
9. Ikegami T, Narayanan K, Won S, Kamitani W, Peters CJ, Makino S (2009) Rift Valley fever virus NSs protein promotes post-transcriptional downregulation of protein kinase PKR and inhibits eIF2alpha phosphorylation. PLoS Pathog 5(2):e1000287. https://doi.org/10.1371/journal.ppat.1000287
10. Ikegami T, Narayanan K, Won S, Kamitani W, Peters CJ, Makino S (2009) Dual functions of Rift Valley fever virus NSs protein: inhibition of host mRNA transcription and post-transcriptional downregulation of protein kinase PKR. Ann N Y Acad Sci 1171(Suppl 1):E75–E85. https://doi.org/10.1111/j.1749-6632.2009.05054.x
11. Cecilia H, Drouin A, Metras R, Balenghien T, Durand B, Chevalier V et al (2022) Mechanistic models of Rift Valley fever virus transmission: a systematic review. PLoS Negl Trop Dis 16(11):e0010339. https://doi.org/10.1371/journal.pntd.0010339
12. Lapa D, Pauciullo S, Ricci I, Garbuglia AR, Maggi F, Scicluna MT et al (2024) Rift Valley fever virus: an overview of the current status of diagnostics. Biomedicines 12(3). https://doi.org/10.3390/biomedicines12030540
13. Poungou N, Sevidzem SL, Koumba AA, Koumba CRZ, Mbehang P, Onanga R et al (2023) Mosquito-borne arboviruses occurrence and distribution in the last three decades in Central Africa: a systematic literature review. Microorganisms 12(1). https://doi.org/10.3390/microorganisms12010004
14. Tercero B, Makino S (2024) Reverse genetics system for Rift Valley fever virus. Methods Mol Biol 2733:101–113. https://doi.org/10.1007/978-1-0716-3533-9_7
15. Borrego B, Moreno S, de la Losa N, Weber F, Brun A (2021) The change P82L in the Rift Valley fever virus NSs protein confers attenuation in mice. Viruses 13(4). https://doi.org/10.3390/v13040542
16. Smith DR, Johnston SC, Piper A, Botto M, Donnelly G, Shamblin J et al (2018) Attenuation and efficacy of live-attenuated Rift Valley fever virus vaccine candidates in non-human primates. PLoS Negl Trop Dis 12(5):e0006474. https://doi.org/10.1371/journal.pntd.0006474
17. Wichgers Schreur PJ, Oreshkova N, Moormann RJ, Kortekaas J (2014) Creation of Rift Valley fever viruses with four-segmented genomes reveals flexibility in bunyavirus genome packaging. J Virol 88(18):10883–10893. https://doi.org/10.1128/JVI.00961-14
18. Kading RC, Crabtree MB, Bird BH, Nichol ST, Erickson BR, Horiuchi K et al (2014) Deletion of the NSm virulence gene of Rift Valley fever virus inhibits virus replication in and dissemination from the midgut of Aedes aegypti mosquitoes. PLoS Negl Trop Dis 8(2):e2670. https://doi.org/10.1371/journal.pntd.0002670
19. Ikegami T (2021) Development of a simian RNA polymerase I promoter-driven reverse genetics for the rescue of recombinant Rift Valley fever virus from vero cells. J Virol 95(7). https://doi.org/10.1128/JVI.02004-20
20. Nogales A, Alonso C, Moreno S, Lorenzo G, Borrego B, Martinez-Sobrido L, Brun A (2024) Novel replication-competent reporter-expressing Rift Valley fever viruses for molecular studies. bioRxiv 2024.08.06.606778; https://doi.org/10.1101/2024.08.06.606778 (preprint)
21. Reed LJ, Muench H (1938) A simple method of estimating fifty per cent endpoints. Am J Epidemiol 27(3):493–497. https://doi.org/10.1093/oxfordjournals.aje.a118408
22. Moreno S, Lorenzo G, Lopez-Valinas A, de la Losa N, Alonso C, Charro E et al (2024) Safety and efficacy upon infection in sheep with Rift Valley fever virus ZH548-rA2, a triple mutant rescued virus. Viruses 16(1). https://doi.org/10.3390/v16010087
23. Borrego B, Brun A (2020) A hyper-attenuated variant of Rift Valley fever virus generated by a mutagenic drug (Favipiravir) unveils potential virulence markers. Front Microbiol 11:621463. https://doi.org/10.3389/fmicb.2020.621463
24. Moreno S, Borrego B, Brun A (2022) Using RVFV as a vector platform for the expression of ruminant disease antigens. Methods Mol Biol 2465:209–225. https://doi.org/10.1007/978-1-0716-2168-4_12

Chapter 3

A Serological Assay for the Detection of Antibodies Generated Against California Serogroup Virus Infection

Alyssa Stulberg, Christina Barry, Alyssia Robinson, Nicole Mueller, Alexander Bello, and Heidi Wood

Abstract

The California serogroup (CSG) viruses are orthobunyaviruses endemic in North America and responsible for the second most common cause of mosquito-borne viral encephalitis in the United States. As the CSG viruses have been neglected and are poorly studied, there are no commercial diagnostic serological assays or reagents available for detection. Therefore, diagnostic laboratories have had to rely on the development of their own in-house serological assays. To develop serological assays, antigenic materials such as recombinant protein, virus-like particles (VLP), or inactivated virus can be used in the IgM antibody capture enzyme-linked immunosorbent assay (MAC-ELISA) to detect CSG virus-specific IgM antibodies. All positive MAC-ELISA results should be confirmed by a well-defined gold standard test method such as the plaque-reduction neutralization test (PRNT).

Key words MAC-ELISA, Virus antigen purification, California serogroup virus, P/N ratio

1 Introduction

CSG viruses belong to the orthobunyavirus genus and include members such as La Crosse virus (LACV), Jamestown Canyon virus (JCV), and snowshoe hare virus (SSHV), with LACV being the most medically significant orthobunyavirus in the United States. The medically important orthobunyaviruses are transmitted during mosquito season, with a reservoir found in mosquitoes and amplifying hosts ranging from small mammals such as rodents, rabbits, and hares to large animals such as deer, depending on the CSG virus. Humans are considered dead-end hosts and there is no evidence to suggest that human-to-human transmission can occur. Following infection, the incubation period is 3–7 days and disease can range from asymptomatic to mild febrile symptoms and

[$]Alyssa Stulberg and Christina Barry have contributed equally to this chapter.

Hani Boshra (ed.), *Bunyaviruses: Methods and Protocols*, Methods in Molecular Biology, vol. 2893,
https://doi.org/10.1007/978-1-0716-4338-9_3,

encephalitis. For LACV infection, up to 1% of patients experiencing acute encephalitis may die. There are no effective treatments or vaccines available for CSG virus infection and the best preventative method is reducing risk for mosquito bites [1, 2].

Diagnostic testing for CSG viruses typically consists of an IgM antibody test followed by a confirmatory PRNT. Molecular assays can also be performed on serum and cerebrospinal fluid (CSF) during acute phase sample collection. Serological testing exhibits a high degree of cross-reactivity between the CSG viruses; therefore, a fourfold higher neutralizing antibody titer must be demonstrated between the various CSG viruses tested to determine the causative agent [3]. Another caveat to serological testing is that IgM antibodies can persist for a long period, in which case the presence of IgM antibodies might be indicative of past infection. Therefore, it is important to collect paired serum samples at least 2–3 weeks apart to distinguish between past or recent infection. Appropriate samples from acutely infected individuals include whole blood, serum, or CSF. For LACV there is a Food and Drug Administration (FDA)-licensed indirect fluorescent antibody test for detection of IgM and IgG antibodies [4]. However, for members of the CSG viruses for which serological assays are not commercially available, the protocol below can be used to generate capture antigen for the MAC-ELISA [5, 6]. The protocol presented below has been adapted from a CDC WNV IgM capture ELISA protocol and is similar to the CDC Zika IgM capture ELISA protocol [7].

2 Materials, Equipment, and Reagents

2.1 Viral Antigen Production

1. CSG stock virus
2. BHK-21 [C-13] cells
3. Corning cell culture flasks, 175 cm^2
4. DMEM, high glucose
5. Fetal bovine serum (FBS), Premium
6. L-Glutamine, 200 mM
7. Penicillin-Streptomycin-Glutamine (100×)
8. DMEM2: DMEM, high glucose, 2% FBS, 1× Penicillin-Streptomycin-Glutamine
9. DMEM5: DMEM, high glucose, 5% FBS, 2 mM L-Glutamine
10. Trypsin-EDTA (0.25%), phenol red
11. DPBS, no calcium, no magnesium
12. BA-1 diluent: 1× Medium 199—Earle's Salts, 0.05 M Tris base buffer pH 7.6, 1% bovine Serum Albumin Fraction V, 1% Penicillin-Streptomycin-Glutamine

13. Sodium bicarbonate 7.5% solution
14. Beta-propiolactone (BPL)
15. Disinfectant (5% Microchem, 1:128 PreEmpt, etc.)
16. 0.45 μM bottle top filter and receiver bottle unit
17. Centricon Plus-70 centrifugal filters, 100 kDa.
18. Conical tubes, 15 and 50 mL
19. Storage Buffer: 1 M Tris buffer, pH 10.5 + 0.85% NaCl +0.002% phenol red
20. Bench-top centrifuge
21. Vacuum pump

2.2 MAC-ELISA

1. Goat Anti-Human IgM antibody
2. California Serogroup Horseradish Peroxidase (HRP) Conjugate 10G5–4/HRP (CDC)
3. SureBlue™ TMB-1 Component Microwell Peroxidase Substrate
4. 96 Well Plate, Non-Treated Surface
5. 2 mL screw cap micro tubes
6. Clear Flat-Bottom Immuno 96-Well Plates
7. 96-well plate lid or sealing film
8. Polyethylene food wrap
9. Skim Milk
10. Polysorbate 20
11. 70% Ethanol
12. 1× PBS, pH 7.2
13. ELISA Carbonate/Bicarbonate Coating Buffer, pH 9.6
14. Wash Buffer: 0.05% Tween-20 in PBS
15. Blocking Buffer: 5% Skim Milk in PBS, 0.5% Tween-20
16. Stop Solution (1 N Sulfuric acid (H_2SO_4))
17. Microplate spectrophotometer @ 450 nm

3 Methods

3.1 Viral Antigen Production

3.1.1 Cell Culture

The BHK-21 [C-13] cell line is ideal for orthobunyavirus propagation. BHK-21 cells are maintained with DMEM5 (*see* **Note 1** on cell maintenance). A mock infected batch of BHK-21 cells should also be prepared as negative antigen control with every batch of viral antigen (*see* **Note 2**).

1. Cells should reach approximately 80% confluency the day of infection in a 175 cm^2 cell culture flask. Usually when BHK-21 cells are approximately 100% confluent the day before infection, an 80% confluency on the infection day can be reached by subculturing the cells at a 1:6 split 24 h before infection. Use 0.25% trypsin to trypsinize the cells.

3.1.2 Infection

1. For infection, calculate the amount of virus required to achieve a Multiplicity of Infection (MOI) of 0.01, using the formula MOI = plaque forming units (PFU)/number of cells. Prepare virus for infection by diluting virus in BA-1 diluent for a total volume of 3 mL per flask. BHK-21 cells at 80% confluency typically represents approximately 1.6×10^7 cells per 175 cm^2 culture flask.
2. Prepare cells for infection by aspirating media and washing monolayer with 7 mL DPBS. Aspirate DPBS.
3. Infect each cell culture flask with 3 mL solution of virus in BA-1 diluent. Incubate culture flasks at 37 °C for 1 h. Rock the flasks every 15–20 min to ensure the cell monolayer is completely covered with virus solution to promote virus adsorption.
4. Following the 1 h incubation, add 27 mL of pre-warmed DMEM2 to bring the final volume up to 30 mL. Incubate the cells at 37 °C for 2–4 days (incubation time depends on virus strain and appearance of cytopathic effect [CPE]). For SSHV, titers peak at approximately 48 h post-infection, whereas JCV peaks at approximately 96 h post-infection.
5. Flasks should be checked twice per day until they reach 80–90% CPE with a microscope. CPE is characterized by destruction of the cell monolayer, with cell rounding, detachment, and clumping. Infected cells should be compared to mock-infected cells as a control for CPE.

3.1.3 Harvest

1. When ideal CPE has been reached, freeze the flasks at −80 °C for at least 1 h to lyse the BHK-21 cells (*see* **Note 3** for extending the freezing time). Thaw flasks at room temperature before continuing with the harvest.
2. Combine flasks from the same infection batch to make a homogeneous mixture of virus prep. Decant mixture into 50 mL conical tubes and clarify by spinning at 4000 rpm for 20 min at 4 °C.
3. Decant supernatant into a sterile 0.45 μM bottle top membrane filter unit with collection bottle and connect to vacuum pump to remove cell debris, crude proteins, and lipids. Turn on vacuum pump to collect filtered virus supernatant in receiver bottle. During supernatant filtration, keep receiver bottle on ice to keep the prep cool.

4. Prepare Centricon Plus-70 filters following the manufacturer's protocols. Briefly, add 70 mL Milli-Q water to the filter and centrifuge at 3500×*g* for 5 min at room temperature (RT) to pre-rinse glycerine residues. Remove water from filter cores just before use by running a recovery spin for 2 min at 1000×*g*, 4 °C.
5. Load virus solution from Subheading 3.1.3, **step 3** to a maximum of 70 mL per Centricon Plus-70 filter. Centrifuge at 3500×*g* for 45 min at 4 °C and remove the flow-through. Recover concentrate by inverting the sample filter cup and placing it on top of the collection cup, then centrifuging at 1000×*g* for 2 min at 4 °C.
6. Mix concentrated virus by pipetting up and down with a micropipette. Transfer concentrated virus to a 15 mL conical tube and record the volume.
7. For every 500 μL of concentrated virus, add 60 μL of Storage Buffer, pH 10.5. Orthobunyaviruse*s* are sensitive to conformational changes at acidic pH, therefore a basic pH of 8.0–9.0 must be maintained [8].
8. Concentrated virus can be stored at 4 °C until inactivation step.

3.1.4 Inactivation

Viral preps can be inactivated with 3 Megarads of gamma irradiation while on dry ice. However, gamma irradiation can negatively affect antigenicity of the viral prep, making it unusable. BPL is the preferred inactivation method for CSG viruses; however, BPL is highly toxic (*see* **Note 4**).

1. Prepare appropriate disinfectant to disinfect contaminated consumables, reagents and media. Place 25 mL water in a 50 mL conical tube, which will later be used to decontaminate pipette tips exposed to BPL.
2. Add 7.5% sodium bicarbonate solution to concentrated virus for a final concentration of 0.2625% to buffer for acidic BPL by-products.
3. For JCV, add BPL to a final concentration of 0.025%. For SSHV, add BPL to a final concentration of 0.1%. Incubate at 4 °C on a shaker plate for 24 h for JCV or 48 h for SSHV.
4. Place tips contaminated with BPL into the 50 mL conical tube containing water for 48 h. This step ensures that BPL is fully hydrolyzed and no longer toxic. After 48 h, place everything into an appropriate disinfectant for 24 h.

3.1.5 Antigen Qualification

A working dilution of viral antigen and CSG-HRP Conjugate (10G5-4/HRP, CDC) should be established by checkerboard qualification, each time new viral antigen and negative antigen stocks are generated or a new conjugate antibody lot is used. A

			Antibody Conjugate											
			1:2000						1:5000					
			1	2	3	4	5	6	7	8	9	10	11	12
SSHV Antigen	1:80	A	SSHV Pos	JCV Pos	Neg	SSHV Pos	JCV Pos	Neg	SSHV Pos	JCV Pos	Neg	SSHV Pos	JCV Pos	Neg
	1:120	B	SSHV Pos	JCV Pos	Neg	SSHV Pos	JCV Pos	Neg	SSHV Pos	JCV Pos	Neg	SSHV Pos	JCV Pos	Neg
	1:250	C	SSHV Pos	JCV Pos	Neg	SSHV Pos	JCV Pos	Neg	SSHV Pos	JCV Pos	Neg	SSHV Pos	JCV Pos	Neg
	1:500	D	SSHV Pos	JCV Pos	Neg	SSHV Pos	JCV Pos	Neg	SSHV Pos	JCV Pos	Neg	SSHV Pos	JCV Pos	Neg
JCV Antigen	1:80	E	SSHV Pos	JCV Pos	Neg	SSHV Pos	JCV Pos	Neg	SSHV Pos	JCV Pos	Neg	SSHV Pos	JCV Pos	Neg
	1:120	F	SSHV Pos	JCV Pos	Neg	SSHV Pos	JCV Pos	Neg	SSHV Pos	JCV Pos	Neg	SSHV Pos	JCV Pos	Neg
	1:250	G	SSHV Pos	JCV Pos	Neg	SSHV Pos	JCV Pos	Neg	SSHV Pos	JCV Pos	Neg	SSHV Pos	JCV Pos	Neg
	1:500	H	SSHV Pos	JCV Pos	Neg	SSHV Pos	JCV Pos	Neg	SSHV Pos	JCV Pos	Neg	SSHV Pos	JCV Pos	Neg

Fig. 1 Example plate layout for antigen/conjugate antibody titration checkerboard qualification. Typically, the antibody conjugate is diluted at 1:2000 and 1:5000 and the antigen is diluted at 1:80, 1:120, 1:250, and 1: 500. The titrations will determine the optimal antigen/antibody conjugate dilution combination for the MAC-ELISA based on previously established positive/negative (P/N) ratios and OD_{450} values for the controls. This figure gives an example of the plate layout for testing SSHV and JCV antigen against the corresponding SSHV, JCV, and negative control antibodies, all performed in duplicate

checkerboard assay will test the antigen at various dilutions against various dilutions of the conjugate antibody using previously established positive and negative control antibody (*see* **Note 5** for selecting positive and negative controls).

1. Prepare positive control (example JCV or SSHV positive antibodies or serum samples), negative control, and negative antigen. Perform a 1 in 400 dilution of each control.
2. Prepare dilutions for conjugate antibody (1:2000 and 1:5000) and antigen (1:80, 1:120, 1:250 and 1:500) and run various combinations for the checkboard qualification (*see* Fig. 1). Methodology for performing the antigen qualification is the same as the MAC-ELISA and can be found in Subheading 3.2.
3. The positive/negative ratios (P/N) can be calculated using the following formula: P/N Ratio = OD_{450} positive control/ OD_{450} negative control against the antigen being tested (*see* **Note 6** for determining acceptable control results based on historical trending data).
4. A background calculation is required against the negative antigen and is calculated using the following formula: background = OD_{450} positive control antigen/OD_{450} negative antigen. Ensure the same positive control is used to test against the virus and the negative antigen. If the background result is <2.0, then the background against the negative antigen is high and the results are invalid.
5. To determine the optimal antigen and antibody conjugate dilution, calculate the mean P/N ratio from historical trending results for the controls and determine the standard deviation (SD).

6. Add the SD to the mean P/N ratio of the positive control to establish the acceptable ranges. The acceptable range of the positive control is the mean historical P/N ratio to mean historical P/N + SD.
7. For the negative control calculate the mean OD_{450} value from historical trending results and determine the SD.
8. Add the SD to the mean OD_{450} of the negative control to establish the target cutoff OD_{450} (negative control mean historical OD_{450} + SD). The acceptable OD_{450} results for the antigen and antibody conjugate qualification should be less than or equal to this target for the negative control.
9. Select an antibody conjugate dilution that gives an acceptable P/N results for your positive controls against the appropriate negative controls (*see* **Note** 7 about selecting a single antibody conjugate dilution for testing against multiple CSG viruses).
10. Perform the antigen/conjugate antibody checkerboard qualification using your CSG positive controls that fall within the acceptable ranges defined in Subheading 3.1.5, **step 6** and the negative control that meets the target cutoff criteria defined in Subheading 3.1.5, **step 8** against the CSG antigen and negative antigen to ensure the following criteria are met at the dilutions selected: CSG virus positive control $P/N > 4.0$ against the appropriate antigen and CSG virus positive control $P/N \leq 4.0$ against the other CSG virus antigens. Ensure there is no background against the negative antigen as defined in Subheading 3.1.5, **step 4**. Test each control at least ten times to establish an expected result for the CSG MAC-ELISA.

3.2 MAC-ELISA

3.2.1 Coating (Day 1)

1. Dilute the capture antibody (AffiniPure™ Goat Anti-Human IgM, $Fc_{5\mu}$) at a 1:2000 dilution by adding 4 μL of antibody into 8 mL of ELISA carbonate/bicarbonate Coating Buffer, pH 9.6 and mix well (*see* **Note 8** regarding qualification of new lots of capture antibody). This is enough to fill each well of a microtiter plate.
2. Add 75 μL of diluted capture antibody to each well of an Immuno 96-well plate. Cover with polyethylene food wrap and store up to 7 days at 4 °C.

3.2.2 Blocking (Day 2)

1. If Blocking Buffer is refrigerated, allow it to reach RT at least 30 min before use (*see* **Note 9** about Blocking Buffer storage). Remove the coated plates from 4 °C and wash plates five times with 300 μL Wash Buffer per well. Blot plates on paper towels to remove excess Wash Buffer.
2. Block the plates with 200 μL Blocking Buffer per well. Incubate at RT for 30 min. Wash plate five times with 300 μL Wash Buffer. Remove Wash Buffer and blot dry before the addition of samples and controls.

3.2.3 Samples and Controls Preparation and Addition

Serum samples and controls are to be prepared to a final dilution of 1:400 in a dilution plate, and cerebral spinal fluid (CSF) can be processed singly as neat.

1. Perform an initial dilution for sample/control by adding 1.5 μL of sample/control to 300 μL Wash Buffer. Mix by pipetting up and down. Transfer 125 μL of 1:200 diluted sample/control to 125 μL Wash Buffer to achieve a final dilution of 1:400.
2. Transfer 50 μL of 1:400 diluted sample/control (or 50 μL neat CSF) to coated well. Cover the plate with a plate sealer and place it in a sealed, humidified plastic container at 37 °C for 1 h.

3.2.4 Antigen Preparation and Addition

Thaw CSG virus antigen and negative antigen before the sample incubation is complete.

1. Remove the assay plate from the 37 °C incubator and wash plates five times with 300 μL Wash Buffer. Blot excess Wash Buffer onto paper towel.
2. Use previously determined CSG and negative antigen dilution from Subheading 3.1.5, **step 10** and use Wash Buffer to dilute antigens.
3. Add 50 μL of diluted antigen to appropriate wells of the coated assay plate.
4. Seal the plate with plastic wrap and place at 4 °C for 10–14 h (overnight).

3.2.5 Conjugate Antibody Addition

Allow Blocking Buffer and SureBlue TMB-1 substrate to reach RT 30 min prior to use.

1. Remove assay plate from 4 °C and wash five times with Wash Buffer. Blot the plate to remove excess Wash Buffer.
2. Use previously determined antibody conjugate dilution from Subheading 3.1.5, **step 10** and use Wash Buffer to dilute antibody conjugate.
3. Add 50 μL of diluted antibody conjugate to each well.
4. Cover the plate with plate sealer and place into a sealed, humidified container at 37 °C for 1 h.

3.2.6 Substrate Addition

1. Remove the assay plate from the 37 °C incubator and wash five times with 300 μL Wash Buffer. Remove excess Wash Buffer by blotting the plate onto paper towel.
2. Add 75 μL SureBlue TMB-1 substrate to each well. Protect from light and incubate at RT for 10 min.
3. Add 50 μL 1 N Sulfuric acid Stop Solution to stop the reaction.
4. Use a microplate spectrophotometer reader to record the OD_{450} values.

Table 1
P/N ratio cut-off values

P/N Ratio	Result
Positive	>4.0
Equivocal	≥3.0<4.0
Negative	<3.0

3.2.7 Interpretation of Results

Calculate the P/N ratios for all samples and controls. The P/N ratios presented in Table 1 are our own pre-determined cutoff values and users must establish their own cutoffs (*see* **Note 10**). For our assay plate to be valid, we determined that the P/N ratio of the positive control must be >4.0. Ensure there is no background against the negative antigen for each positive control based on the background calculation described in Subheading 3.1.5, **step 4**. If the plate is valid, proceed with test result interpretation. It is recommended that paired samples should be submitted, especially for those that yield equivocal results. Positive samples should be confirmed by PRNT.

1. For test sample results, a P/N ratio of >4.0 is positive, P/N ratio between ≥3.0 to <4.0 is equivocal, and a negative result yields a P/N ratio <3.0 for CSG IgM antibodies (*see* Table 1).
2. If a sample is positive, a background evaluation must be performed. See Subheading 3.1.5, **step 4** for background calculation. The sample should have a background value of ≥2.0 to be considered valid.

3.2.8 Limitations

1. IgM can persist in patient serum after initial infection, and the presence of IgM may not be indicative of acute infection but of an exposure at some point.
2. Cross-reactivity may be observed between members of the CSG viruses, therefore a confirmatory PRNT should be performed to confirm the presence of virus-specific neutralizing antibodies.
3. Samples might be collected before the formation of IgM antibodies resulting in a false negative, therefore paired samples should be submitted with the convalescent sample submitted 2–3 weeks following the acute sample.

4 Notes

1. BHK-21 cells require washing with PBS before trypsinization. With trypsinization at RT, BHK-21 cells will begin lifting off the surface after 2 min. BHK-21 cells grow fast and may triple

in numbers under 24 h. Therefore, using a flask that is approximately 100% confluent the day before infection and splitting that at a ratio of 1:6 will result in a confluency of about 70–80%.

2. A batch of BHK-21 cells inoculated with BA-1 (no virus), also known as "mock infected cells," needs to be prepared as a negative antigen to assess background in the CSG MAC-ELISA.
3. The virus culture in flasks can be left frozen at −80 °C for extended periods of time if you wish to process at a later time point.
4. Ensure appropriate respiratory protection training and a respirator fit-testing has been completed before proceeding. Refer to the Material Safety Data Sheet (MSDS) for BPL for additional safety precautions. Refer to your institutional guidelines when working with BPL and working with live RG2 viruses. Include signage indicating that BPL is in use and identify the type of virus you are working with to prevent others from entering the room.
5. Choose controls for the CSG MAC-ELISA based on confirmed PRNT titers. Negative controls used in the CSG MAC-ELISA must show negative results by PRNT. Positive controls used in the CSG MAC-ELISA must show positive results by PRNT. It is important to select positive controls for each CSG virus that are not cross-reactive or show at least a fourfold difference in PRNT titer between CSG viruses. For example, a sample with a 1:20 PRNT titer to JCV and a 1:20 PRNT titer to SSHV would not be a suitable control because of cross-reactivity. An example of a suitable SSHV positive control would be a sample with a Negative result or a 1:20 PRNT titer to JCV and a high PRNT titer to SSHV (1:160–1:640). Likewise, an example of a suitable JCV positive control would be a sample with a Negative result or a 1:20 PRNT titer to SSHV and a high PRNT titer to JCV (1:160–1:640).
6. Historical trending of the OD_{450} values and P/N ratios should be performed for the positive control and historical trending of the OD_{450} values should be performed for the negative controls throughout the history of the method to ensure the assay remains consistent over time.
7. It is important to select one antibody conjugate dilution to test against all CSG antigens generated and to select for a dilution that exhibits lower cross-reactivity.
8. New lots of capture antibody should be quality checked. In our experience, we found no variation between different lots of this antibody.
9. Blocking buffer can be stored at 4 °C for up to 10 days.

10. The positive and negative cutoff values presented in this document were determined in-house and users of this document must establish their own cutoff values. Each lab should perform their own characterization of positive and negative control results based on the selection of positive controls described in **Note 5** and historical trending of the OD_{450nm} results for control testing. Based on the in-house characterization and determination of P/N values, the cut-off criteria for a positive, equivocal and negative result may differ between laboratories. Users must take into consideration seroprevalence in the population being tested and persistent IgM levels.

References

1. Webster D, Dimitrova K, Holloway K et al (2017) California serogroup virus infection associated with encephalitis and cognitive decline, Canada, 2015. Emerg Infect Dis 23: 1423–1424
2. Colebunders RL (2009) Control of communicable diseases manual, 19th edition control of communicable diseases manual, 19th edition edited by David L. Heymann Washington, DC: American Public Health Association, 2008. 746 pp. $45.00 (hardcover). Clin Infect Dis 49:1292–1293
3. Evans AB, Peterson KE (2021) Cross reactivity of neutralizing antibodies to the encephalitic California serogroup orthobunyaviruses varies by virus and genetic relatedness. Sci Rep 11: 16424
4. Jones TF, Erwin PC, Craig AS et al (2000) Serological survey and active surveillance for La Crosse virus infections among children in Tennessee. Clin Infect Dis 31:1284–1287
5. Mincer J, Materniak S, Dimitrova K et al (2021) Jamestown canyon and snowshoe hare virus seroprevalence in New Brunswick. Off J Assoc Med Microbiol Infect Dis Canada 6:213–220
6. Martin DA, Muth DA, Brown T et al (2000) Standardization of immunoglobulin M capture enzyme-linked immunosorbent assays for routine diagnosis of arboviral infections. J Clin Microbiol 38:1823–1826
7. EUA-CDC-ZikaMAC-ifu.pdf. https://www.fda.gov/media/96373/download
8. Hover S, Charlton FW, Hellert J et al (2023) Organisation of the orthobunyavirus tripodal spike and the structural changes induced by low pH and K+ during entry. Nat Commun 14:5885

Chapter 4

Molecular Diagnosis of *Phlebovirus riftense* Infection Using an L-Segment-Based Real-Time RT-PCR Method

Aurélie Pédarrieu and Catherine Cêtre-Sossah

Abstract

Rift Valley fever (RVF) is one of the main vector-borne zoonotic diseases that affects a wide range of ruminants and humans in Africa and the Arabian Peninsula.

Several techniques involving cell culture and molecular biology methods have been developed to diagnose RVF infection. Success partly relies on sending samples to a national or reference laboratory in good conditions and having the capacity to perform the appropriate diagnostic test in the matrix of interest during the period of viremia where high loads of viral particles are present. Conventional and duplex/multiplex real-time reverse transcriptase polymerase chain reaction (RT-PCR) assays are currently the most rapid and sensitive molecular tests for the detection and quantification of Rift Valley fever virus (RVFV) during outbreaks. The one-step real-time duplex RT-PCR technique that is detailed in this chapter is based on the L segment of RVFV. Results must be carefully validated and interpreted, taking into account the results of both positive and negative controls.

Key words Molecular diagnosis, *Phlebovirus riftsense*, Cell culture

1 Introduction

Rift Valley fever is an emerging mosquito-borne disease usually detected in its epizootic form over large areas of an affected country after heavy rainfall events and flooding [1]. The host range is wide with several animal species of livestock infected (caprine, ovine, bovine, cameline), wildlife, and humans [2, 3]. The health, socio-economic and dietary consequences of these epidemics and/or epizootics are direct and indirect, with mass abortion and high mortality rates of neonates in the ruminant population, morbidity in human populations, and the development of food insecurity [4]. Rift Valley fever is caused by *Phlebovirus riftsense*, which belongs to the genus *Phlebovirus* in the family *Phenuiviridae* of the order *Hareavirales* [5]. First identified in the Great Rift Valley in Kenya in 1930 and initially confined to the African continent, it subsequently spreads to Madagascar, the archipelago of Comoros,

Hani Boshra (ed.), *Bunyaviruses: Methods and Protocols*, Methods in Molecular Biology, vol. 2893,
https://doi.org/10.1007/978-1-0716-4338-9_4,

and the Arabian Peninsula [6]. There is a growing concern that Rift Valley fever virus (RVFV) will extend its current range due to the wide variety of mosquito species able to transmit to several mammal hosts. This includes species distributed in countries outside Africa and Arabian Peninsula where RVFV is not yet known to circulate despite the environmental factors driving and favoring its circulation [7–10]. Recent outbreaks of RVF in Madagascar, Mayotte, Niger, Uganda, and Sudan raised international interests in improving diagnostic and vaccine control strategies [11–14].

The earlier the clinical signs of disease are recognized by the farmer and the earlier the clinical diagnosis is confirmed by laboratory tests, along with rapid reporting to the relevant veterinary authorities, the better the disease will be managed and controlled. Success partly relies on sending samples to a regional, national, or reference laboratory that has the capacity to test for the presence of the genome of RVFV in the serum during the acute phase of the disease, i.e., the period of viremia with high loads of viral particles, which lasts 1–6 days after infection. Indeed RVF virus isolation and detection of RVFV-specific IgM antibodies or nucleic acids are conventional methods that can be routinely used [3, 15]. Virus isolation is one of the most reliable source of evidence of infection, considered as a gold standard for establishing the viral aetiology of a disease. Once the virus has been isolated, RT-PCR and sequencing will help confirm the identity of the isolated virus.

Conventional and duplex/multiplex real-time reverse transcriptase polymerase chain reaction (RT-PCR) (2 to 5 fluorescent channels for detection) and loop-mediated isothermal amplification (LAMP) assays, are currently the most rapid and sensitive molecular tests for the detection and quantification of RVFV during outbreaks [16–20] (Table 1).

Next generation sequencing (NGS) approaches have been recently developed but most of these techniques are expensive and require dedicated trained personnel and equipped biosafety level 3 laboratories that are often not available in the areas where the disease occurs [21, 22].

The one-step real-time PCR technique that is detailed in this chapter is based on the L segment of RVFV and beta-actin adapted from Bird et al., 2007 [19] and Toussaint et al., 2007 [23] respectively.

2 Materials

2.1 Equipment

1. Plastic gloves, gowns, and eye protection and, if available, use of a fitted N95 respirator to filter air-borne particles to ensure the biosafety of the laboratory staff, PPE (personal protective equipment) for field collection,

Table 1
Developed RT-PCR and RT-LAMP techniques able to detect *Phlebovirus riftsense*

Method	Segment	Additonal information	References
Conventional RT-nested PCR	S (NSs)	Specific, sensitive tool in the early phase of infection No significant difference compared to virus isolation	[18]
Conventional RT-nested PCR	S, L	Use of a generic RT- nested-PCR, followed by sequencing of the amplified fragment	[24]
One-step real-time RT-PCR	M (G_N)	Use of primers and 5′ nuclease detection probe	[17]
One-step real-time RT-PCR	L	High analytical sensitivity (5 RNA copies of in vitro-transcribed RNA/reaction equivalent to 0.1 PFU of infectious virus/reaction) Use of armored enterovirus RNA as an internal, exogenous control	[19, 25]
One-step real-time RT-PCR	S		[26]
One-step real-time RT-PCR TaqMan	L		[27]
Two-step real-time RT-PCR	S (NSs)	Use of synthetic RNA transcribed in vitro	[16]
Real-time RT-PCR TaqMan	S, L	Sequence-based primer/fluorogenic	[28]
Multiplex real-time RT-PCR	S (NSs), M, L	Use of an exogenous RNA control Overall sensitivity of 92.5%, specificity of 95% DIVA tool (Differentiating Infected from Vaccinated Animals)	[20]
Real-time RT-LAMP (loop-mediated isothermal amplification)	L	Run at a constant temperature (63 °C)	[29]
Real-time RT-LAMP	L	No cross-reactivity with other genetically related and unrelated arboviruses Use in less-well-equipped laboratories in Africa	[30]
RT-LAMP	S	Vertical flow visualization strip	[31]
One-step RT-LAMP	M		[32]
One-step RT-RPA (recombinase polymerase amplification)	S		[33]
Field-deployable reverse transcription insulated isothermal PCR (RT-iiPCR)	M, L		[34]

(continued)

Table 1
(continued)

Method	Segment	Additonal information	References
Multiplex PCR-based macroarray	M	Hybridization of PCR amplicons to probes on a macroarray chip, followed by colorimetric detection	[35]
Multiplex PCR and oligonucleotide microarray	S	Multipathogen and high-throughput method	[36]
High-resolution melting (HRM) multiplex RT-PCR	S	Pan-arbovirus detection assay based on highly multiplexed assays to differentiate the broad diversity of arboviruses	[37]

S stands for Small, M for Medium, NS for Nonstructural

2. Cool box for storage at +4 °C for field collections,
3. Biosafety level 3 laboratory (BSL-3),
4. Microbiological safety hood,
5. Refrigerator set at +4 °C,
6. Freezer set at −80 °C,
7. Thermocycler,
8. Vortex,
9. Pipetman single channel
   (i) P1000 (100–1000 μL)
   (ii) P200 (20–200 μL)
   (iii) P20 (1–20 μL)
   (iv) P10 (1–10 μL)

2.2 Plasticware and Reagents

1. Eppendorf-type microtubes 1.5/2 mL, DNase/RNase free
2. Strips of 8 PCR microtubes (0.2 mL) or 96-well PCR plates (0.2 mL) with their caps
3. Combination of 10, 20, 200, 1000 μL filter tips
4. RNA extraction reagents

   Several types of kits are worldwide commercially available, either based on silica membranes or on magnetic beads with adapted protocols depending on the type of materials used.

   Several examples are given below:

   – RNA extraction kit based on silica membranes (Nucleospin RNA virus from Macherey Nagel/SV total RNA isolation system from Promega/RNA Tissue or Rneasy Mini kit from Qiagen)

Table 2
List of primers and probes used for RT-PCR

Primes and probe names	Sequence (5′-3′)	Amplified gene
2912F	TGAAAATTCCTGAGACACATGG	Segment L
2981R	ACTTCCTTGCATCATCTGATG	Segment L
2950P-FAM	FAM-CAATGTAAGGGGCCTGTGTGGACTTGTG-BHQ1	Segment L
ACT-F 1005	CAGCACAATGAAGATCAAGATCATC	Exon 4–5 β-actin
ACT-R 1135	CGGACTCATC GTACTCCTG CTT	Exon 4–5 β-actin
ACT-P 1081-VIC	VIC-TCGCTGTCCACCTTCCAGCAGATGT-BHQ1	Exon 4–5 β-actin

- RNA extraction automated kits based on magnetic beads (RNA/DNA) (ID Gene Mag Universal Extraction Kit from ID IndiMag Pathogen kit from Indical using the robot KingFisher (Thermofisher))
- RNAse-free H_2O

5. RT-PCR reagents
   - AgPath One-step RT-PCR kit (reference: 4387424, Ambion, ThermoFisher)
   - Primers and probes at a concentration of 100 μM (Table 2)
   - Three RT-PCR controls: (i) a positive extraction control: RVFV Internal Reference Material-Method (IRM-Method) consisting of extracted RNA from RVFV culture supernatant prepared at a concentration corresponding to a correct amplification of the target gene with a known expected Ct value, (ii) a negative PCR control with H_2O RNAse-free abbreviated NTC (H_2O; NTC, Negative Template Control) to test for the absence of contamination during the PCR reaction, (iii) a negative extraction control with H_2O RNAse-free, Cext (H_2O C ext., External Control), to ensure the absence of contamination during the various stages of the method, from the RNA extraction step to the real-time RT-PCR one.

2.3 Types of Samples, Storage Conditions, and Preparation of the Samples

Because of the high-titer viremia after *Phlebovirus riftsense* infection in animals, several specimens can be used for diagnostic testing such as serum, whole blood, uterine and vaginal excretion fluids, and tissues including those from aborted fetuses (spleen, liver, brain). Immediate transport in appropriate containers of the specimen to the laboratory (under controlled temperature, fresh on ice/4 °C or frozen at −80 °C in liquid nitrogen) has to be considered as a key step for the accurate diagnosis of the virus from clinical specimens, as viruses are described as heat labile. Specimens must be collected during the early stage when the clinical signs are acute, before fever subsides, as viremia is only transient. In the case of live animals, specimens should include heparinized blood and serum, while the tissues of choice from dead animals or aborted fetuses are the liver, spleen, and brain [1], as they are the main sites of viral replication.

2.3.1 In the Case of Live Animals

1. Wear appropriate PPE to ensure biosafety (gloves, a lab coat, eye protection, fitted N95 respirator to filter air-borne particles).
2. From the jugular vein collect 5 mL of whole blood with anticoagulant (preferably ethylene diamine tetra-acetic acid (EDTA)) or without anticoagulant (in the case of serum) during the febrile stage of the disease in Vacutainer tubes (5/10 mL tubes). After collection, the samples should be kept in sealed containers. The surfaces of all specimen containers should be decontaminated with 10% sodium hypochlorite at 0–4 °C if the specimens are transported to the laboratory in less than 48 h. If transport to the lab requires more than 48 h, the samples should be frozen at −80 °C.

2.3.2 In the Case of Postmortem Samples

1. Wear appropriate PPE to ensure biosafety (gloves, a lab coat, eye protection, and fitted N95 respirator to filter air-borne particles).
2. Collect several aliquots of approximately 1 cm^3 of liver, spleen, brain, or abortion products. After collection, the samples should be kept in sealed containers. The surfaces of all specimen containers should be decontaminated with 10% sodium hypochlorite at 0–4 °C if the specimens are transported to the laboratory in less than 48 h. If transport to the lab requires more than 48 h, the samples should be frozen at −80 °C.
3. Prepare and label a sterile Eppendorf tube containing 1 mL of MEM medium per piece of organ to be treated with the addition of two beads (diameter 2 mm) to each Eppendorf tube;
4. Add one piece of each organ to one of the previously prepared 1.5 mL Eppendorf microtubes;
5. Vortex twice for 2 min until the piece is completely dissolved, keep the sample on ice until the following stage;
6. Centrifuge at 200×*g* for 10 min at 4 °C;
7. Pipette the supernatant and keep it at exactly 4 °C; this is the starting material for further RNA extraction

3 Methods

Phlebovirus riftsense can be detected preferably from serum collected without anticoagulant or plasma/blood collected with anticoagulant, during the febrile stage of the disease in live animals, or from liver, spleen and brain of animals that have died, or from aborted fetuses. This procedure describes the two steps (step 1: RNA extraction followed by step 2: RVFV-specific amplification by PCR) for the detection of the L segment (Large segment) of the RVF virus, using real-time PCR in duplex format (simultaneous amplification in a single reaction of the target gene and the β-actin gene (constitutive gene), from serum (field samples).

The genome of the RVFV is made up of tri-segmented RNA (three segments L: large; M: medium; S: small). As illustrated in Fig. 1 below, the L segment codes for the RNA-dependent RNA polymerase, the M segment for two surface glycoproteins GN and GC and for the nonstructural protein (NSm), and the S segment for the nucleoprotein (N) and for a nonstructural protein (NSs).

3.1 RNA Extraction

Several kits are commercially available. Manufacturer's protocol has to be followed. Eluted RNA has to be kept at 4 °C if RT-PCR is performed the day of the RNA extraction, or kept at −80 °C if RT-PCR is postponed.

3.2 RT-PCR

3.2.1 Preparation of Working Solution of Primers and Probes

Probes are photosensitive, so they must be consequently protected from light and their exposure time must be minimized. Each primer and probe stock solution are delivered at a concentration of 100 μM, and this must be checked on each tube. On receipt, the primer stock solution is aliquoted to 20 μL, the probe solution is aliquoted to 2 μL and stored at <−16 °C.

1. Prepare the working solution of 10 μM of each primer (2912F, 2981R, ACT-F1005, ACT-R1135) and 5 μM of each probe (2950 RVF, 1081 ACT) following Table 3 below;

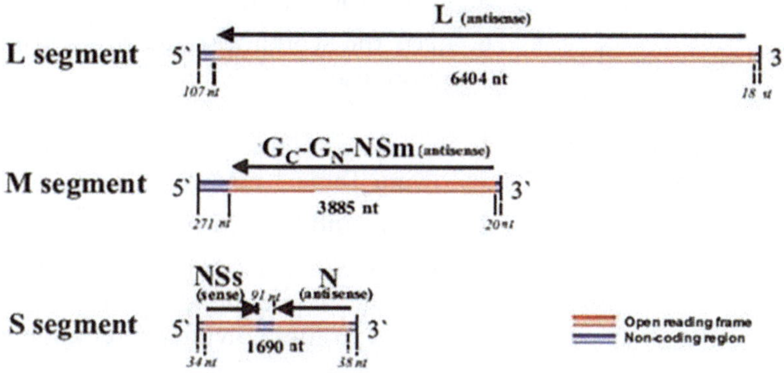

Fig. 1 Schematic representation of the RVF virus genome (MP12 strain) [38]

Table 3
Preparation of the working solutions of primers and probes

Primer/probe name-final concentration	H_2O (μL), x	Stock solution 100 μM (μL), y	Final volume to get the final concentration
Primer 2912 F-10 mM	180	20	200
Primer 2981 R-10 mM	180	20	200
Primer ACT 1005 F-10 mM	180	20	200
Primer ACT 1135 R-10 mM	180	20	200
Probe 2950 RVF-5 mM	38	2	40
Probe 1081 ACT –5 mM	38	2	40

Table 4
Preparation of the RT-PCR Mix

Components of the master mix	Volume for 1 reaction (μL)
2× Mix AgPath	10
25× RT-PCR	0.8
2912 F 10 μM	1
2981 R 10 μM	1
ACT 1005 F 10 μM	1
ACT 1135 R 10 μM	1
Probe 2950 RVF P 5 μM	0.75
Probe 1081 ACT P 5 μM	0.75
H_2O	3.7
Total	20

2. Prepare and label a sterile 0.5- or 1.5-mL Eppendorf tube;
3. Add the x volume of H_2O to each of the tube;
4. Add the y volume of stock of primers or probe to each of the appropriate Eppendorf tube;
5. Vortex for 1 min;
6. Keep the working solutions of primers/probes on ice/4 °C before use.

3.2.2 Preparation of the RT-PCR Enzyme Mix

1. Prepare a volume of mix equivalent to X reactions +10% following Table 4 below;
2. Prepare and label "Mix" a sterile 1.5-mL Eppendorf tube;
3. Add the volume of H_2O and then add the volume of each of the other components starting with the next coming bigger volume, using individual filtered tip between each component.
4. Vortex for 1 min;

5. Keep the mix solution on ice/4 °C;
6. Prepare and label the number of individual tube, strips of 8 microtubes or a 96-well PCR microplate needed;
7. Keep the working solutions of primers/probes on ice/4 °C before use;
8. Distribute the amount of 20 μL in each PCR tube;
9. Add 5 μL of RNA (sample or control) starting from the two negative controls, and ending with the positive control;
10. Close tubes with appropriate caps starting from the two negative controls and ending with the positive control;
11. Centrifuge briefly to remove any bubbles and ensure thorough homogenization;
12. Switch ON the computer and the thermocycler;
13. Run the following program:

	Temperature	Time	Number of cycles
PCR	45 °C	10 min	1 cycle
	95 °C	10 min	1 cycle
	95 °C	15 sec	40 cycles
	60 °C	1 min[a]	

[a]Measurement of the fluorescence

14. Place the tube/strip/microplate in the thermocycler;
15. Start the run;
16. Tested samples are stored at <−65 °C.

3.2.3 Validation of the Assay

1. For each of the three controls, individually check the amplification curve using the thermal cycler software and the Ct value displayed.
2. The assay is validated if the following decision rules are met (Table 5):

Table 5
Validation of the positive and negative controls

		β actin gene—VIC		Target gene—FAM	
Controls	Name	Ct values	Amplification curve	Ct values	Amplification curve
Positive	IRM-method	a < IRM-method VIC < b	Characteristic	c < IRM-method FAM < d	Characteristic
Negative	NTC	No Ct	Not characteristic	No Ct	Not characteristic
	Cext	No Ct	Not characteristic	No Ct	Not characteristic

(a) The positive extraction control (IRM method) has the status "positive" with an average Ct value within the expected range in β actin (VIC) and in the target gene (FAM) that each laboratory defines depending upon its IRM with confidence intervals (a, b for IRM-Method VIC and c, d for IRM-method FAM) of their positive controls;

- a < IRM-method VIC < b
- c < MRI-method FAM < d

(b) The negative PCR control (NTC) must be "negative" for the target and β actin genes;

(c) The negative extraction control (Cext) must be "negative" for the target and β actin genes

3.2.4 Interpretation of the Data

The values obtained for the samples are interpretable only if the expected values for the controls are obtained (Fig. 2).

1. For each sample, individually check the amplification curve given by the thermal cycler software and the Ct value displayed, and interpret the "positive, negative, or inconclusive" status.

   Several statuses are defined:

- Positive: the sample shows a fluorescence curve characteristic of an exponential amplification phase which crosses the fluorescence threshold (=threshold) for the target gene and the β actin gene. A Ct value corresponding to the crossing of the y-axis and the line corresponding to the detection threshold is given. This value for the target gene must be below the threshold value of 40 Ct, with a β actin value below the threshold value of 40 Ct. In that case, « the detection of RVFV genome is positive. »
- Negative: three possibilities:
  - either the sample shows no amplification curve for the target gene (No Ct), regardless of the β actin value obtained;
  - or the sample shows an amplification curve for the target gene with a Ct value greater than 40 Ct or No Ct, with a β actin value below the threshold value of 40 Ct;
  - or the sample shows a correct amplification curve for the target gene, but with a β actin value outside >40 Ct.

"No Ct" is then displayed for the sample. In that case, "the detection of the RVFV genome is negative."

- Inconclusive (*see* **Note 1**): the sample has a Ct value for the target gene in FAM that is <40 Ct but the fluorescence curve is not characteristic of an exponential amplification

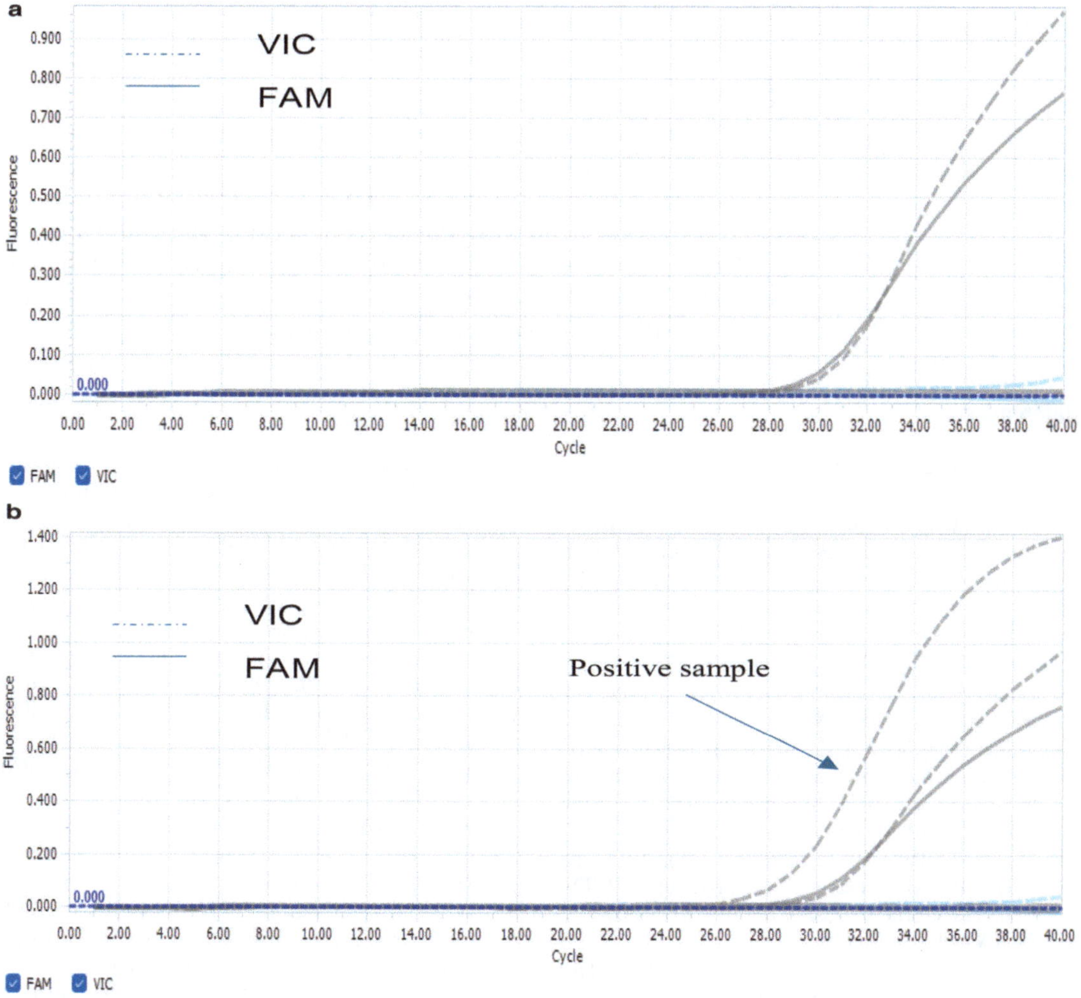

Fig. 2 RVFV-specific RTqPCR amplification curves. (**a**) Validation of the assay (negative and positive controls). (**b**). Example of a positive serum sample

	β actin gene—VIC		Target gene—FAM	
Status	**Ct values**	**Amplification curve**	**Ct values**	**Amplification curve**
Positive	<40 Ct	Characteristic	<40 Ct	Characteristic
Negative	<40 Ct	Characteristic	No Ct	Not characteristic
	<40 Ct	Characteristic	>40 Ct	Characteristic
	>40 Ct	Characteristic	<40 Ct	Characteristic
Inconclusive	No Ct	Not characteristic	<40 Ct	Not characteristic

4 Notes

1. Procedure for inconclusive samples: Dilute the sample RNA to 1:10 and repeat the PCR using the same technique based on this 1:10 dilution.

   If the diluted inconclusive RNA is detected or not detected in the target gene with a detected β actin gene result (z +/-zz Ct), the sample status will be returned detected. If the diluted inconclusive RNA is undetected with an undetected β actin result, the sample status is still inconclusive.

References

1. Gerdes T (2004) Rift Valley fever. Rev Sci Tech Off Int Epiz 23:613–623
2. Javelle E, Lesueur A, Pommier de Santi V (2020) The challenging management of Rift Valley fever in humans: literature review of the clinical disease and algorithm proposal. Ann Clin Microbiol Antimicrob 19:4
3. Bird BH, Ksiazek TG, Nichol ST, MacLachlan NJ (2009) Rift Valley fever virus. J Am Vet Med Assoc 234:883–893
4. Wanyoike F, Rich KM (2010) An assessment of the regional and national socio-economic impacts of the 2007 Rift Valley fever outbreak in Kenya. Am J Trop Med Hyg 83:52–57
5. Sasaya T, Palacios G, Briese T, Di Serio F, Groschup MH, Neriya Y et al (2023) ICTV virus taxonomy profile: Phenuiviridae 2023. J Gen Virol 104:001893
6. Linthicum KJ, Britch SC, Anyamba A (2016) Rift Valley fever: an emerging mosquito-borne disease. Annu Rev Entomol 61:395–415
7. Moutailler S, Krida G, Schaffner F, Vazeille M, Failloux AB (2008) Potential vectors of Rift Valley fever virus in the Mediterranean region. Vector Borne Zoonotic Dis 8:749–753
8. Turell MJ, Wilson WC, Bennett KE (2010) Potential for North American mosquitoes (Diptera: Culicidae) to transmit Rift Valley fever virus. J Med Entomol 47:884–889
9. Brustolin M, Talavera S, Nuñez A, Santamaria C, Rivas R, Pujol N et al (2017) Rift Valley fever virus and European mosquitoes: vector competence of *Culex pipiens* and *Stegomyia albopicta* (=Aedes albopictus). Med Vet Entomol 31:365–372
10. Tantely LM, Boyer S, Fontenille D (2015) A review of mosquitoes associated with Rift Valley fever virus in Madagascar. Am J Trop Med Hyg 92:722–729
11. Baudin M, Jumaa AM, Jomma HJE, Karsany MS, Bucht G et al (2016) Association of Rift Valley fever virus infection with miscarriage in Sudanese women: a cross-sectional study. Lancet Glob Health 4:e864–e871
12. De St Maurice A, Nyakarahuka L, Purpura L, Ervin E, Tumusiime A, Balinandi S et al (2016) Notes from the field: Rift Valley fever response—Kabale District, Uganda, March 2016. MMWR Morb Mortal Wkly Rep 65:1200–1201
13. Tambo E, Olalubi OA, Sacko M (2016) Rift Valley fever epidemic in Niger near border with Mali. Lancet Infect Dis 16:1319–1320
14. Promed Archive Number: 20190222.6331808
15. WOAH terrestrial manual 2023. Chapter 3.1.10. Rift Valley fever. https://www.woah.org/fileadmin/Home/eng/Health_standards/tahm/3.01.19_RVF.pdf
16. Garcia S, Crance JM, Billecocq A, Peinnequin A, Jouan A, Bouloy M et al (2001) Quantitative real-time PCR detection of Rift Valley fever virus and its application to evaluation of antiviral compounds. J Clin Microbiol 39:4456–4461
17. Drosten C, Gottig S, Schilling S, Asper M, Panning M, Schmitz H et al (2002) Rapid detection and quantification of RNA of Ebola and Marburg viruses, Lassa virus, Crimean-Congo hemorrhagic fever virus, Rift Valley fever virus, dengue virus, and yellow fever virus by real-time reverse transcription-PCR. J Clin Microbiol 40:2323–2330
18. Sall AA, Macondo EA, Sène OK, Diagne M, Sylla R, Mondo M et al (2002) Use of reverse transcriptase PCR in early diagnosis of Rift Valley fever. Clin Diagn Lab Immunol 9:713–715

19. Bird BH, Bawiec DA, Ksiazek TG, Shoemaker TR, Nichol ST (2007) Highly sensitive and broadly reactive quantitative reverse transcription-PCR assay for high-throughput detection of Rift Valley fever virus. J Clin Microbiol 45:3506–3513
20. Wilson WC, Romito M, Jasperson DC, Weingartl H, Binepal YS, Maluleke MR et al (2013) Development of a Rift Valley fever real-time RT-PCR assay that can detect all three genome segments. J Vir Methods 193:426–431
21. Brinkmann A, Ergunay K, Radonić A, Kocak Tufan Z, Domingo C, Nitsche A et al (2017) Development and preliminary evaluation of a multiplexed amplification and next generation sequencing method for viral hemorrhagic fever diagnostics. PLoS Negl Trop Dis 11:e0006075
22. Schulz A, Sadeghi B, Stoek F, King J, Fischer K, Pohlmann A et al (2022) Whole-genome sequencing of six neglected arboviruses circulating in Africa using sequence-independent single primer amplification (SISPA) and MinION nanopore technologies. Pathogens 11(12):1502
23. Toussaint JF, Sailleau C, Breard E, Zientara S, De Clercq K (2007) Bluetongue virus detection by two real-time RT-qPCRs targeting two different genomic segments. J Virol Methods 140:115–123
24. Sánchez-Seco MP, Echevarría JM, Hernández L, Estévez D, Navarro-Marí JM, Tenorio A (2003) Detection and identification of Toscana and other phleboviruses by RT-nested-PCR assays with degenerated primers. J Med Virol 71:140–149
25. Drolet BS, Weingartl HM, Jiang J, Neufeld J, Marszal P, Lindsay R et al (2012) Development and evaluation of one-step rRT-PCR and immunohistochemical methods for detection of Rift Valley fever virus in biosafety level 2 diagnostic laboratories. J Virol Methods 179:373–382
26. Weidmann M, Sanchez-Seco MP, Sall AA, Ly PO, Thiongane Y, Lô MM et al (2008) Rapid detection of important human pathogenic Phleboviruses. J Clin Virol 41:138–142
27. Busquets N, Xavier F, Martín-Folgar R, Lorenzo G, Galindo-Cardiel I, Pérez del Val B et al (2010) Experimental infection of young adult European breed sheep with Rift Valley fever virus field isolates. Vector Borne Zoonotic Dis 10:689–696
28. Mwaengo D, Lorenzo G, Iglesias J, Warigia M, Sang R, Bishop RP et al (2012) Detection and identification of Rift Valley fever virus in mosquito vectors by quantitative real-time PCR. Virus Res 169:137–143
29. Peyrefitte CN, Boubis L, Coudrier D, Bouloy M, Grandadam M, Tolou HJ et al (2008) Real-time reverse-transcription loop-mediated isothermal amplification for rapid detection of Rift Valley fever virus. J Clin Microbiol 46:3653–3659
30. Le Roux CA, Kubo T, Grobbelaar AA, van Vuren PJ, Weyer J, Nel LH et al (2009) Development and evaluation of a real-time reverse transcription-loop-mediated isothermal amplification assay for rapid detection of Rift Valley fever virus in clinical specimens. J Clin Microbiol 47:645–651
31. Han Q, Zhang S, Liu D, Yan F, Wang H, Huang P et al (2020) Development of a visible reverse transcription-loop-mediated isothermal amplification assay for the detection of Rift Valley fever virus. Front Microbiol 11:590732
32. Wekesa F, Wamalwa M, Oduor R, Binepal Y, Ateya L, Okumu N et al (2023) Development and validation of rapid colorimetric reverse transcription loop-mediated isothermal amplification for detection of Rift Valley fever virus. Adv Virol 2023:1863980
33. Euler M, Wang Y, Nentwich O, Piepenburg O, Hufert FT, Weidmann M (2012) Recombinase polymerase amplification assay for rapid detection of Rift Valley fever virus. J Clin Virol 54:308–312
34. Trujillo JD, Wilson WC, Craig A, Van den Bergh C, Wang T, Thompson P et al (2024) Rift Valley fever virus M and L genome segment detection: a comparison of field-deployable reverse transcription insulated isothermal PCR (RT-iiPCR) and laboratory-based multiplex reverse transcription real-time PCR. J Clin Microbiol 62:e0043023
35. Venter M, Zaayman D, van Niekerk S, Stivaktas V, Goolab S, Weyer J et al (2014) Macroarray assay for differential diagnosis of meningoencephalitis in southern Africa. J Clin Virol 60:50–56
36. Yao W, Yang Z, Lou X, Mao H, Yan H, Zhang Y (2021) Simultaneous detection of Ebola virus and pathogens associated with hemorrhagic fever by an oligonucleotide microarray. Front Microbiol 12:713372
37. Villinger J, Mbaya MK, Ouso D, Kipanga PN, Lutomiah J, Masiga DK (2017) Arbovirus and insect-specific virus discovery in Kenya by novel six genera multiplex high-resolution melting analysis. Mol Ecol Resour 17:466–480
38. Flick R, Bouloy M (2005) Rift Valley fever virus. Curr Mol Med 5:827–834

Chapter 5

Ultracentrifugation Purification of Cache Valley Virus Using Iodixanol

Akeel Baig

Abstract

Ultracentrifugation is an invaluable technique frequently used for either crude or highly purified viral preparations. Virus produced by cells is first cleared of cellular debris through benchtop centrifugation, then purified by ultracentrifugation in a stepwise iodixanol gradient solution. Here we describe the purification of Cache Valley virus using iodixanol, a nonionic, nontoxic, iso-osmotic solution.

Key words Cache Valley virus, Ultracentrifugation, Iodixanol, Optiprep

1 Introduction

Cache Valley virus (CVV) is a negative-sense, single-stranded RNA virus with a genome consisting of three RNA segments [1]. CVV is an arthropod-borne bunyavirus classified within the class *Bunyavir-icetes*, genus *Orthobunyavirus*, and Bunyamwera group [2]. Figure 1 shows a schematic of the virus (A) along with a depiction of the segmented genome consisting of large (RNA-dependent RNA polymerase), medium (two glycoproteins: Gn, Gc), and small segments (NSs and N) (B). Unlike a majority of other *Bunyaviricetes* members, serologic studies demonstrate that vector-borne CVV also infect larger mammals, with virus being isolated from horses, cattle, and sheep. CVV has been shown to cause reproductive losses in small ruminants [3–6]. The transmission cycle of Bunyaviruses may involve a number of vertebrate hosts, with a propensity for mosquito vectors, including several species of *Aedes*. Specifically, CVV has been isolated from various species of the *Anopheles* or *Culicine* genera of mosquitoes, including *Anopheles quadrimaculatus* and *Culiseta inornate* [7]. Most prominently, CVV is of concern in animal husbandry due to its ability to cause abortion, stillbirth, and congenital defects of the musculoskeletal and central nervous systems in affected ruminants [3, 8].

Hani Boshra (ed.), *Bunyaviruses: Methods and Protocols*, Methods in Molecular Biology, vol. 2893,
https://doi.org/10.1007/978-1-0716-4338-9_5,

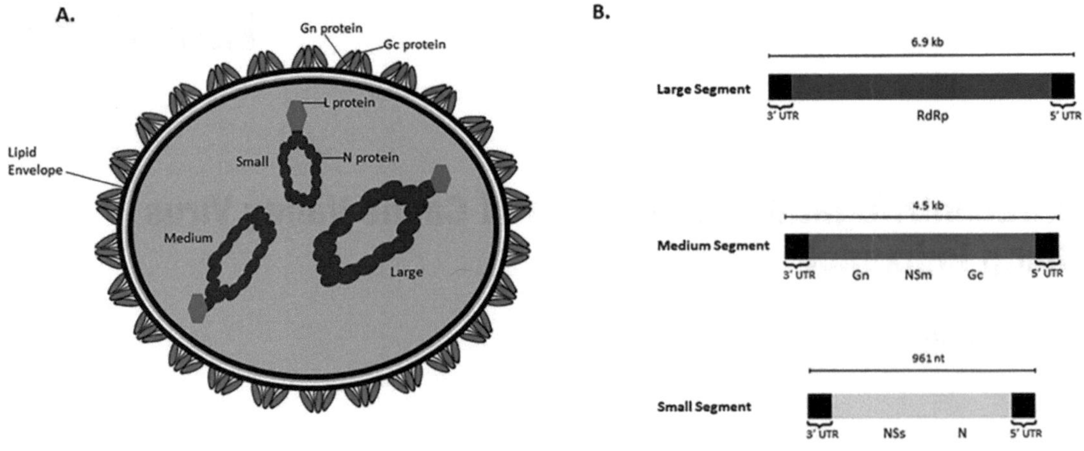

Fig. 1 (**a**) Schematic depiction of Cache Valley virus structure. (**b**) Schematic depiction of Cache Valley virus negative sense RNA genome consisting of three segments, large, medium, and small. Large segment encodes a single RNA-dependent RNA polymerase gene. Medium segment encodes two glycoproteins (Gn and GC) along with a nonstructural protein NSm. The small segment encodes the two proteins NSs and N

Common ultracentrifugation density gradient compounds used for virus purification are CsCl, sucrose, and iodixanol. The latter is an iodinated density gradient compound that is nontoxic to biological material [9, 10]. It is nonionic, water soluble, iso-osmotic at all concentrations and available as a 60% (w/v) solution with a density of 1.32 g/mL [11].

Iodixanol has been reported to increased transduction, both in vitro and in vivo, when compared with CsCl gradient-purified vectors [12]. Due to the low viscosity and iso-osmotic properties, iodixanol solutions are found to preserve virus functionality and particle integrity [13]. In addition, compared with CsCl, the low toxicity allows for the direct use of purified viruses in vitro and in vivo without removal of iodixanol [13].

Sucrose can also prove problematic as gradient compound when purifying viruses as it has been shown to cause loss of surface glycoproteins in enveloped viruses [14]. In addition, Kol et al. demonstrated that iodixanol preserved mechanical and morphological properties compared with sucrose [15].

Iodixanol, commercially available as OptiPrep™ has been used to purify various viruses from the class *Bunyaviricetes*, specifically, Bunyamwera virus [16, 17], Rift Valley Fever virus [18–20], Hantavirus [21–24], Hazara virus [25], and Crimean-Congo Hemorrhagic Fever virus [26]. Currently, there are no reports of Cache Valley virus purification using iodixanol. Here we describe a method for Cache Valley virus purification using OptiPrep™ and ultracentrifugation.

2 Materials

2.1 Cell Culture

1. A549 cells.
2. Dulbecco's Modified Eagles's Medium (DMEM) complete media containing 10% Fetal Bovine Serum (FBS) and 1% L-glutamine.
3. Complete Opti-MEM media with 10% KnockOut Serum Replacement and 1%L-glutamine.

2.2 Virus Purification

Required solutions:

1. Solution A: OptiPrep™.
2. Solution B, diluent: 0.02 M Tris-HCl, 0.2 M NaCl, 2 mM EDTA, pH 7.4.
3. Solution C, working solution (30% w/v iodixanol): Mix 1:1 equal volumes of Solution A (OptiPrep™) and Solution B.
4. Solution D, buffer (50% v/v Solution B): 0.01 M Tris-HCl, 0.1 M NaCl, 1 mM EDTA, pH 7.4.

   Protease inhibitors in Solutions B and D optional.
5. Rotor requirement: Swinging-bucket rotor with approx 38 mL (e.g. Beckman SW28).

3 Methods

3.1 Cell Culture

1. Culture A549 cells in DMEM complete media until total cell count reaches 2.5×10^7 (*see* **Notes 1** and **2**).
2. Following the subsequent split, culture cells in complete Opti-MEM media until a total cell count of 2×10^8 is reached.

3.2 Viral Infection and Cytopathetic Effect

1. Seed 1×10^7 A549 cells on 20×150 mm plates in 10 mL complete Opti-MEM media and incubate for 18 h at 37 °C in 5% CO_2.
2. Infect cells with a 1:100 dilution of 4.5×10^6 pfu/mL preparation of CVV.
3. Cytopathetic effect is clearly visible after 48 h.

3.3 Supernatant Clearance

1. Pool supernatant from all 20 plates into 4×50 mL siliconized screw cap tubes (*see* **Note 3**).
2. Centrifuged tubes at $3300{\times}g$ for 30 min at 4 °C to pellet cellular debris.
3. Transfer cleared supernatants to fresh 4×50 mL siliconized screw cap tubes.

3.4 Ultracentrifugation Cushion

1. Partially purify the virus by sedimentation through a 20% (w/v) iodixanol cushion. Prepare the cushion by diluting Solution C with Solution D for a total volume of 240 mL (*see* Subheading 2.2).
2. Place 19 mL of 20% cushion in each of 6 × 38 mL ultracentrifuge tubes.
3. Add 19 mL of cleared supernatants on top of cushion for all six tubes from **step 2**.
4. Ultracentrifuge the cleared supernatants at 60,000×*g* for 1.5 h at 4 °C.
5. Repeat **steps 2–4** to partially purify the entire 200 mL volume of cleared supernatant.
6. Resuspend virus containing pellets in Solution D (*see* Subheading 2.2) by adding 3 mL to one tube, resuspending pellet, then transferring the resuspended pellet into a second tube, and repeating the process for all 12 pellet containing tubes.

3.5 Virus Ultracentrifugation

1. During **step 2**, Subheading 3.4 prepare 50 mL each of iodixanol solutions of 10%, 15%, 20%, 25%, and 30% (w/v) by diluting Solution C with Solution D (*see* Subheading 2.2).
2. In each ultracentrifugation tube for the 38 mL swinging-bucket rotor, create a discontinuous gradient by adding 7.5 mL of 30% iodixanol, then carefully pipette 7.5 mL 25% iodixanol on top of the 30%.
3. Repeat this layering until the 10% concentration is finally added. Figure 2a shows a schematic of the discontinuous gradient.
4. Repeat for all six tubes.
5. Top-load 0.5 mL of resuspended viral pellet to each of the 6 × 38 mL tubes.
6. Centrifuge the gradients at 250,000×*g* for 1.5 h at 4 °C. Use a slow deceleration program.
7. The virus band is in the 25% iodixanol fraction. Figure 2b shows the virus band found within the 30% iodixanol concentration. Extract the virus band using a syringe.
8. The final virus concentration will be ~2.7 × 10^7 pfu/mL.

4 Notes

1. A complete cell culture description is beyond the scope of this chapter. Only the key points are mentioned here.

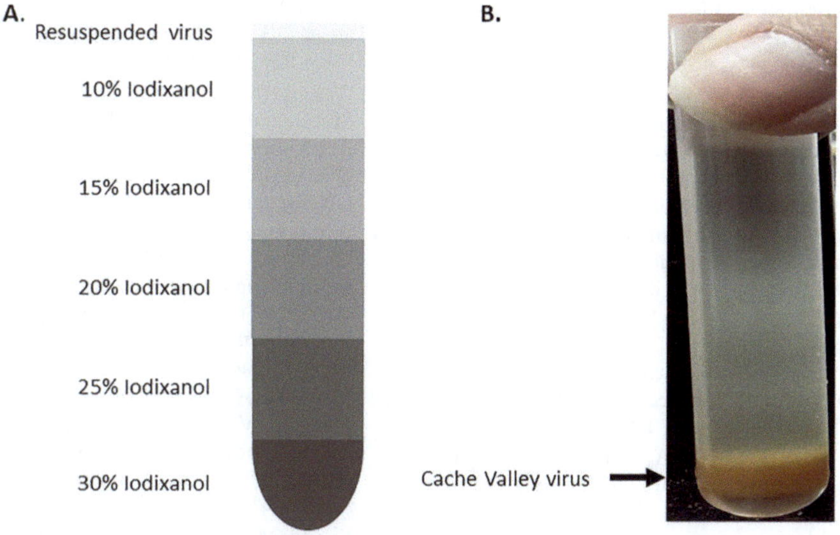

Fig. 2 (**a**) Illustrated representation of iodixanol gradient. (**b**) Representative image of purified Cache Valley virus. The virus band is found within the 30% iodixanol concentration

2. Although A549 cells were infected for virus propagation, other cell lines are also possible. However, the time to CPE may vary.
3. Cells are not lysed to avoid contaminating the supernatants with intracellular and viral proteins.

Acknowledgments

Teresa Cabral and Areeb Khan for technical assistance.

References

1. Edwards JF (1994) Cache Valley virus. Vet Clin N Am Food Anim Pract 10(3):515–524
2. Holden P, Hess AD (1959) Cache Valley virus, a previously undescribed mosquito-borne agent. Science (1979) 130(3383):1187–1188
3. Chung SI, Livingston CW, Edwards JF, Gauer BB, Collisson EW (1990) Congenital malformations in sheep resulting from in utero inoculation of Cache Valley virus. Am J Vet Res 51(10):1645–1648
4. Chung SI, Livingston CW, Jones CW, Collisson EW (1991) Cache Valley virus infection in Texas sheep flocks. J Am Vet Med Assoc 199(3):337–340
5. Edwards JF, Angulo AB, Pannill EC (2003) Theriogenology question of the month. J Am Vet Med Assoc 222(10):1361–1362
6. Edwards JF, Livingston CW, Chung SI, Collisson EC (1989) Ovine arthrogryposis and central nervous system malformations associated with in utero Cache Valley virus infection: spontaneous disease. Vet Pathol 26(1):33–39
7. Calisher CH, Francy DB, Smith GC, Muth DJ, Lazuick JS et al (1986) Distribution of Bunyamwera serogroup viruses in North America, 1956–1984. Am J Trop Med Hyg 35(2):429–443
8. Crandell RA, Livingston CW, Shelton MJ (1989) Laboratory investigation of a naturally occurring outbreak of arthrogryposis-hydranencephaly in Texas sheep. J Vet Diagn Invest 1(1):62–65
9. Svaland MG, Haider T, Langseth-Manrique K, Andrew E, Hals PA (1992) Human

pharmacokinetics of iodixanol. Investig Radiol 27(2):130–133
10. Nossen JØ, Aakhus T, Berg KJ, Jorgensen NP, Andrew E (1990) Experience with iodixanol, a new nonionic dimeric contrast medium preliminary results from the human phase I study. Investig Radiol 25:S113–S114
11. Ford T, Graham J, Rickwood D (1994) Iodixanol: a nonionic iso-osmotic centrifugation medium for the formation of self-generated gradients. Anal Biochem 220:360–366
12. Lock M, Alvira M, Vandenberghe LH, Samanta A, Toelen J et al (2010) Rapid, simple, and versatile manufacturing of recombinant adeno-associated viral vectors at scale. Hum Gene Ther 21(10):1259–1271
13. Segura MM, Kamen AA, Garnier A (2011) Overview of current scalable methods for purification of viral vectors. Methods Mol Biol 737:89–116
14. Palker TJ (1990) Mapping of epitopes on human T-cell leukemia virus type 1 envelope glycoprotein. In: Blattner WA (ed) Human retrovirology: HTLV. Raven Press, New York, pp 435–445
15. Kol N, Tsvitov M, Hevroni L, Wolf SG, Pang H-B et al (2010) The effect of purification method on the completeness of the immature HIV-1 gag shell. J Virol Methods 169(1):244–247
16. Novoa RR, Calderita G, Cabezas P, Elliott RM, Risco C (2005) Key golgi factors for structural and functional maturation of Bunyamwera virus. J Virol 79(17):10852–10863
17. Cabezas P, Risco C (2006) Studying cellular architecture in three dimensions with improved resolution: ta replicas revisited. Cell Biol Int 30(9):747–754
18. Wolf MC, Freiberg AN, Zhang T, Akyol-Ataman Z, Grock A et al (2010) A broad-spectrum antiviral targeting entry of enveloped viruses. Proc Natl Acad Sci 107(7):3157–3162
19. Freiberg AN, Sherman MB, Morais MC, Holbrook MR, Watowich SJ (2008) Three-dimensional organization of Rift Valley fever virus revealed by cryoelectron tomography. J Virol 82(21):10341–10348
20. Weingartl HM, Zhang S, Marszal P, McGreevy A, Burton L, Wilson WC (2014) Rift Valley fever virus incorporates the 78 kDa glycoprotein into virions matured in mosquito C6/36 cells. PLoS One 9(1):e87385
21. Buranda T, Wu Y, Perez D, Jett SD, Bondu-Hawkins V et al (2010) Recognition of decay accelerating factor and αvβ3 by inactivated hantaviruses: toward the development of high-throughput screening flow cytometry assays. Anal Biochem 402(2):151–160
22. Hall PR, Hjelle B, Brown DC, Ye C, Bondu-Hawkins V et al (2008) Multivalent presentation of antihantavirus peptides on nanoparticles enhances infection blockade. Antimicrob Agents Chemother 52(6):2079–2088
23. Bisoffi M, Hjelle B, Brown DC, Branch DW, Edwards TL et al (2008) Detection of viral bioagents using a shear horizontal surface acoustic wave biosensor. Biosens Bioelectron 23(9):1397–1403
24. Prescott JB, Hall PR, Bondu-Hawkins VS, Ye C, Hjelle B (2007) Early innate immune responses to Sin Nombre hantavirus occur independently of IFN regulatory factor 3, characterized pattern recognition receptors, and viral entry. J Immunol 179(3):1796–1802
25. Surtees R, Dowall SD, Shaw A, Armstrong S, Hewson R et al (2016) Heat shock protein 70 family members interact with Crimean-Congo hemorrhagic fever virus and Hazara virus nucleocapsid proteins and perform a functional role in the Nairovirus replication cycle. J Virol 90(20):9305–9316
26. Wang X, Li B, Guo Y, Shen S, Zhao L et al (2016) Molecular basis for the formation of ribonucleoprotein complex of Crimean-Congo hemorrhagic fever virus. J Struct Biol 196(3):455–465

Chapter 6

Cryogenic Electron Microscopy of Rift Valley Fever Virus

Michael B. Sherman and Thomas J. Smith

Abstract

Rift Valley fever virus (RVFV) is an important livestock and human pathogen. It is also a potential bioweapon owing to its ability to spread by aerosols. It is an enveloped virus containing surface protrusions composed of two viral glycoproteins, G_c and G_n; the viral core contains ribonucleoprotein complexes. We describe the use of cryogenic electron microscopy (Cryo-EM) of single particles to visualize the three-dimensional (3D) organization of RVFV. The spherical RVFV virions have a diameter of 100 nm with icosahedral symmetry ($T = 12$) and 720 prominent protrusions on their surface.

In this chapter we describe the general protocols to prepare virus suspensions for Cryo-EM, image acquisition of virus particles embedded in vitreous water, and their image processing. Because of RVFV being classified as a BSL3 and potentially as a select agent, we also describe the protocol of performing Cryo-EM grid preparation and imaging in our Cryo-EM BSL3 containment, along with the general etiquette of using BSL3 containment for research, including entry into containment, dressing up for work there, manipulation with the agent, grid preparation, transfer of the frozen grids into a microscope, etc., waste disposal, and finally exit from the containment. The microscope decontamination with ClO_2 is briefly described as well because that technique is not widely used in the Cryo-EM field.

Key words Rift Valley fever virus, RVFV, Cryogenic electron microscopy, Cryo-EM, Image processing, Biological safety level 3, Cryo-EM decontamination

1 Introduction

Cryo-EM has become an invaluable tool for studying the 3D organization of biological macromolecules/complexes, including viruses and other species [1, 2]. It allows one to get information on the structure of whole virions at the atomic level with particles preserved in a thin layer of vitrified (amorphous) water. Vitrification is performed by ultrafast freezing of the suspension of the virions in a buffer solution [3–5]. The freezing is so fast (~10^6 °C/s) that water molecules do not have time to form ice crystals that would damage the integrity of the sample. Therefore, the virions are preserved in a near-native state in biologically relevant buffers; their vitrified suspensions on cryo-EM grids represent a snapshot

Hani Boshra (ed.), *Bunyaviruses: Methods and Protocols*, Methods in Molecular Biology, vol. 2893,
https://doi.org/10.1007/978-1-0716-4338-9_6,

of the structure at the time of freezing. Unfortunately, the frozen sample is very sensitive to electron irradiation in the electron microscope and can only be briefly exposed to the electron beam. Therefore, the images are underexposed and hence noisy with a low signal/noise ratio. To compensate for the noise in the individual images, multiple particle images are collected, and the images are averaged to boost the signal and to reduce the noise in the combined reconstruction. In modern cryo-EM, thousands of micrographs are typically collected that yield up to several hundred thousand images of individual particles or more. The information from all the images is combined and used to extract the structural data, resulting in high-resolution 3D volumes reconstructed from the image set.

Many viruses have an icosahedral shape/structure that is characterized by very high symmetry. In this case, a single virion image is equivalent to 60 particle images for asymmetrical particles. That in turn allows one to collect a lower number of virion images in Cryo-EM to obtain the same resolution in the 3D reconstructions. That is why icosahedral particles are so popular in Cryo-EM projects. Not every sample amenable to Cryo-EM has this kind of organization and many specimens do not have any symmetry at all, such as ribosomes. But at least all ribosomes have the same 3D structure and therefore their images could be combined to enhance the signal to noise ratio. In an even more general case, specimen particles of the same nature (e.g. pleiomorphic viruses or cell organelles, whole cells) would not even have the same size or shape, and therefore will have to be reconstructed without combination of the information from other particles. In those cases, another cryo-EM technique, cryogenic electron tomography, Cryo-ET, [6–8] could be used, although in most cases it has yielded lower resolution reconstructions.

Bunyaviruses, in general, and phleboviruses in particular, were thought to have a pleiomorphic organization where every particle has a unique shape/size. This would prevent averaging and consequently thwart attempts at obtaining high-resolution 3D reconstructions. Fortunately, RVFV is an exception to this generalization and has a regular icosahedral shape, making structural analyses possible. Another example of an icosahedral phlebovirus was reported for Uukuniemi virus [9]. In our case all the RVFV virions looked the same except for the differences in their orientations in frozen water. Since vitrification happens so fast (on a microseconds scale), the particles do not have time to change during freezing leading to nearly perfect preservation of them in the native conformation. Differences in orientations of the particles in vitrified ice present views from random directions in images, allowing for the reconstruction of the original 3D structure.

The quality of the reconstructions depends on the number of particles in the image set, the quality of the preservation in vitreous ice, and the distribution of orientations of the individual particle images. The highest quality image reconstructions come from datasets with diverse orientations, a large number of particles, and the high quality of the images themselves. Image recording is also important; modern digital cameras, especially direct electron detectors (DED) introduced into Cryo-EM several years ago [1, 10] drastically improved the quality of the reconstructions. The higher the resolution of the final reconstruction, the finer the details can be discerned in it. Even in the case of lower resolution reconstructions, the known structures of individual components/domains can be fitted (docked) [11] into the cryo-EM maps for near-atomic representations of the virus structure.

For RVFV we had a limited dataset of less than 300 particles [12] and therefore the resolution of virus features in the final 3D map was limited to 27 Å. Nonetheless, we were able to clearly resolve surface protrusions from the viral envelope and their mutual arrangement in the virus with some covered "communication channels" connecting individual protrusions. The resolution was not high enough to unambiguously delineate which proteins comprised the various parts of protrusions, to resolve contacts within the protrusions, or their attachment to the virion shell. The shell structure was not resolved either. Similar results were obtained in [13] that were directly comparable to ours. Regardless of the lack of distinct protein features in the maps, the reconstructions provided valuable information about virion organization and led to further experiments to probe the structure/function of the capsid.

2 Materials

1. VeroE6 cells
2. Cell growth media
3. Serum free media.
4. Table-top centrifuge, A-4-62 rotor
5. Amicon Ultra concentrators, 15 mL, 100 kDa cut-off
6. OptiPrep (iodixanol, ~12 mL/tube)
7. T75 flasks
8. Buffer: 10 mM Tris·HCl, 100 mM NaCl, 1 mM EDTA, pH 7.2
9. SW-41 rotor and compatible ultracentrifuge.
10. Polyclonal mouse hyper-immune ascitic fluid against RVFV

2.1 EM Supplies

1. Quantifoil or C-flat grids 2 × 2 or 2 × 1 or larger size holes.
2. Glow discharge device (e.g. Solarus 950) or plasma cleaner (e.g. PELCO easiGlow) for EM grids.

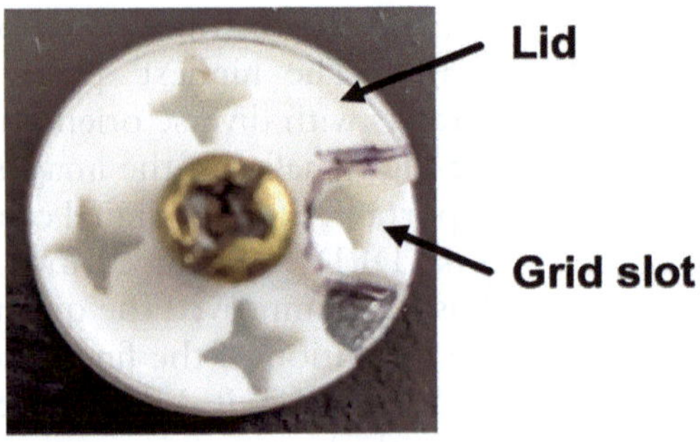

Fig. 1 Cryo-button for storage of vitrified Cryo-EM grids. Plastic base has four slots for grids, there is rotating lid on top and a holding screw

3. Cryo-EM plunger (Vitrobot IV, EM GP/GP2) (*see* **Note 1**).
4. Cryo-EM storage box ("button", Fig. 1).
5. Liquid nitrogen Dewars (1 L for specimen transfer, 30–45 L for vitrified grids storage, 4 L for auxiliary operations with liquid nitrogen (LN_2)).
6. Ethane or mixture of ethane/propane gas cylinder.
7. Cavicide.
8. 70% v/v ethyl alcohol solution.
9. Respirator (PAPR or N95 face mask).

3 Methods

3.1 Virus Preparation (See Note 2)

1. Grow VeroE6 cells to congruency in T75 flasks.
2. Infect VeroE6 cells with RVFV (multiplicity factor of 1).
3. About 6–8 h post-infection (p.i.), discard media from the flask and replace cell culture with fresh serum-free media.
4. Monitor the cells for CPE and harvest virus in 36–48 h p.i.
5. Remove the cell debris from the supernatant by centrifugation at 2000 rpm for 15 min at 4 °C.
6. Purify the virus suspension further using iodixanol density gradient centrifugation (13–22% in 10 mM Tris·HCl, 100 mM NaCl, 1 mM EDTA, pH 7.2) in a SW-41 rotor at 210,000×*g* for 1.5 h at 4 °C.
7. Collect fractions and test for the presence of RVFV by Western blotting using a polyclonal mouse hyper-immune ascitic fluid against RVFV (*see* **Note 3**).
8. Concentrate the secreted virus by ultrafiltration (Amicon Ultra concentrators) of the initial 5–10 mL (per flask) to sufficient concentration for EM analysis.

3.2 Cryo-EM Grids Preparation

1. Take four holey grids, clean their surface with either Plasma Cleaner (e.g. Solarus) or in a Glow Discharge unit (e.g. EasiGlow). Grids should become hydrophilic.
2. Cleaning: Solarus, 30–60 s, H_2–O_2 mixture; EasiGlow, 30 s, 20 mA, 04 mBar. Wear gloves when handling grids/reaching inside the units. Use EM grade tweezers (*see* **Note 4**).
3. Start a cryo-plunger (Vitrobot IV, or EM-GP2), let it stabilize the temperature/humidity in the chamber. Put fresh blotting paper in the blotter(s).
4. Setup blotting program.
   (a) Vitrobot IV:
      (i) Turn on the plunger using power switch on the back of the unit.
      (ii) Wait until the program has started. At that point the screen above the environmental chamber lights up and shows the graphical user interface window with temperature, humidity, and blotting settings.
      (iii) Adjust temperature and humidity in the chamber by moving sliders using a stylus. In most cases humidity is set to 100%, temperature varies from room temperature to +4 °C. Turn on the humidifier by clicking the "ON" checkbox.
      (iv) Adjust the blot time, blot force, and wait time by clicking on "Options," which brings up a settings page.
   (b) EM-GP/GP2:
      (i) Turn ON the unit using the switch on the back.
      (ii) Select from the list, or setup a program; check and edit settings for the current session including chamber temperature and humidity (usually 85%). Adjust blotting time and setup front or back blotting by setting rotations before specimen application and grid blotting. Setup a pre-blot waiting time if needed.
5. Cool down the LN_2 box (Vitrobot), or internal Dewar (EM-GP), fill up the ethane cup. In the case of Vitrobot, put the cold box into its holder.
6. Pick up a pre-cleaned grid with plunger tweezers, attach the tweezers to the rod in the plunger.
7. Initiate the blotting process. Usually, the rod with tweezers would be drawn into the environmental chamber of the plunger and will be positioned for specimen injection.

8. Inject the sample using 3–4 μL of the viral particle suspension. The sample should be sufficiently concentrated (a titer of 10^9 or higher).
9. Let it sit in the chamber for 10–30 s before blotting.
10. Blot the grid for 1–15 s depending on the chamber temperature/humidity and drop the grid held in tweezers into the cryogen (ethane or ethane/pentane mixture).
11. Carefully transfer frozen grid into a precooled (LN_2) cryo-EM storage box (button, Fig. 1).
12. Repeat until all grids are frozen (*see* **Note 5**).
13. Store the button in a LN_2 Dewar for subsequent use in a cryo-EM for imaging.
14. Clean working space and collect the waste generated during grid vitrification (*see* **Notes 3**, **7**, and **8**).

3.3 Cryo-EM in BSL3 Containment

Transfer purified RVFV to a BSL3 cryo-EM containment facility described in [14–16].

The containment is designed following the *Biosafety in Microbiological and Biomedical Laboratories* (BMBL) [17] recommendations for BSL3 facilities, including an anteroom with secured interlocking doors and physical separation from the rest of the Center, and negative air pressure. A downstream clean change room with personal protective equipment (PPE) storage is included in the design. In the change room there is a hands-free operated sink with eye washing station and a chemical safety shower. Access to the BSL3 containment is restricted to authorized researchers and support personnel. Required PPE includes a seamless-front gown, double gloves, and a respirator. Persons exiting the laboratory leave their gowns and respirators in the change room and are required to thoroughly wash their hands with soap before leaving containment.

Purified and concentrated virus suspensions are vitrified as reported previously:

1. Don a pair of gloves, respiratory protection (PAPR or N95 face mask depending on the risk assessment for the project), and long-sleeve wrap around cover gown.
2. Place a cryo-plunger (Fig. 2a, b) in the BSC, and cool down with liquid nitrogen in order to prepare the EM grids for vitrification before the vial with the agent is inserted into the cabinet (*see* **Note 6**).
3. Once the necessary number of grids are frozen in the BSC, transfer them into cryo-buttons. Transfer the cryo-buttons into a 1 L transfer Dewar filled with LN_2; seal the Dewar inside the BSC, and surface-decontaminate from outside with Cavicide or 70% v/v ethanol. The cryo-buttons can then be taken out for grid transfer into the Cryo-EM. Under no

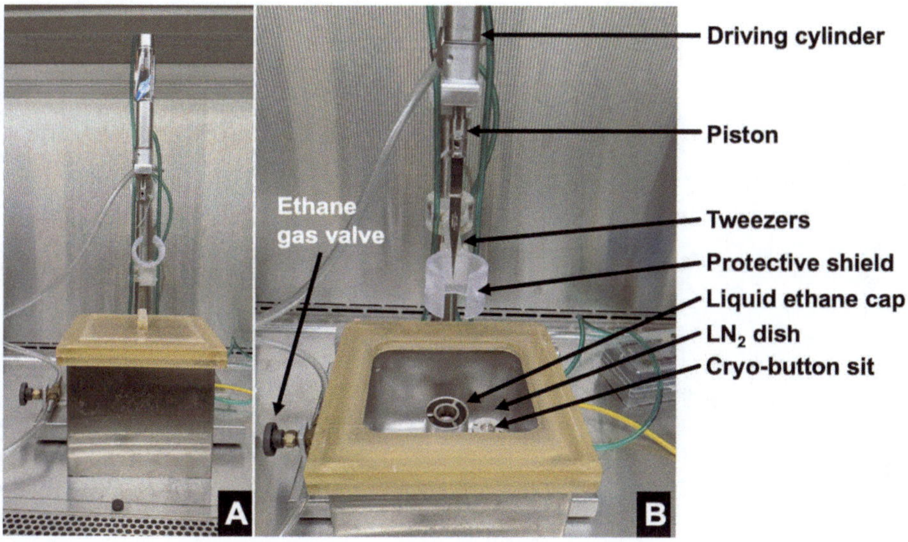

Fig. 2 (**a**) In-house-built pneumatic cryo-plunger, an overview; (**b**) Detailed view of the plunger with various parts shown

circumstances should cryo-buttons with grids be thawed or warmed to a temperature higher than −150 °C; in the case of an accidental warm-up, decontaminate the grids and transfer buttons in Cavicide, and discard the grids.

4. Discard the used LN_2 from the plunger in the BSC into a Styrofoam box and decontaminate with Cavicide after the LN_2 has evaporated.
5. After completing all work in the lab, store and secure the RVFV before exiting.
6. In the anteroom, decontaminate the PAPR, gloves and cover gown with 70% v/v ethyl alcohol.
7. Wash your hands thoroughly with soap and water at the sink located in anteroom (*see* **Notes 3**, **7**, and **8**).

3.4 Specimen Imaging

1. The Cryo-EM should have an anti-contamination device with a cryo-box surrounding the specimen in the microscope column. That box protects the sample in the microscope column from ice-contamination while the grid is being inspected and imaged. Cool down the microscope anti-contamination device with LN_2 at least a half hour before starting specimen transfer.
2. Precool the cryo-transfer station for specimen cryo-holders (located at the microscope console) with LN_2 to at least −150 °C. Insert a cryo-specimen holder (Gatan 626) into the station and transfer a cryo-button with vitrified RVFV from the 1 L Dewar into it as well.

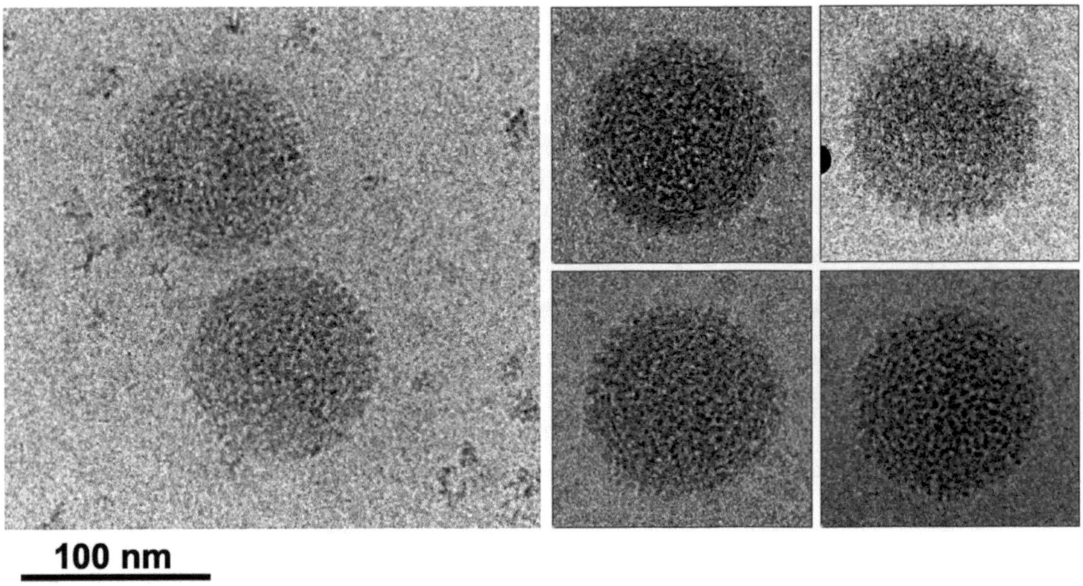

Fig. 3 RVFV virus particles embedded in vitreous ice. Field of view with two virus particles on the left. Four more separate particles shown on the right

3. Loosen the lid of the button and insert a grid with frozen RVFV into the holder, secure it with a clip ring, and close the holder's cryo-shield to protect the grid from humid air. Transfer the station to the microscope countertop adjacent to the column and insert the holder into the specimen airlock, pre-evacuate the airlock with the rotary and diffusion vacuum pumps, then insert the holder inside the microscope column for imaging (*see* **Note 9**).
4. Inspect the grid in the microscope at low magnification, define areas suitable for imaging (*see* **Note 10**). Nowadays imaging is controlled by a program (SerialEM, Leginon, etc.) allowing for massive data collection without operator intervention.
5. Use −0.7 to −2.5 μm defocus range for imaging. If the Cryo-EM is equipped with an electron energy filter (Omega, Bio-Quantum, Selectrix, etc) use 10–20 eV slit width for imaging. The best results are achieved with imaging using DED cameras. Acquire images with ~40–80 electrons/$Å^2$ electron irradiation dose. Figure 3 presents an example of RVFV particles embedded in vitreous ice. Because of an overwhelming amount of noise in the images, the protrusions are not obvious in the raw images but are clearly resolved in the 3D Cryo-EM map (Fig. 4b).
6. Acquire necessary number of images to be able to extract several tens of thousands of virus particle images from them.

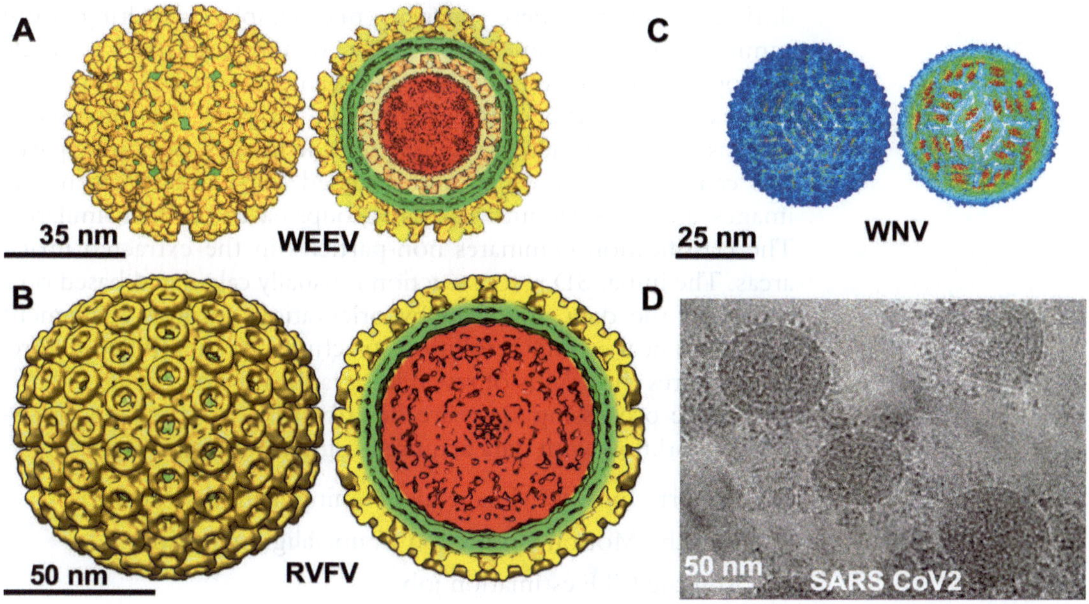

Fig. 4 Examples of infectious agents studied in the UTMB Center for Virus Imaging in Sealy Center for Structural Biology and Molecular Biophysics. (**a**) Western Equine Encephalitis Alphavirus. (**b**) Rift Valley Fever Phlebovirus MP-12. (**c**) West Nile Flavivirus. Left sides of the panels show surface rendering of the whole particles, the right sides show front half of the maps removed to reveal inside organization of the virions. (**d**) SARS CoV2 particles in vitreous ice. Surface spikes and nucleocapsids are visible. Scale bars indicate dimensions of the viruses in individual panels

7. When imaging of the grid is completed, transfer the holder from the microscope either into the precooled cryo-transfer station for another grid transfer, or into a vacuum pumping station where the holder is heated to 100 °C for several hours to decontaminate the holder and the grid. Dispose the grid by placing it in Cavicide or a 10% bleach solution; keep the holder in the pumping station to be ready for the next imaging session.
8. Keep the used, empty cryo-boxes in 70% v/v ethanol solution for at least 24 hours, and dry before the next use (*see* **Notes 3**, 7, and **8**).

4 Processing Of The Data

Image processing is done using readily available program packages like cryoSPARC [18] and Relion [19]. Modern DEDs (Falcon4, K3, Apollo, etc.) produce series of image frames ("movies") that are aligned relative to each other, which allows to reduce specimen drift during imaging. These images (micrographs) resulting from summed "movie" frames have higher quality compared to single micrographs produced by older style cameras or by photographic film exposures. After frame alignment, estimate the microscope contrast transfer function (CTF) from the images. Specimen particles can then be automatically searched for, extracted from the images, and classified into separate groups based on their similarity. The classification eliminates non-particles in the extracted image areas. The initial 3D reconstruction is usually calculated based on a subset of the data, followed by orientations/origins refinement leading to near atomic resolution structures. As mentioned before, this requires thousands of raw micrographs for processing.

Image processing steps are described below based on cryoSPARC workflow with data acquired using a DED camera:

1. Import "movies" (image files coming from the camera).
2. Run the Motion Correction (frame alignment) job.
3. Run the CTF estimation job.
4. Inspect images, remove bad ones based on insufficient resolution, excessive motion during exposure, etc.
5. Start particles search/selection. Run Blob picker job after setting the particle diameter (1000 Å for RVFV).
6. Check results using the Inspect Particle Picks job.
7. Extract RVFV particles from the images using 768 size boxes (2 Å/pxl).
8. Run a 2D Classification job to eliminate debris in the extracted boxes.
9. Select good classes using a Select 2D Classes job.
10. Repeat steps 8 and 9 several times until there are no bad classes in the 2D Classification results.
11. Run an Ab initio 3D reconstruction job to obtain an initial model for subsequent refinement. Use particles from the 5 to 10 best classes from **step 8**.
12. Use the model obtained in **step 11** as input for 3D refinement of particles stack from **step 9**.
13. Repeat the 3D refinement **step 12** until the final resolution and quality of the map stops improving.

This concludes the image processing phase. If the resolution of the final EM map is better than ~3.5 Å, ab initio atomic model building is possible. Otherwise, atomic resolution structures of individual proteins/domains (or homologs) from the world-wide PDB/PDBe database [20] can be fitted into cryo-EM 3D map. Often X-ray crystallography tools are used for structure fitting/refinement, e.g. Coot [21] and/or Phenix [22].

5 Notes

1. For BSL3 agents use manual plunger contained in a BSC (Fig. 2a, b).
2. Specimen Preparation in BSL3. All operations including virus production and its storage/handling for CryoEM preparation, imaging, and disposal of used materials are performed in a BSL3 containment following general BSL3 practices, and researchers are required to wear appropriate PPE while working with BSL3 agents. CryoEM grids are prepared in a ducted BSC. Secondary containers are opened at least 20 min after bringing them into the BSC to allow for recovery of the laminar air flow in the BSC. Also, all the necessary tools and chemicals are placed inside the BSC in advance to minimize air disturbances. The cryo-plunger is precooled with LN_2 before the vial with the sample is placed into the BSC.
3. Waste Disposal. All waste generated in containment is either collected into biohazard containers, autoclaved, and discarded into regulated medical waste containers, that are then incinerated as part of UTMB's waste management process, or is decontaminated chemically using Cavicide, 70% v/v alcohol, or freshly prepared 10% bleach. Autoclaving is performed using standard protocols; autoclave tape is used as an indicator of proper temperature during the process, and biological bacterial spore inactivation tests are used with every autoclave cycle to ensure full decontamination.
4. Clean grids no longer than 30 min before using them in cryo-plunger.
5. Specimen vitrification. To preserve biological samples for CryoEM, they must be flash-frozen in a cryogen (typically liquid ethane or propane) with an extremely high cooling rate ($>10^5$ to 10^6 °C/s). A special, pneumatically operated cryo-plunger is developed (Fig. 2a, b) for that purpose. A small volume, approximately 2–3 μL, of a purified agent (RVFV) suspended in buffer is pipetted to an EM grid. The excess liquid is wicked off with filter paper, and the grid is plunged into the cryogen. This procedure leads to water vitrification, a process in which water hardens like glass, preventing the formation of ice crystals and thereby preserving near-native structures of the agents. This method leaves the agents immobilized in a solid water matrix ("vitreous ice"), reducing the risk of researcher's aerosol exposure as long as the grids remain frozen.
6. With manual cryo-plunger, the person doing the vitrification controls the blotting time. The whole vitrification process is performed manually including specimen application to the

grid, blotting, and transfer of the frozen grids into precooled cryo-storage boxes (buttons, Fig. 1).

7. Post-imaging BSL3 operation, decontamination process. Once the imaging session is complete, the cryo-holder is removed from the microscope and placed into precooled cryo-transfer station. The grid (still frozen) is removed and immediately immersed in Cavicide. A new grid is then inserted into the holder and transferred to the microscope for a new imaging session.

   For decontamination, the cryo-holder is transferred while still cold into a vacuum pumping station where it is warmed under high vacuum and maintained at ~100 °C for at least 1–6 h to ensure proper decontamination of the tip that contacted the grids. In some cases, the grids remain in the holder for decontamination, followed by discarding them in a bottle with Cavicide.

   The cryo-transfer station and all tools used during the transfer of frozen grids to and from the cryo-holder are decontaminated by spraying and wiping with 70% v/v ethyl alcohol solution or immersing in Cavicide. Used cryo-grid storage buttons are stored in 70% v/v ethyl alcohol solution for at least 24 h and thoroughly dried before the next use. The cryo-transfer station and other tools used for specimen transfer are also sprayed with Cavicide or 70% v/v ethanol and held in an oven at 60 °C for at least 1 hour before the next use. Based on thermal stability assays of all agents used in the facility, this treatment should be capable of inactivating any residual virus.

8. Potential incidents affecting safety operation. One of the major biosafety concerns with Cryo-EM in BSL3 is microscope contamination in case of an accident. In case of thawing of a grid inside the microscope column or in the specimen chamber during specimen transfer and pre-pumping, the agent could be released into the column and spread over a large area. Liquids cannot be used for decontamination inside the Cryo-EM because of the ultra-high vacuum inside. Therefore, a special decontamination protocol had to be developed to be compatible with Cryo-EM operation. Although the cryo-grid is clamped in the holder by a clip-ring, it could potentially be dislodged resulting in poor thermal contact with the holder leading to its warmup. If the grid fell from the holder, it would contact warm parts of the microscope column, and instantly thaw. Although the volume of the agent suspension is small, it would be difficult to locate a grid or droplet inside the microscope and perform spot decontamination using recommended protocols for infectious agents.

   We have performed extensive development and testing such a protocol together with the microscope manufacturer

(JEOL) and ClorDiSys company specializing in chlorine dioxide decontamination equipment. We decided to use ClorDiSys Minidox ClO_2 generator and coupled it to a Cryo-EM JEM 2200FS. A mixture of nitrogen/ClO_2 gas is prepared in a mix box with a preset humidity level and then injected into the microscope. The Minidox generator maintains the ClO_2 concentration and humidity while running a continuous closed loop gas flow including the microscope section being decontaminated and the mix box. Once a preset exposure value of ClO_2 gas is reached, Minidox initiates purging of the system with nitrogen gas and when the system reaches 0.1 ppm ClO_2 concentration level, stops the cycle. The microscope at that point is considered to be devoid of infectious contaminants and safe for use and maintenance. All decontamination runs are validated using *Bacillus Atrophaeus* biological spore indicators. That protocol allows us to work not only with viruses that become unstable at elevated temperatures (~60 °C) but also with thermally resistant infectious bacteria, and even with bacterial spores.

9. Potential exposure risk. This procedure is the only manipulation step with a frozen sample that is done outside of a BSC, so a respirator is required.
10. Grids quality. Ideally, a vitrified grid would have even ice thickness across. However, some cryo-plungers by design produce vitrified grids with ice gradient spanning from very thin ice in the holes to too thick for imaging (Vitrobot, wedge geometry of the blotters). Small size specimens need thinner ice for high resolution reconstructions while larger samples such as viruses require thicker ice so they can be fully embedded. During grid vitrification and subsequent handling grids can get damaged and bent, which leads to reduction of useful imaging area. Examples are shown in Fig. 5. To get good quality vitrified grids, the user needs to try different blotting parameters, blotting time, strength of blotting (e.g. "blot force" in the Vitrobot), temperature, and humidity in the plunger's chamber, etc.

6 Structures of Agents Studied in the Center

Several infectious agents have been studied in the Center, including western equine encephalitis virus (WEEV), an alphavirus that causes highly debilitating and often fatal disease in humans and horses [23]. WEEV is classified as a NIAID high priority Category B biothreat agent. This virus was successfully imaged in the Center and reconstructed its structure in 3D at 1.5 nm resolution (Fig. 4a, [24]). To our knowledge, this was the first BSL3 virus ever imaged in BSL3 containment using Cryo-EM.

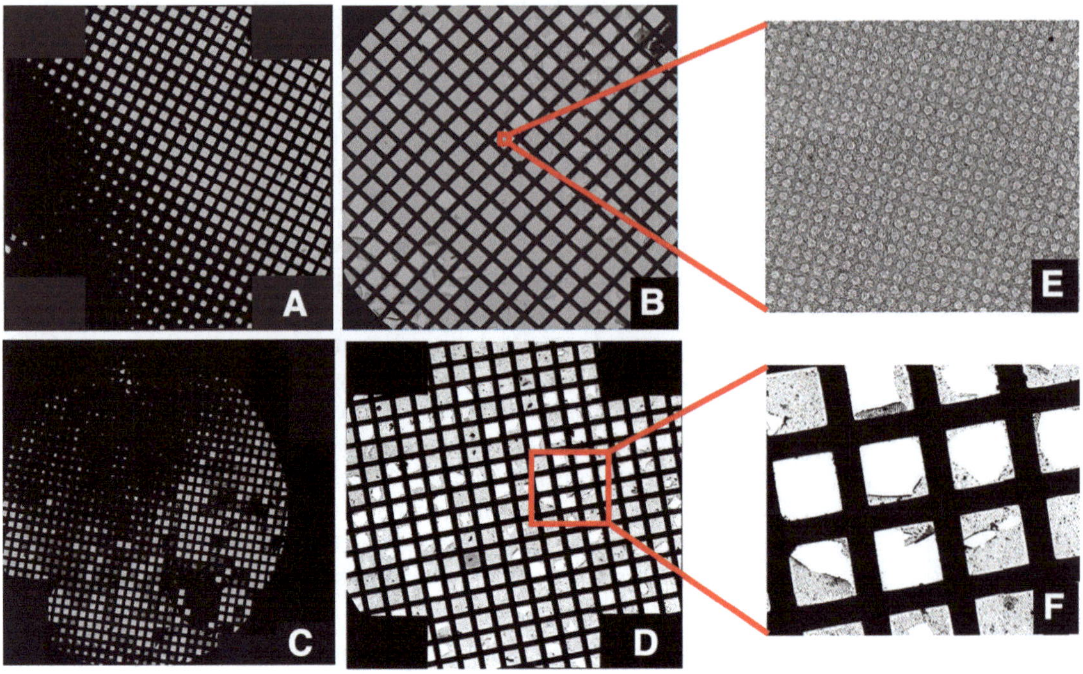

Fig. 5 (**a**) Vitrified grid showing ice thickness gradient. Could be good for various size samples. Lower left part of the grid might have good ice thickness for larger size specimens (e.g. viruses). (**b**) Good grid with uniformly thin ice across the grid. Useful for high resolution imaging of smaller size particles (ribosomes, proteins, nucleic acids, etc.). (**c**) Uneven ice coverage of the grid; may still be used for larger size specimens (viruses). (**d**) Broken and practically dry grid. Unusable for microscopy. (**e**) Magnified fragment of (**b**) showing holes with even thickness of ice inside. (**f**) Fragment of (**d**) showing broken area of the grid. Remaining holes are dry and devoid of ice

Another example of a 3D reconstruction obtained in the Center is that of RVFV, a prototypical bunyavirus. RVFV is classified as a NIAID priority Category A biothreat agent. Work from the W. M. Keck Virus Imaging Center produced the first three-dimensional reconstruction of RVFV from electron microscopy images (Fig. 4b) [12, 25]. This structure provides a model for many bunyaviruses and could aid in the design of antiviral drugs, diagnostics, and effective vaccines.

West Nile virus (WNV) is a flavivirus spread by Culex mosquitoes. It is a leading cause of mosquito borne human infections in the continental USA. The virus is causing disease with various manifestations (chills, rash, malaise, headache, backache, arthralgia, myalgia, and eye pain) [26]. There is no vaccine available for treatment, that is why studying the structure of the virus is important. Wild type WNV along with several mutants were imaged in the Center. WNV was recently reclassified into BSL2 category and was imaged in a Titan Krios microscope with 3D map reaching resolution of 3 Å (Fig. 4c, unpublished data).

SARS-CoV2 is a coronavirus that caused a world-wide outbreak in 2020 [27]. We imaged the wild type of SARS-CoV2 viruses in our BSL3 containment as single particles and as a tilt series in the microscope for electron tomography. A fragment of a micrograph with SARS-CoV2 particles is shown in Fig. 4d. Several different projects required wild type SARS-CoV2 imaging (e.g [28].).

Acknowledgements

The W. M. Keck Virus Imaging Center is part of the Sealy Center for Structural Biology and Molecular Biophysics. It was established through generous gifts from W. M. Keck Foundation, and the Kleberg Foundation, and by a grant from Health Resources and Services Administration.

We are thankful to A. N. Freiberg, S. C. Weaver, D. Beasley, and G. Diamond for providing virus specimens.

References

1. Vinothkumar KR, Henderson R (2016) Single particle electron cryomicroscopy: trends, issues and future perspective. Q Rev Biophys 49:e13. https://doi.org/10.1017/s0033583516000068
2. Subramaniam S, Kühlbrandt W, Henderson R (2016) CryoEM at IUCrJ: a new era. IUCrJ 3 (Pt 1):3–7. https://doi.org/10.1107/s2052252515023738
3. Dubochet J, McDowall AW (1981) Vitrification of pure water for electron microscopy. J Microsc 124(3):RP3–RP4
4. McDowall AW, Chang JJ, Freeman R, Lepault J, Walter CA, Dubochet J (1983) Electron microscopy of frozen hydrated sections of vitreous ice and vitrified biological samples. J Microsc 131(Pt 1):1–9
5. Dubochet J (2012) Cryo-EM--the first thirty years. J Microsc 245(3):221–224. https://doi.org/10.1111/j.1365-2818.2011.03569.x
6. Kurner J, Medalia O, Linaroudis AA, Baumeister W (2004) New insights into the structural organization of eukaryotic and prokaryotic cytoskeletons using cryo-electron tomography. Exp Cell Res 301(1):38–42
7. Chreifi G, Chen S, Jensen GJ (2021) Rapid tilt-series method for cryo-electron tomography: characterizing stage behavior during FISE acquisition. J Struct Biol 213(2):107716. https://doi.org/10.1016/j.jsb.2021.107716
8. Zimmermann L, Chlanda P (2023) Cryo-electron tomography of viral infection—from applications to biosafety. Curr Opin Virol 61:101338. https://doi.org/10.1016/j.coviro.2023.101338
9. Overby AK, Pettersson RF, Grunewald K, Huiskonen JT (2008) Insights into bunyavirus architecture from electron cryotomography of Uukuniemi virus. Proc Natl Acad Sci USA 105(7):2375–2379. https://doi.org/10.1073/pnas.0708738105
10. Hanske J, Sadian Y, Muller CW (2018) The cryo-EM resolution revolution and transcription complexes. Curr Opin Struct Biol 52:8–15. https://doi.org/10.1016/j.sbi.2018.07.002
11. Trellet M, van Zundert G, Bonvin A (2020) Protein-protein modeling using Cryo-EM restraints. Methods Mol Biol 2112:145–162. https://doi.org/10.1007/978-1-0716-0270-6_11
12. Sherman MB, Freiberg AN, Holbrook MR, Watowich SJ (2009) Single-particle cryo-electron microscopy of Rift Valley fever virus. Virology 387(1):11–15. https://doi.org/10.1016/j.virol.2009.02.038
13. Huiskonen Juha T, Överby Anna K, Weber F, Grünewald K (2009) Electron Cryo-microscopy and single-particle averaging of Rift Valley fever virus: evidence for GN-GC glycoprotein heterodimers. J Virol 83(8):3762–3769. https://doi.org/10.1128/jvi.02483-08
14. Sherman MB, Freiberg AN, Razmus D, Yazuka S, Koht C, Hilser VJ, Lemon SJ, Brocard A, Zimmerman D, Chiu W, Watowich SJ, Weaver SC (2010) A unique BSL-3 cryo-electron microscopy laboratory at UTMB. Appl Biosaf 15(3):130–136
15. Sherman MB, Trujillo J, Leahy I, Razmus D, Dehate R, Lorcheim P, Czarneski MA,

Zimmerman D, Newton JT, Haddow AD, Weaver SC (2013) Construction and organization of a BSL-3 cryo-electron microscopy laboratory at UTMB. J Struct Biol 181(3):223–233. https://doi.org/10.1016/j.jsb.2012.12.007

16. Sherman MB, Trujillo J, Bammes BE, Jin L, Stumpf MW, Weaver SC (2015) Decontamination of digital image sensors and assessment of electron microscope performance in a BSL-3 containment. AIMS Biophys 2:153–162. https://doi.org/10.3934/biophy.2015.2.153
17. Biosafety in microbiological and biomedical laboratories, 6th edn (2020) https://www.cdc.gov/labs/pdf/SF__19_308133-A_BMBL6_00-BOOK-WEB-final-3.pdf
18. Punjani A, Rubinstein JL, Fleet DJ, Brubaker MA (2017) cryoSPARC: algorithms for rapid unsupervised cryo-EM structure determination. Nat Methods 14(3):290–296. https://doi.org/10.1038/nmeth.4169
19. Kimanius D, Dong L, Sharov G, Nakane T, Scheres SHW (2021) New tools for automated cryo-EM single-particle analysis in RELION-4.0. Biochem J 478(24):4169–4185. https://doi.org/10.1042/bcj20210708
20. Lawson CL, Baker ML, Best C, Bi C, Dougherty M, Feng P, van Ginkel G, Devkota B, Lagerstedt I, Ludtke SJ, Newman RH, Oldfield TJ, Rees I, Sahni G, Sala R, Velankar S, Warren J, Westbrook JD, Henrick K, Kleywegt GJ, Berman HM, Chiu W (2011) EMDataBank.org: unified data resource for CryoEM. Nucleic Acids Res 39 (Database issue):D456–D464. https://doi.org/10.1093/nar/gkq880
21. Emsley P, Lohkamp B, Scott WG, Cowtan K (2010) Features and development of coot. Acta Crystallogr D Biol Crystallogr 66 (Pt 4):486–501. https://doi.org/10.1107/S0907444910007493
22. Liebschner D, Afonine PV, Baker ML, Bunkoczi G, Chen VB, Croll TI, Hintze B, Hung LW, Jain S, McCoy AJ, Moriarty NW, Oeffner RD, Poon BK, Prisant MG, Read RJ, Richardson JS, Richardson DC, Sammito MD, Sobolev OV, Stockwell DH, Terwilliger TC, Urzhumtsev AG, Videau LL, Williams CJ, Adams PD (2019) Macromolecular structure determination using X-rays, neutrons and electrons: recent developments in Phenix. Acta Crystallogr D Struct Biol 75(Pt 10):861–877. https://doi.org/10.1107/S2059798319011471
23. Smith DW, Mackenzie JS, Weaver SC (2009) Alphaviruses. In: Richman DD, Whitley RJ, Hayden FG (eds) Clinical virology. ASM Press, Washington, DC, pp 1241–1274
24. Sherman MB, Weaver SC (2010) Structure of the recombinant alphavirus Western equine encephalitis virus revealed by cryoelectron microscopy. J Virol 84(19):9775–9782. https://doi.org/10.1128/JVI.00876-10
25. Freiberg AN, Sherman MB, Morais MC, Holbrook MR, Watowich SJ (2008) Three-dimensional organization of Rift Valley fever virus revealed by cryoelectron tomography. J Virol 82(21):10341–10348. https://doi.org/10.1128/JVI.01191-08
26. Watts DM, Granwehr BP, Shope RE, Solomon TOM, Tesh RB (2006) Japanese encephalitis and West Nile and other flavivirus infections. In: Tropical infectious diseases, vol 1. Elsevier Inc., pp 823–830. https://doi.org/10.1016/b978-0-443-06668-9.50078-8
27. Wu D, Wu T, Liu Q, Yang Z (2020) The SARS-CoV-2 outbreak: what we know. Int J Infect Dis 94:44–48. https://doi.org/10.1016/j.ijid.2020.03.004
28. Diamond G, Molchanova N, Herlan C, Fortkort JA, Lin JS, Figgins E, Bopp N, Ryan LK, Chung D, Adcock RS, Sherman M, Barron AE (2021) Potent antiviral activity against HSV-1 and SARS-CoV-2 by antimicrobial Peptoids. Pharmaceuticals 14(4):304

Chapter 7

Detection and Characterization of Antiviral Proteins Against Bunyaviruses in Mosquito-Derived Cells

P. Böhmer, M. Reuter, M. Altinli, and E. Schnettler

Abstract

Viruses of the class *Bunyaviricetes* are often transmitted by arthropods, including mosquitoes. The innate immune response in mosquitoes comprises several pathways, including sequence-specific degradation through RNA interference (RNAi). It is known that bunyavirus infections are targeted by the innate immune response in mosquitoes and derived cells; however, detailed information is often still missing. In this chapter, we describe a detailed protocol to determine if a protein of interest acts antiviral against bunyavirus infection in mosquito-derived cells. In line, we also describe a detailed protocol to investigate the effect of overexpression of a mosquito-encoded protein or analysis of mutant proteins on bunyavirus infection.

Key words dsRNA transfection, Mosquito-derived cells, *Bunyaviricetes*, Antiviral, RNAi

1 Introduction

Investigating the antiviral activity of specific mosquito proteins in mosquito-derived cell culture by either dsRNA / siRNA-based silencing approach or overexpression has been a very important method in the past for various mosquito infecting viruses, including viruses of the class *Bunyaviricetes* [1–3]. The dsRNA-based silencing approach involves the production of sequence-specific dsRNA molecules by in vitro transcription with T7 RNA polymerase from PCR products, followed by their transfection to mosquito-derived cells with the help of lipocomplexes. These dsRNA molecules are recognized as foreign in the cell and enter the exogenous siRNA pathway, resulting in the production of 21 nt-long small interfering (si)RNAs, which in turn are incorporated into the RNA-induced silencing complex, specifically the Argonaute-2-protein. Following, complementary mRNA is recognized and

Authors Böhmer and Reuter have equally contributed to this chapter.

Hani Boshra (ed.), *Bunyaviruses: Methods and Protocols*, Methods in Molecular Biology, vol. 2893,
https://doi.org/10.1007/978-1-0716-4338-9_7,

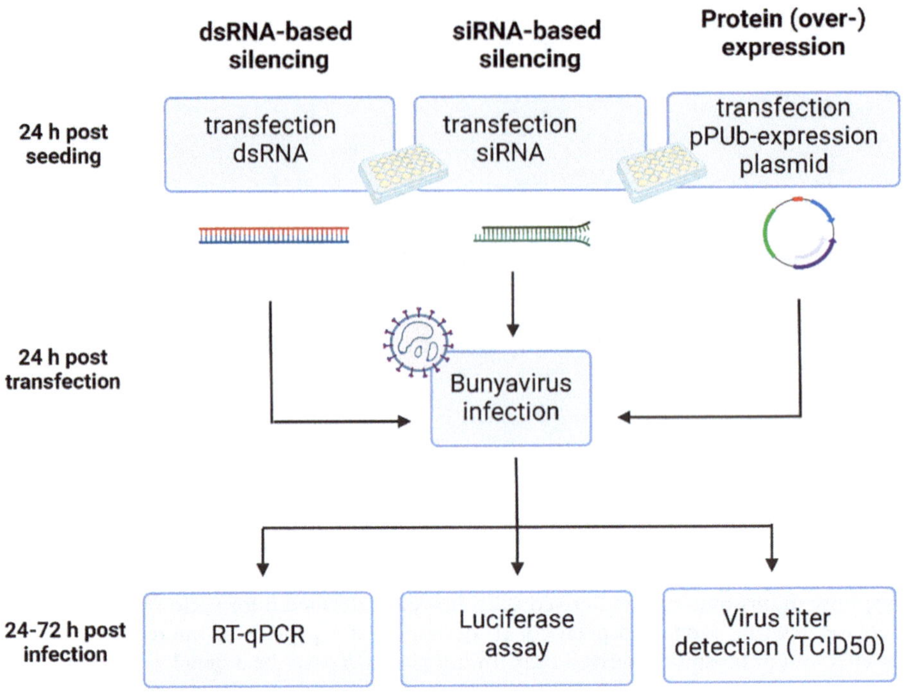

Fig. 1 Schematic outline of experiments to detect and characterize antiviral proteins against bunyaviruses in mosquito-derived cells using silencing and overexpression assays. For details, see the text

degraded, resulting in reduced protein production [4]. Due to the lack of antibodies for most mosquito proteins, successful transcript silencing is often determined by quantitative real-time PCR. The effect of protein silencing on virus infection is investigated following virus infection of dsRNA/ siRNA transfected cells (Fig. 1), either detecting virus production (TCID50 or plaque assays), viral RNA (qRTPCR), or reporter detection in case of reporter viruses (e.g. luciferase).

On the other hand, overexpression of proteins of interest and their effect or interaction with bunyavirus infection are used to investigate more detailed molecular interactions (Fig. 1). This can either be done in cells already endogenously expressing the protein of interest or resubstituting the protein of interest or mutants in cells not expressing the protein of interest (e.g., knockout cell lines produced by CRISPR/Cas9 or TALENS). Such an investigation is specifically useful to understand the importance of various domains or even specific amino acids of, for example, key antiviral proteins, like the Dicer-2 or Argonaute-2 protein of the siRNA pathway in mosquitoes [5, 6]. Until now, most knowledge about RNAi proteins still comes from the model organism Drosophila. Although this knowledge can often be extrapolated, it has been shown that this is not always the case, for example, looking at the encoded Piwi, Argonaute, and Aubergine proteins [7]. In addition, overexpression of wild-type target proteins, which were identified to be downregulated upon infection, can be used to determine its effect and possible antiviral activity on virus infection.

2 Materials

2.1 Cell Culture

1. *Aedes aegypti*-derived Aag2-AF5 (ECACC 19022601), Aag2-AF319 (CRISPR/Cas9 produced Dicer-2 knock out cells; ECACC 19022602), and BHK-21 (ECACC 85011433) cells.
2. Medium*:* Aag2-derived cells: Leibovitz L15 medium supplemented with 10% heat-inactivated Fetal Calf Serum (FCS), 10% Tryptose phosphate broth (TPB) and 1× Penicillin/ Streptomycin. BHK-21 cells: GMEM supplemented with 10% heat-inactivated Fetal Calf Serum (FCS) and 1× Penicillin/ Streptomycin. DPBS (sterile), Trypsin-EDTA. Store at 4 °C.
3. Cell scraper.
4. 25 cm^2 culture flask
5. 24-well and 96-well culture plates
6. Neubauer counting chamber.
7. Cell incubator 28 °C without CO_2 supply.
8. Cell incubator 37 °C with 5% CO_2.

2.2 Design of Primers and Generation of Templates for In Vitro Transcription of dsRNA Production

Any RNA isolation method (either for total RNA or mRNAs) can be used. We report here how it works for total RNA isolation using NucleoZOL®.

2.2.1 RNA Isolation

1. NucleoZOL® (Machery&Nagel).
2. 70% EtOH.
3. Isopropanol.
4. RNase-free water.

2.2.2 cDNA Production

1. Rnase-free water.
2. Random hexamer primers.
3. Oligo(dT) primers.
4. M-MLV 5× Buffer.
5. M-MLV (200 U/μL).
6. RNase inhibitor.
7. dNTPs (10 mM).

2.2.3 PCR with T7-Primers

1. Kod Hot Start polymerase Kit (Thermofisher, Waltham USA), or any other proof-reading DNA polymerase.
2. gene-specific primers with attached T7 sites.

3. 2% Agarose Gel.
4. Gel and PCR Clean-up kit.

2.3 In Vitro Transcription and dsRNA Formation

1. Ambion™ MEGAscript™ RNAi kit (Thermofisher).

2.4 Transfection and Infection

1. *Aedes aegypti*-derived cells (AF5 or AF319) in a 24-well plate, seeded 1 day prior to transfection (2.8*10^5 cells per well), to have a confluence of 50–70% on the day of transfection.
2. DharmaFECT2 (Horizon, Waterbeach, United Kingdom).
3. Reduced Serum medium OptiMEM.
4. Double-stranded RNA (specific for the investigated target and a nontargeting control).
5. Commercially bought siRNAs (specific for the investigated target and a nontargeting control).
6. Plasmids pPUb, expressing the protein of interest (e.g. Dicer-2 wt or mutants) and a control plasmid (e.g. expressing eGFP) from the ubiquitous promoter of *Ae. aegypti*.
7. Virus stock (with a known sufficient titer of the Bunyavirus to be investigated, either wildtype or expressing a luciferase reporter protein).

2.5 Effect on Virus Infection

2.5.1 qRT-PCR

1. 96-well plate.
2. PCR foil.
3. QuantiTect SYBR Green PCR Mastermix (Qiagen, Hilden, Germany), or any other quantitative PCR Mastermix.
4. Target-specific primer pair.
5. Housekeeping gene-specific primer pair (e.g. ribosomal S7).
6. RNase-free H20.

2.5.2 TCID50

1. 96-well plate.
2. Formaldehyde.
3. Crystal violet.

2.5.3 Luciferase Measurement

1. Passive Lysis Buffer (Promega).
2. Luciferase kit, according to the reporter luciferase used (Promega).
3. Luminometer and corresponding plates.

3 Methods

Carry out all procedures at room temperature unless otherwise specified. dsRNA and siRNA molecules, as well as RNA, should be kept on ice.

3.1 Cell Culture

1. AF cells and BHK-21 cells are cultured in a 25 cm^2 flask either at 28 °C without CO_2 (AF cells) or 37 °C with 5% CO_2 (BHK-21 cells); respectively. Cells are subcultured when reaching confluency.
2. For subculturing the AF cells, they are detached using the cell scraper and resuspended in their culture medium by slowly pipetting up and down several times. In the case of BHK-21 cells, the media is removed and the cells are shortly washed with PBS. After removing the PBS, 1 mL Trypsin-EDTA is added to detach the cells. After detaching, 4 mL of complete culture media is added. Transfer the desired amount to a new 25 cm^2 flask, previously filled with fresh culture media. For routine passaging, use a splitting ratio of 1:5 for AF5 cells and 1:10 for BHK-21 cells.
3. For silencing or overexpression assays, AF cells are counted using a Neubauer counting chamber and 2.8*10^5 cells are seeded in 1 mL culture medium per well in a 24-well plate. Experiments are performed in technical triplicates, therefore seed three wells for each construct/dsRNA/siRNA to be tested for a specific bunyavirus.
4. For TCID50 assays, BHK-21 cells are counted, and 4*10^4 cells are seeded in 180 μL per well in a 96-well plate.

3.2 Design of Primers and Generation of Templates for In Vitro Transcription of dsRNA Production

After identifying a target mRNA for silencing, a specific target nucleotide sequence should be chosen to produce dsRNA from this sequence.

1. Choose a dsRNA target sequence that is between 150 and 500 bp long and shows no similarity with other host mRNA regions for at least a 19 nt stretch.
2. Design primers for the last 20–22 nt of the ends of the target side. Add to the 5′ end of the designed primers the sequence for the T7 RNA polymerase promoter (5'gta ata cga ctc act ata ggg 3′).

3.2.1 RNA Isolation

For total RNA isolation (either for dsRNA production or to quantify viral infection, see 3.6.1), seed 2.8*10^5 cells in 1 mL culture medium per well in a 24-well plate 3 days prior to RNA isolation.

1. Remove cell medium from AF cells.
2. Add 500 μL Nucleozol®.

3. Ensure complete lysis by pipetting up and down (*see* **Note 1**).
4. Add 200 μL RNase-free water per 500 μL Nucleozol®.
5. Shake the sample vigorously for 15 s.
6. Incubate at room temperature for 5 min.
7. Centrifuge for 15 min at 12.000 × g at RT (*see* **Note 2**).
8. Transfer 500 μL of the supernatant to a new tube.
9. Add 500 μL isopropanol per 500 μL supernatant to precipitate RNA.
10. Incubate 10 min at room temperature.
11. Centrifuge the sample for 10 min at 12.000 g.
12. Discard supernatant (*see* **Note 3**).
13. Add 500 μL 70% EtOH for washing.
14. Centrifuge for 3 min at 8000 g.
15. Remove ethanol by pipetting (*see* **Note 4**).
16. Repeat the washing step (Add 500 μL 70%EtOh, centrifuge 3 min at 800 g).
17. Remove the supernatant (*see* **Note 5**).
18. Dissolve the pellet in 25 μL RNAse-free water (*see* **Note 6**).
19. Measure RNA concentration (e.g. Nanodrop).

3.2.2 cDNA Production and T7-Based PCR Product

1. Mix 1.5 μg total RNA with 1 μL of oligo(dT) primers in a total volume of 15 μL filled up with ddH20.
2. Incubate for 5 min at 70 °C.
3. Place 1 min on ice.
4. Add 5 μl 5× M-MLV buffer, 1.25 μL dNTPs (10 mM), 1 μL RNAse inhibitor, 1.75 μL water (molecular grade, RNAse-free), 1 μL M-MLV (200 U/μL).
5. Incubate 1 h at 42 °C.
6. Run a KOD hot start PCR reaction with the generated cDNA as a template using your designed primers (*see* **Note 7**).
7. Run the PCR product on a 2% TAE agarose gel and purify the PCR product with a kit according to the manufacturer's instructions (*see* **Note 8**).

3.3 In Vitro Transcription and dsRNA Production

Use the purified PCR products as templates for in vitro transcription and dsRNA production

1. Generate dsRNA according to the manufacturer's instruction using the RNAi Megascript kit (Thermofisher) (*see* **Note 9**).
2. Measure the dsRNA concentration (e.g. Nanodrop), make aliquots, and store the dsRNA at −80 °C or use it directly for transfection experiments.

3.4 Transfection of Cells with dsRNA/siRNA or pPUb-Expression Plasmid

dsRNA or siRNA-based silencing or overexpression of a protein of interest is done 24 h prior to infection (Fig. 1). The following procedure exemplifies one well of a 24-well plate:

1. *On Day 1*: 24 h prior to transfection, seed 2.8 * 10^5 AF cells/mL in 1 mL in a 24-well plate.
2. Incubate the cells at 28 °C without CO_2.
3. *On Day 2*: Prepare either (i) the in vitro transcribed dsRNA, (ii) the commercially bought siRNAs (sequence-specific and control), or (iii) the pPUb-based expression plasmid.
4. Check your seeded cells for confluency under a microscope. They should be evenly distributed at the bottom of the well at about 50–80% confluency.
5. Prepare two 1.5 mL reaction tubes.
6. Prepare Tube 1: (i) 200 ng dsRNA, (ii) 20 nM siRNA or (iii) 500 ng pP-Ub-expression plasmid+ reduced serum medium (OptiMEM) in a total volume of 50 μL (*see* **Note 10**).
7. Prepare Tube 2: 2 μL Dharmafect 2 + 48 μL reduced serum medium (OptiMEM) (*see* **Note 11**).
8. Incubate for 5–10 min.
9. Add the mix of tube 2 to tube 1. The total transfection volume will be 100 μL/well. (*see* **Notes 12** and **13**).
10. Mix by tapping or vortexing shortly and centrifuge briefly.
11. Incubate the mix for 30 min at room temperature. The incubation time should be a minimum of 20 min and should not exceed 1 h.
12. After the incubation time, add the mix drop by drop to the cells.
13. Incubate cells at 28 °C without CO_2 for 24 h.

3.5 Infection

1. *On Day 3*: Check your cells under a microscope. (*see* **Note 14**).
2. Remove the medium from the well and replace it with fresh medium (0.5 mL) (*see* **Note 15**).
3. For the infection, calculate the amount of virus needed per well according to the desired multiplicity of infection (MOI) in 100 μL growth medium.

$$\frac{\text{cell number}\left(2.8 * 10^4\text{cells}\right) * \text{desired MOI}}{\text{Virustiter}} * 1000$$

amount of virus added per well(μL)

4. Add the virus onto the cells drop by drop in a circular manner. Leave the virus on the cells for 2 h at 28 °C without CO_2.

Afterwards, remove the inoculum from the well and replace it with fresh medium (1 mL).

5. Remove the medium from the well and replace it with fresh medium (1 mL).
6. Incubate the cells at 28 °C without CO_2 for 48 h.

3.6 Effect on Virus Infection

Depending on the used virus (wt or expressing a luciferase reporter protein) different assays are performed to quantify virus infection. Generally, effects on the viral RNA in the cells are detected by qRT-PCR and are combined with the detection of infectious virus particles in the cell culture media ($TCID_{50}$). In the case of luciferase reporter expressing viruses, the effect on virus protein production by luciferase assays is performed. Again, it can be combined with the detection of infectious virus particles in the cell culture media ($TCID_{50}$). In addition, successful silencing of the target mosquito mRNA in the case of dsRNA/ siRNA-based silencing needs to be checked by qRT-PCR. In case of (over-)expression, the presence of the protein can be detected by SDS-PAGE and Western blot analysis or immunofluorescent staining.

1. Check your cells under the microscope (*see* **Note 16**).
2. Take off the supernatant and either proceed with 3.6.1 RNA isolation and RT-qPCR, 3.6.2 $TCID_{50}$ or 3.6.3. Luciferase measurement.

3.6.1 RT-qPCR

1. Isolate total RNA (*see* 3.2.1).
2. Produce cDNA from your isolated RNA (see 3.2.2) (*see* **Notes 17, 18, 19**, and **20**).
3. Use your produced cDNA to run two Sybr Green qRT-PCRs (Qiagen) reactions with 1 μL cDNA template each:

   PCR 1 to check the silencing efficiency of your transcript: qRT-PCR1 Reaction 1 with a primer pair targeting your gene of interest (*see* **Note 21**) and a housekeeper gene (e.g. ribosomal S7 [8]).

PCR 2 to check the effect of your silencing upon viral replication. qRT-PCR 2 with a primer pair targeting your virus and a housekeeper (ribosomal S7).

3.6.2 TCID50

1. Take off the supernatant and either use it directly or freeze it at −80 °C until use.
2. Seed 4 *10^4 BHK-21 cells in 180 μL per well in a 96-well plate and incubate overnight at 37 °C and 5% CO_2.
3. Add 20 μL of supernatant in the first row (use four columns per sample).
4. Gently mix and take 20 μL of row 1 and add to row 2 (*see* **Note 22**).

5. Repeat this process, creating tenfold dilutions. Leave one row as uninfected control.
6. Incubate the plate at 37 °C, 5% CO_2, for the appropriate time, depending on the virus (check the plate for CPE every day).
7. After the incubation is complete, fix the plate for 30–60 min in 8% formaldehyde in a fume cupboard.
8. Gently tap off excess formaldehyde before adding 100 μL crystal violet per well.
9. Gently wash off the dye using tap water and let the plate air-dry.

3.6.3 Luciferase Measurement

1. Dilute 5× passive lysis buffer (Promega) with water to get 1× passive lysis buffer.
2. Remove cell medium.
3. Lyse cells using 100 μL 1× passive lysis buffer (Promega) (**Notes 23** and **24**).
4. Incubate the cells 15–20 min at room temperature on a rocking plate.
5. Measure luminescence according to the manufacturer's protocol.

3.7 Data Analysis

3.7.1 Data Analysis of qRT-PCR Data

1. For analysis of qRT-PCR data the $2^{-\Delta\Delta CT}$ method is used with the ribosomal S7 gene as housekeeping gene.
2. After calculating the relative quantification, each sample is normalized to its control knockdown. To normalize it calculate: *RQ knockdown target/RQ knockdown control.*
3. Present your data in a graph (see example below; Figs. 2 and 3).

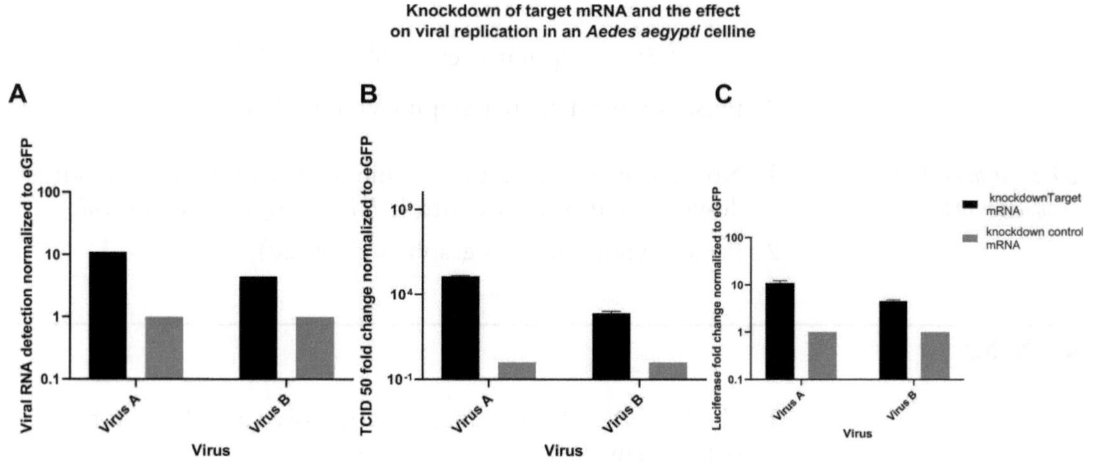

Fig. 2 Exemplary graph to show the effects on bunyavirus infection upon silencing a target mosquito mRNA. Regarding the method the results should be represented in different graphs. The graph should present the used virus on the x-axis and the fold change of either (**a**) viral RNA detection (**b**) TCID50 fold change or (**c**) Luciferase fold change. The data is normalized to the control treatment (nontargeting dsRNA/ siRNA)

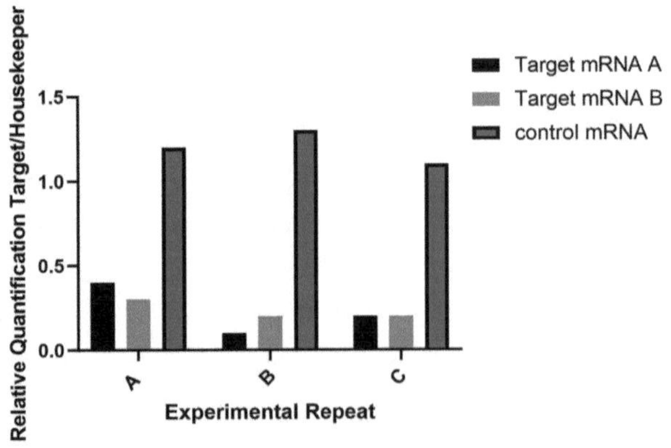

Fig. 3 Exemplary graph showing the qRT-PCR data to determine the success of silencing the target transcript. The Y axis shows the relative Quantification determined by the $2^{-\Delta\Delta CT}$ method. The x-axis shows the different experimental repeats

3.7.2 Data Analysis of TCID 50 Data

1. Determine the number of positive (no staining; no cells due to CPE) and negative wells (blue staining of the well, due to cells).
2. Calculate the 50% Tissue Culture Infectious Dose ($TCID_{50}$) using the Reed-Muench, Spearman-Kärber or Weil method. For example, the Reed-Muench formula is as follows, with I being the interpolated value of the 50% endpoint and h being the dilution factor:

$$I = \frac{(\%\text{of wells infected at dilution above } 50\% - 50\%)}{(\%\text{of wells infected at dilution above } 50\% - \%\text{of wells infected at dilution below } 50\%)}$$

$$50\%\text{endpoint titer} = 10^{\text{logtotaldilutionabove}50\% -}$$

3. Present your data in a graph (see Fig. 2b).

3.7.3 Data Analysis of Luciferase Data

1. Normalize your measured luminescence of your target knockdown to that of your control: *RLU target/RLU control.*
2. Present your data in a graph (*see* Fig. 2c).

4 Notes

1. NucleoZOL® homogenate can be stored at −80 °C for up to a year for later use.
2. A blue pellet should be seen at the bottom. Sometimes, no pellet is visible, and the fluid looks bluish. In both cases, you can continue.

3. The RNA appears as a gel-like pellet at the bottom of the tube. White RNA can be seen as contamination. Sometimes, you don't see a pellet, but continue anyway.
4. Remove the supernatant carefully since the pellet is not as adherent anymore.
5. Drying of the pellet is not necessary and might even reduce solubility.
6. Leave the sample on ice for 20–60 min for efficient solubilization.
7. Calculate the primer annealing temperature according to the binding sequence.
8. To verify the amplified sequence, the PCR product should be sequenced afterwards.
9. To increase the amount of perfect dsRNA molecules, the in vitro transcript is heated up and cools down slowly before adding DNAse I and RNAseA.
10. Check your concentration of dsRNA prior to use! When handling RNA always keep RNA on ice!
11. Tube 2 can be prepared as a mastermix for all wells.
12. Pipette Dharmafect slowly to prevent shearing forces, which might destroy the Dharmafect micelles and reduce transfection efficiency.
13. Make sure to change the tips between each dsRNA-Mix and pipette directly into the liquid as to create as little droplets on the side of the tube as possible.
14. The cell shape and appearance might have changed due to the transfection. Clumped and/or detached cells are normal cell behaviour after transfection.
15. Replace the medium carefully, as you work with semi-adherent and transfected cells, which can be washed off easily. An additional flask with growth medium for infection is advised to prevent cross-contamination with viruses.
16. Depending on which antiviral protein was silenced, cells can show a severe reaction and detach and/or die.
17. You can use less RNA but you have to adjust the total amount of RNA in the other samples accordingly for comparing samples.
18. For viruses with a poly-A tail use oligo(dT) primers for other viruses' random hexamers.
19. For checking silencing efficiency, use oligo(dT) primers since the dsRNA can sometimes be detected in case of cDNA production with random hexamers.

20. When using random Hexamers, incubate 1 h at 37 °C and inactivate the MMLV by 10 min at 70 °C.
21. Primer pairs that are used to check knockdown efficiency should not overlap with the region of the gene that is targeted by the dsRNA.
22. Change tips for each time to ensure accurate dilutions.
23. Passive Lysis Buffer is provided as 5× buffer. Dilute first in ddH20 .
24. You can store the lysed cells in Passive Lysis Buffer at −20 °C for 1 month.

Acknowledgements

This work is supported by the German Centre for Infection Research (DZIF; TTU 01.708, MA; ES) and the DFG (project number 497659464; MR and 500072465; PB).

References

1. Walsh E, Torres T Z B Rückert C (2022) Culex mosquito Piwi4 is antiviral against two negative-sense RNA viruses. Viruses 14(12)
2. Dietrich I, Shi X, McFarlane M, Watson M, Blomström AL, Skelton JK, Kohl A, Elliott RM, Schnettler E (2017) The antiviral RNAi response in vector and non-vector cells against orthobunyaviruses. PLoS Negl Trop Dis 11(1): e0005272
3. Blair CD (2023) A brief history of the discovery of RNA-mediated antiviral immune defenses in vector mosquitos. Microbiol Mol Biol Rev 87(1):e0019121
4. Tikhe CV, Dimopoulos G (2021) Mosquito antiviral immune pathways. Dev Comp Immunol 116:103964
5. Gestuveo RJ, Parry R, Dickson LB, Lequime S, Sreenu VS, Arnold MJ, Khromykh AA, Schnettler E, Lambrechts L, Varjak M, Kohl A (2022) Mutational analysis of Aedes aegypti dicer 2 provides insights into the biogenesis of antiviral exogenous small interfering RNAs. PLoS Pathog 18(1):e1010202
6. Varjak M, Maringer K, Watson M, Sreenu VB, Fredericks AC, Pondeville E, Donald CL, Sterk J, Kean J, Vazeille M, Failloux AB, Kohl A, Schnettler E (2017) Aedes aegypti Piwi4 is a noncanonical PIWI protein involved in antiviral responses. mSphere 2(3)
7. Lewis SH, Salmela H, Obbard DJ (2016) Duplication and diversification of dipteran Argonaute genes, and the evolutionary divergence of Piwi and aubergine. Genome Biol Evol 8(3): 507–518
8. Leggewie M, Cherer C, Altinli A, Gestuveo RJ, Sreenu VB, Fuss J, Vazeille M, Mousson L, Badusche M, Kohl A, Failloux AB, Schnettler E (2023) The Aedes aegypti RNA interference response against Zika virus in the context of co-infection with dengue and chikungunya viruses. PLoS Negl Trop Dis 17(7):e0011456

Chapter 8

Vector Competence Assays for RVFV in Mosquitoes

Lotty Birnberg and Núria Busquets

Abstract

Rift Valley fever virus (RVFV) is the causative agent of Rift Valley fever, a mosquito-borne zoonotic febrile illness that primarily affects ruminants, camels, and humans. It is endemic in sub-Saharan Africa and the Arabian Peninsula, albeit with the potential to spread to and establish in non-endemic countries where mosquito vectors are present. Assessing the vector competence (VC) of local mosquito populations is essential for risk evaluation and to develop more accurate control and surveillance strategies. Herein, parameters for VC experimental design, as well as protocols for mosquito rearing and laboratory techniques used for RVFV VC assays are comprehensively described.

Key words Rift Valley fever phlebovirus, Vector competence, Experimental infection, Artificial blood feeding, Laboratory colony, Field-caught mosquito

1 Introduction

Rift Valley fever phlebovirus (RVFV) belongs to the family *Phenuiviridae* (formerly the order *Bunyaviriales*), now the class *Bunyaviricetes.* Within this family, phleboviruses of medical relevance such as Sandfly fever Toscana (TOSV), Naples (SFNV), and Sicilian (SFSV) viruses are primarily transmitted by sandflies, nonetheless, RVFV is predominantly a mosquito-borne pathogen [1]. RVFV virions consist of a host cell-derived glycoprotein envelope with icosahedral symmetry and a single-stranded, predominantly negative-sense tripartite RNA genome comprised of large (L; 6.4 kb), medium (M; 3.2 kb), and small (S; 1.7 kb) segments [2]. The L segment encodes the viral RNA-dependent RNA polymerase (RdRp; L protein). The M segment contains five in-frame start codons used for the expression of nonstructural proteins (NSm, and 78 kDa protein) and two structural glycoproteins Gn and Gc [3]. The S segment is ambisense, encoding the nucleoprotein N in the genomic sense and the nonstructural protein (NSs) in the antigenomic sense [2]. L and N proteins are essential for viral RNA replication and transcription, glycoproteins are important for attachment and entry into the host

Hani Boshra (ed.), *Bunyaviruses: Methods and Protocols*, Methods in Molecular Biology, vol. 2893,
https://doi.org/10.1007/978-1-0716-4338-9_8,

cells, while nonstructural proteins are involved in pathogenesis [4]. It is assumed that the 78 kDa protein may be involved in the dissemination and transmission of the virus within the mosquito [5].

RVFV is a zoonotic arbovirus that primarily affects ruminants, camels, and humans. It is endemic to sub-Saharan Africa and the Arabian Peninsula [6, 7] albeit, with the potential to spread to, and establish in non-endemic countries through transport of infected animals and/or mosquitoes [8–11]. Since its discovery in Kenya 1930, periodic outbreaks of RVFV have been reported throughout its endemic geographical area. Outbreaks are characterized by high morbidity and increased abortion incidence in adult ruminants (e.g. sheep, goats and cattle) [12], and high mortalities in neonates and young animals [13, 14]. In humans, RVFV infections mostly result in a mild febrile illness, but a small percentage of infections may progress to a life-threatening hemorrhagic fever, meningoencephalitis, and ocular complications [13]. RVFV is primarily transmitted among ruminants by the bite of a broad spectrum of mosquitoes from the genera *Aedes* and *Culex* [15]. Epidemic transmission has been associated with heavy rain periods when RVFV-infected *Aedes* females transmit RVFV horizontally to domestic ruminants (amplifying vertebrates), and high density *Culex* populations increase the transmission to humans and other animals [13]. During inter-epidemic periods, there is evidence of a low-level circulation of the virus in wild animals and livestock [16], and it is hypothesized that RVFV could be maintained in *Aedes* mosquitoes by vertical transmission (infected mosquitoes to the eggs/progeny) [5], even though, this remains uncertain.

Since RVFV is considered as a category A priority pathogen by the National Institute of Allergy and Infectious Diseases (NIAID) and belongs to the World Organization for Animal Health (WOAH)'s list of notifiable animal diseases of concern [17], and owing to its dreadful impacts on the economy and public health in endemic countries, determining the ability of local vectors in endemic and non-endemic countries to transmit RVFV (vector competence) is essential to evaluate more accurately the interactions between virus and vector genotypes (G × G) [18] to increase the preparedness for future, potential, introduction/outbreaks. Consequently, the present chapter aims to provide a comprehensive guide of procedures to conduct vector competence assays for RVFV transmission in mosquitoes that have been proven successfully in our laboratory.

2 Materials

2.1 Mosquito Collection and Rearing

1. Ovitrap.
2. Entomological dipper.
3. White plastic trays (22 × 15 × 6 cm).
4. Sterile plastic 3 mL pipettes.
5. Strainer.
6. Tetra Goldfish Sticks.
7. Fish tank oxygen pump.
8. Sterile 100 mL plastic cups.
9. Insect rearing cages 30 × 30 × 30 cm (e.g. Bugdorm).
10. Glass 50 mL bottles.
11. Filter paper.
12. Sucrose 10% solution.
13. Black plastic 10 × 8 cm cups.
14. Hay infusion: Use 700 mg of hay per 100 mL of boiling water. Wait until it cools and preserve at 4 °C.
15. Climatic chambers (T°C, %RH, photoperiod).

2.2 Artificial Blood Feedings

1. Hemotek feeding system.
2. Heparinized bovine blood.
3. Adenosine 5′-triphosphate (ATP) disodium salt hydrate.
4. Pathogen-free (SPF) chicken skins.
5. Parafilm.
6. 1000 μL pipette/sterile tips with filter.

2.3 RVFV Strain Preparation

All the materials should be sterile. All solutions should be warmed up at 37 °C prior use. RVFV should always be handled within Biosafety Level 3 (BSL3) enhanced facilities inside a Class II Biosafety Cabinet (BSC). The use of filtering facepiece (FFP) 3 masks or Purifying Air Powered Respirators (PAPR) are mandatory, based on a previous risk assessment.

2.3.1 Propagation

1. Cell culture 75 cm^2 flasks (T75).
2. African green monkey (Vero) cells.
3. *Aedes albopictus* clone (C6/36) cells.
4. Maintenance medium: Dulbecco's Modified Eagle Medium (DMEM), Fetal bovine serum (FBS) 2%.
5. Post-infection medium: DMEM, FBS 2%, Glutamine 1%, Penicillin-streptomycin (PNS) 2%.

6. Sodium bicarbonate solution 7.5%.
7. Fetal bovine serum.
8. RVFV.
9. 1000 and 200 μL pipettes/sterile tips with filter.
10. 15 mL conical tubes.
11. 50 mL conical tubes.
12. Electronic pipetting pump.
13. 10 mL disposable pipettes
14. Incubator (37 °C, 5% CO_2).
15. Centrifuge (capacity for 50 mL conical tubes, 2000 rpm).
16. 1.5 mL cryovials.
17. Microscope (100×).
18. Water bath set at 37 °C.

2.3.2 Titration by $TCID_{50}$ Dose

1. 96-well cell-culture plates with lid.
2. African green monkey (Vero) cells.
3. RVFVs (one to be titrated and another with known titer).
4. 1.5 mL microcentrifuge tubes.
5. 1000 and 200 μL pipettes/sterile tips with filter.
6. Maintenance medium: DMEM, FBS 2%.
7. Post-infection medium: DMEM, FBS 2%, Glutamine 1%, PNS 2%.
8. Multichannel 200 μL pipette.
9. Incubator (37 °C, 5% CO_2).
10. Microscope (100×).

2.4 Vector Competence Assays

2.4.1 Preparation of Experimental Mosquito Groups for Infections

1. Battery-powered insect aspirator.
2. CO_2 horizontal-flow plate with CO_2 cylinder and flow regulator.
3. Entomological forceps.
4. Plastic containers (15 × 10 cm) with mesh lids.
5. Filter paper cylinders.
6. Cotton.
7. Sucrose 10% solution.

2.4.2 RVFV Infections

1. Class II Biosafety cabinet (BSC).
2. Insulation chambers/glove box.
3. Hemotek feeding system.

4. Feeding membrane (e.g. Pathogen-free (SPF) chicken skins or Parafilm).
5. Adenosine 5′-triphosphate (ATP) disodium salt hydrate.
6. Fresh heparinized bovine blood.
7. RVFV suspension.
8. CO_2 horizontal-flow plate with CO_2 cylinder and flow regulator.
9. Entomological forceps.
10. Cardboard cages (~5 × 5 cm) with mesh lids.
11. 1.5 mL microcentrifuge tubes.
12. DMEM, PNS 2%.

2.4.3 Maintenance of RVFV Infected Mosquitoes

1. Climatic chambers (T°C, %RH, photoperiod).
2. Insect rearing cages 25 × 25 × 25 cm (e.g. Bugdorm).
3. Cotton.
4. Sucrose 10% solution.

2.5 Mosquito Dissection and Saliva Extraction

1. Insulation chamber/glove box.
2. CO_2 horizontal-flow plate with CO_2 cylinder and flow regulator.
3. Glass microscope slides.
4. Entomological forceps.
5. Entomological needles # 4.
6. 1.5 mL microcentrifuge tubes.
7. DMEM, PNS 2%.
8. FBS 2%.
9. Sterile glass beads, 0.2 mM.
10. 20 μL pipette/tips with filter.

2.6 RVFV Detection and Quantification

2.6.1 Viral RNA Extraction

1. Class II Biosafety cabinet (BSC).
2. Mechanical homogenizer (e.g. Tissuelyser).
3. Microcentrifuge.
4. Virus RNA extraction kit.
5. Absolute EtOH.
6. Thermoblock compatible with 1.5 mL tubes.
7. 1000 and 200 μL pipettes/sterile tips with filter.
8. 1.5 mL microcentrifuge tubes.

2.6.2 RT-qPCR for RVFV Detection

1. RVFV oligonucleotides: RVS, AAAGGAACAATG GACTCTGGTCA ; RVAs, CACTTCT TACTACCATGTCCTCCAAT.
2. RVFV probe: RVP, AAAGCTTTGATATCT CTCAGTGCCCCAA.
3. AgPath-ID™ One-Step RT-PCR.
4. 1.5 mL microcentrifuge tubes.
5. RT-qPCR thermocycler and compatible PCR plates.
6. 1000, 200, 20 μL pipettes/sterile tips with filter.

2.7 Titration of RVFV Infected Samples

2.7.1 Titration of Mosquito Head/Body

1. Class II Biosafety cabinet (BSC).
2. 96-well cell-culture plates with lid
3. African green monkey (Vero) cells.
4. Maintenance medium: Dulbecco's Modified Eagle Medium (DMEM), Fetal bovine serum (FBS) 2%.
5. Antibiotic-antimycotic 1× (i.e. penicillin, streptomycin, and amphotericin B).
6. 1.5 mL microcentrifuge tubes.
7. Refrigerated microcentrifuge.
8. 1000, 200 μL pipettes/sterile tips with filter.
9. Incubator (37 °C, 5% CO_2).
10. 200 μL multichannel pipette.
11. Crystal violet fixative: 125 mL formaldehyde 37%, 52.5 mL EtOH 95%, 25 mL glacial acetic acid, 297.5 mL distilled water, 2 g crystal violet powder.
12. Distilled water.
13. Microscope (100×).

2.7.2 Titration of Mosquito Saliva

1. Class II Biosafety cabinet (BSC).
2. 6-well cell-culture plates with lid.
3. African green monkey (Vero) cells.
4. Maintenance medium: Dulbecco's Modified Eagle Medium (DMEM), Fetal bovine serum (FBS) 2%.
5. FBS.
6. Water bath set at 42 °C.
7. Agarose solution: 2 g agarose, 100 mL autoclaved water.
8. Solution B: 94 mL DMEM 2×, 4 mL adipose-derived stromal vascular fraction (SVF), 2 mL anti-anti 100×.
9. Incubator (37 °C, 5% CO_2).
10. Disposable sterile 5 mL pipettes/electronic pipetting pump.
11. 1000, 200 μL pipettes/sterile tips with filter.

12. Crystal violet fixative: 125 mL formaldehyde 37%, 52.5 mL EtOH 95%, 25 mL glacial acetic acid, 297.5 mL distilled water, 2 g crystal violet powder.
13. Spatula.
14. Distilled water.
15. Microscope (100×).

3 Methods

To assess the vector competence for RVFV in mosquitoes, several aspects (described below) should be considered. From the combination of these aspects, multiple experimental setups could be originated, and the most appropriate assay to answer the scientific question should be selected.

3.1 Experimental Design

1. Select the mosquito species and populations (*see* **Note 1**).
2. Select the virus strain and titer (*see* **Note 2**).
3. Decide which environmental conditions (temperature, relative humidity and photoperiod) are going to be tested (*see* **Note 3**).
4. Choose the appropriate infection approach (*see* **Note 4**).
5. Select the time points to be evaluated (*see* **Note 5**).
6. Choose virus detection techniques (*see* **Note 6**).

3.2 F0 and Laboratory Mosquito Rearing

Local mosquito populations are reared in NBS2 facilities. Exotic species should be reared under BSL3 conditions to prevent escapes into nature. Under laboratory conditions, the mosquito's life cycle develops inside climatic chambers.

1. To obtain F0 and/or start a mosquito colony, collect eggs and/or late-instar larvae directly from ovitraps or breeding sites using an entomological dipper (*see* **Note 7**), respectively, and take them to the laboratory.
2. Once in the laboratory, transfer eggs and/or 200 larvae to sterile-plastic trays (22 × 15 × 6 cm) containing 750 mL of water (*see* **Note 8**).
3. When eggs hatch, use a 3 mL plastic pipette to transfer 200 larvae to sterile-plastic trays (22 × 15 × 6 cm) containing 750 mL of water (*see* **Note 8**).
4. Using a strainer filtrate "old" water from larvae trays and place larvae on a new sterile-plastic tray with fresh water (*see* **Note 8**). Change water from larvae trays every 2–3 days.
5. Add fish food pellets (Tetra Goldfish Sticks) daily according to larval developmental stage as shown in Table 1.

Table 1
Larval feeding calculations

Day after hatching	*Aedes albopictus* (factor)	*Culex pipiens* (factor)	Pellet calculation[a]
1	× 0.1	× 0.05	$P = \frac{\text{larvae number} \times \text{factor per day}}{20}$
2			
3	× 0.2	× 0.1	
4			
5	× 0.3	× 0.2	
6			
7	× 0.4	× 0.3	
8			
9	× 0.5	× 0.4	
∞			

[a]The number of fish pellets should be rounded up

6. When pupae appear, cut off the tip of a 3 mL plastic pipette (to prevent hurting pupae) and transfer up to 150 pupae to a sterile 100 mL plastic cup containing water and place it inside 30 × 30 × 30 cm insect rearing cages (BugDorm). Make sure that adult density is between 2000 and 3000 individuals to prevent bottlenecks and do not affect population fitness.
7. Fill a 50 mL glass bottle with 10% sucrose solution and insert a filter paper fan in the bottle for mosquitoes to feed ad libitum. Place the bottle inside the mosquito cage and replace it every 3 days to avoid fungal contamination.
8. F0 7-to-10-days-old females can be blood-fed (*see* Subheading 3.3) to produce the first laboratory generation (F1) and/or can be used for vector competence assays (*see* Subheading 3.5).
9. To obtain F1 mosquitoes, 4 days after blood feeding, add 100 mL of dechlorinated tap water to a black plastic cup (10 × 8 cm, height × diameter) (*see* **Note 9**) and place it into the mosquito cage for females to lay eggs.
10. When eggs are laid (*see* **Note 10**), remove the oviposition cup from the mosquito cage and repeat **steps 2–8**.
11. Repeat **steps 2–9** for subsequent laboratory generations.
12. It is recommendable to properly characterize (genetically) and screen mosquito populations for the presence of endosymbionts (e.g. Wolbachia and /or Asaia) and viruses (arboviruses and ISVs) to better results explanation.

3.3 Artificial Blood Feedings

1. Remove the sucrose solution from adult mosquito cages 24–48 h before blood feeding to induce appetite (*see* **Note 11**).
2. For artificial blood feedings using a Hemotek feeding system (Fig. 1), set the Hemotek power unit at 37 °C (Fig. 1a).

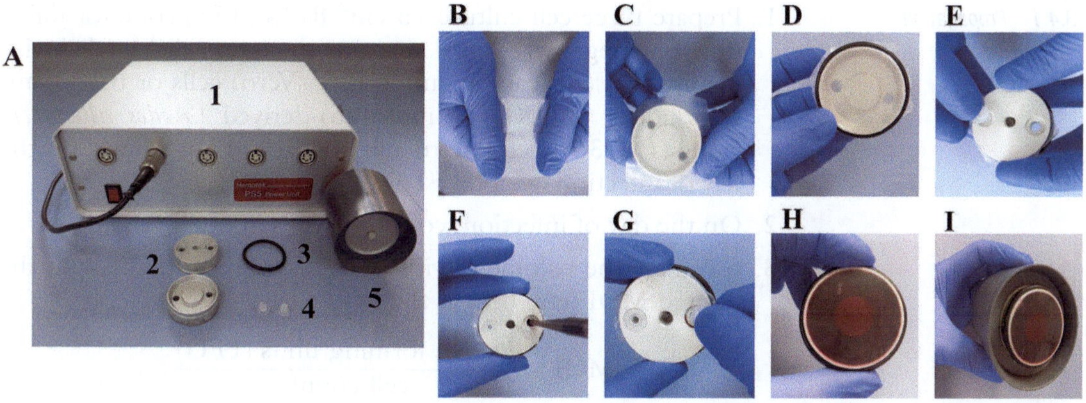

Fig. 1 Hemotek feeding system assembly, (**a**) Hemotek feeding system, 1 power unit, 2 meal reservoir, 3 O-ring, 4 plastic plugs, 5 feeder; (**b**) Feeding membrane (Parafilm); (**c**) Place feeding membrane on the meal reservoir; (**d**) Fixate membrane with the O-ring; (**e**) Place one plastic plug to cover one of the openings at the bottom of the meal rese4rvoir; (**f**) Fill the meal reservoir with blood; (**g**) Cover the second opening at the bottom of the meal reservoir; (**h**) Meal reservoir with blood ready to use; (**i**) screw meal reservoir with blood to the feeder

3. Place the feeding membrane (*see* **Note 12**) on the meal reservoir (Fig. 1b, c), fixate it with an O-ring (Fig. 1d) and place a plastic plug to cover only one of the two openings at the bottom of the meal reservoir (Fig. 1e).
4. Use a 1000 μL pipette to fill the meal reservoir with fresh heparinized bovine blood (*see* **Note 13**) through the uncovered opening (Fig. 1f).
5. Cover the opening with another plastic plug (Fig. 1g), screw the meal reservoir to the feeder (Fig. 1h, i) and connect the feeder to the Hemotek power unit.
6. After few minutes, verify the temperature of the heater and if it is warm, offer it to the mosquitoes for 1 h (*see* **Note 14**).
7. After blood feeding, discard blood remains and Parafilm following the approved biosafety protocols [19].
8. Place the meal reservoir, the O-ring, and plastic plugs in a container with a 10% household bleach solution (around 4000 ppm free chlorine) for 10 min. Scrub them until they are clean, rinse, and air-dry.

3.4 RVFV Preparation

The following protocol is for the propagation of a South African virulent strain (i.e. 56/74 strain) (*see* **Note 15**). RVFV should always be handled within Biosafety Level 3 (BSL3) enhanced facilities inside a Class II Biosafety Cabinet (BSC). The use of filtering facepiece (FFP) 3 masks or Purifying Air Powered Respirators (PAPR) are mandatory, upon previous risk assessments. All the equipment, BSC and work benches should be decontaminated, and disposable material should be discarded following the approved biosafety protocols [19].

3.4.1 Propagation

1. Prepare three cell culture 75 cm^2 flasks (T75) (one for infection, one negative control, and one to count cells) seeded with mammalian (African green monkey (Vero) cells or baby hamster kidney (BHK)) or mosquito-derived (*Aedes albopictus* clone (C6/36)) cell line to achieve 90% of confluency on the day of infection (*see* **Note 16**).
2. On the day of infection, count the cells in one of the flasks.
3. Calculate the volume of virus needed to infect the flask at the desired MOI (e.g. 0.1) as follows:

$$\mathrm{MOI} = \frac{\text{Plaque forming units (PFU)}}{\text{cell count.}}$$

and then,

$$\mathrm{RVFV\ (mL)} = \frac{\mathrm{PFU}}{\text{RVFV strain titer } \frac{\mathrm{PFU}}{\mathrm{mL}}}$$

4. Add the volume of virus required obtained in **step 3** to a sterile 15 mL conical tube and complete this volume to 1 mL with DMEM maintenance medium. Set aside.
5. Remove medium from the flasks and label, one with RVFV and the other as negative control. Include the cell line, date and technician.
6. Add 1 mL of DMEM maintenance medium to flask labeled as the negative control.
7. Add the virus solution to the RVFV propagation flask.
8. Rock gently both flasks to ensure that the virus/medium is evenly distributed through the cell monolayer.
9. Incubate at 37 °C, 5% CO_2 for 1 h 30 min. rocking the flasks every 30 min.
10. Once the adsorption time has passed, slowly add 19 mL of post-infection medium.
11. Check both flasks to verify the cell monolayer status and return flasks to the incubator.
12. Check flasks every day until cytopathic effect (CPE) is observed (*see* **Note 17**).
13. To harvest the virus, collect all the supernatant from RVFV flask using an electronic pipetting pump, and transfer it to a sterile 50 mL conical tube (*see* **Note 18**).
14. Harvest supernatant from the negative control as in **step 13**.
15. Adjust the FBS to 10% and adjust pH by adding sodium bicarbonate 7.5% solution until the color of the media turns into pink. The stock should not be frozen with an acid pH.
16. Centrifuge at 4 °C, 2000 rpm for 10 min.

17. Aliquot supernatant (e.g. 1 mL each) without disrupting pellet and label vials with virus strain, cell line used, passage number, technician, and date of harvest (*see* **Note 19**).
18. Store at −75 °C.

3.4.2 Titration by $TCID_{50}$ Dose

Titration process is performed by duplicate.

1. Seed two 96-well cell-culture plates with Vero cells to achieve 90% of confluency on the day of titration.
2. Prepare tenfold serial dilutions of RVFV being titrated as follows: add 900 μL of DMEM at room temperature to ten 1.5 mL microcentrifuge tubes and label −1 up to −10. Then add 100 μL of RVFV to the first tube and mix well. Transfer 100 μL of this dilution to the second tube performing up and down with the micropipette at least three times. Repeat this process until reaching the tenth tube. It is crucial to change the tip between dilutions.
3. Add 900 μL of DMEM to another 1.5 mL microcentrifuge tube and dilute 100 μL of RVFV stock with known titer as the positive control (*see* **Note 20**).
4. Label the 96-well plate columns (1–12) with the number of the dilutions: from −1 to −10, positive control and negative control.
5. Remove medium with a multichannel pipette without touching the cell monolayer.
6. Add 100 μL of the corresponding dilution to wells A–H.
7. Add 100 μL of the harvested supernatant of the negative control to wells A–H in column 12.
8. Add 100 μL of the positive control to wells A–H in column 11.
9. Place the plate lid and gently rock the plate.
10. Incubate at 37 °C, 5% CO_2 for 1 h 30 min.
11. Using a multichannel pipette add 100 μL of post-infection medium and incubate until CPE is observed (around 7 days).

3.4.3 Titer Calculation by 50% Tissue Culture Infectious Dose ($TCID_{50}$) Assay (Reed-Muench, 1938)

1. At 7 dpi, place the 96-well plate on a microscope. On a plate template register wells showing CPE as positive and wells without CPE as negative. Verify controls. All positive control replicates (A–H) should present CPE and all negative control replicates should have the cell monolayer preserved.
2. Complete Table 2 with the following calculations:
   A. The cumulative number of positive wells beginning from the highest dilution (−10).
   B. The cumulative number of negative wells beginning from the lowest dilution (−1).

Table 2
Data for $TCID_{50}$ titer calculation

Dilution	# Positive wells	# Negative wells	A	B	Ratio $\frac{A}{(A+B)}$	% Positive
−1						
−2						
−3						
−4						
−5						
−6						
−7						
−8						
−9						
−10						

Positive well, CPE
Negative well, no CPE
(A) The cumulative number of positive wells beginning from the highest dilution (−10)
(B) The cumulative number of negative wells beginning from the lowest dilution (−1)
Ratio of A/(A + B) per dilution (same row)
Percentage of the ratio per dilution

Ratio of $\frac{A}{(A+B)}$ per dilution.
Percentage of the ratio per dilution.

3. Using the information on Table 2 calculate the proportional distance with the equation:

$$\text{prop.dist.} = \frac{(\% > 50\%) - 50\%}{(\% > 50\%) - (\% < 50\%)}$$

4. Then, $\log TCID_{50} =$ (log of the dilution >50%) + (prop. dist. × log of the dilution factor).
5. Calculate the reciprocal of this number to obtain the titer in $TCID_{50}$ /0.1 mL and convert it to $TCID_{50}$ /mL.

3.5 Vector Competence Assays

3.5.1 Preparation of Mosquito Groups for Experimental Infections

1. About 2 or 3 days before the infections, aspirate 7–10-days-old females (field or laboratory) directly from adult cages using battery-powered insect aspirators (Fig. 2b).
2. Place the collection container of the aspirator on the surface of a CO_2 horizontal-flow plate (5 L/min CO_2) to anesthetize specimens (Fig. 2c) (*see* **Note 21**).
3. Once they are asleep, open the container and spread specimens through the plate to classify and count them (Fig. 2d, e).

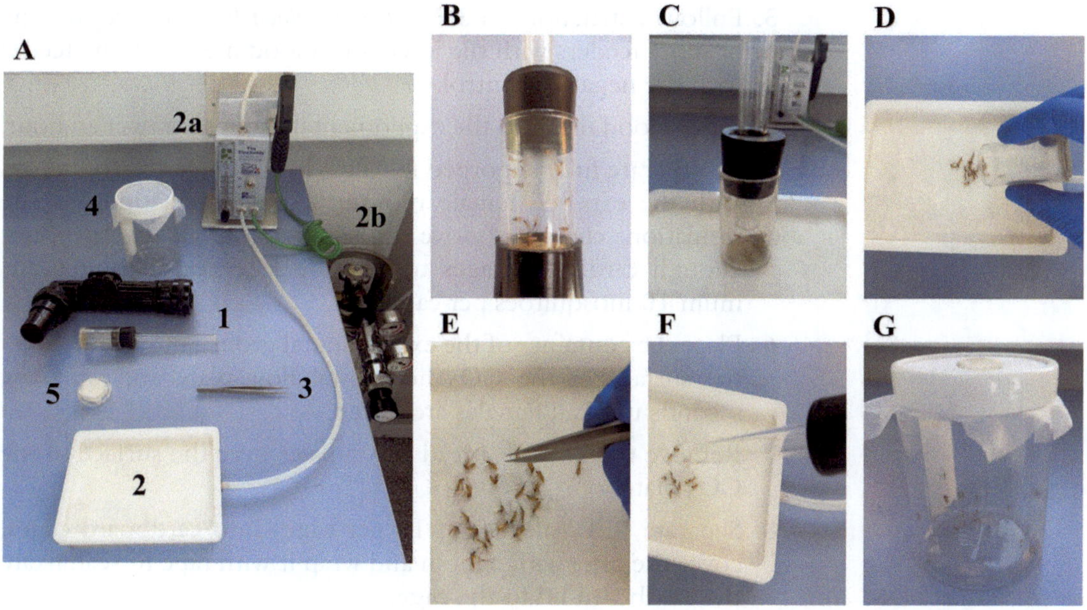

Fig. 2 Preparation of mosquito groups for experimental infections. (**a**) Materials for experimental groups preparation, **1** battery-powered insect aspirator with collection container and tube, **2** CO_2 horizontal-flow plate connected to a flow regulator (**2a**) and CO_2 cylinder (**2b**), **3** entomological forceps, **4** plastic container with mesh lid, **5** cotton pledget soaked in 10% sucrose solution; (**b**) Collection container with aspirated mosquitoes; (**c**) Collection container on the surface of the CO_2 horizontal-flow plate to anesthetize specimens; (**d**) Specimens scattered for female classification; (**e**) female selection; (**f**) females aspiration for experimental groups; (**g**) mosquito experimental group inside plastic container with mesh lid and sucrose pledget

4. Separate around 30 females per group (*see* **Note 22**). Aspirate and place them inside an appropriate container with a mesh on its lid (Fig. 2f, g). Prepare as many groups as needed considering that per time point a minimum of 30 fully engorged females are recommended. The mortality rate of the population used must be considered as well.
5. Place a cotton pledget soaked in sucrose 10% for mosquitoes to feed ad libitum and recover from anesthesia (Fig. 2g).

3.5.2 RVFV Infections

If the mosquitoes are reared at BSL2 facilities, enter experimental groups to BSL3 1 or 2 days before the infection for specimens' acclimatization.

RVFV Oral Exposure

1. Prepare the infectious blood meal, mixing fresh bovine blood with the virus to obtain the desired viral dose concentration (e.g. 7 log $TCID_{50}$/mL).
2. Prepare the negative control, mixing fresh bovine blood with the same volume of virus but instead of virus use the harvested negative control.

3. Follow instructions in 3.3 *artificial blood feeding*. Prepare one or more feeders with the infectious blood meal and one feeder with the negative control.
4. Offer blood meals to the experimental groups for over an hour.
5. To separate fully engorged females (FEF), enter the containers with the exposed female mosquitoes, one at a time, to the insulation chamber/glove box (*see* **Note 23**). Introduce enough cardboard cages to separate FEF considering maximum 10 mosquitoes per cage.
6. Place the container of the experimental groups upside down on the surface of the CO_2 horizontal flow plate and wait few seconds until all females are asleep.
7. Remove the lid and spread the females over the surface of the CO_2 plate.
8. Separate FEF (maximum 10 per cage) into cardboard cages, cover the cage with a mesh and wrap it with tape to seal, fixate the cardboard lid to the cage.
9. Place cotton pledgets soaked in sucrose 10% on top of the mesh for mosquitoes to feed ad libitum and recover from anesthesia.
10. Separate a subset of FEF (e.g. 3) per group (RVFV exposed and not exposed) as day 0 post exposure (0 dpe) controls.
11. Place each control mosquito individually inside 1.5 mL microcentrifuge tubes containing 500 μL of DMEM, PNS 2% and 2 mm glass beads.
12. Count all specimens to ensure that the number of FEF and unfed females agrees with the total number of mosquitoes within the container of each exposed group.
13. Label and number individual FEF cages according to experimental group.
14. Place experimental groups into bugdorms labeled accordingly and place them into the climatic chamber.
15. Decontaminate all the equipment, BSC and work benches, discard all disposable material following the approved biosafety protocols [19].

3.6 Maintenance of RVFV-Infected Mosquitoes

Maintenance to complete the extrinsic incubation period is conducted inside climatic chambers.

1. Soak cotton pledgets with a sucrose 10% solution.
2. Place one sucrose pledget over the mesh lid of each cardboard cage for mosquitoes to feed ad libitum.
3. Change the sucrose pledget daily until the end of the experiment.
4. Record specimen mortality daily.

3.7 Mosquito Dissection and Saliva Extraction

At every selected time point, a subset of females (*see* **Note 24**) will be sacrificed to calculate VC rates. Dissections will be performed inside the insulation chamber provided with the CO_2 horizontal flow plate, sterile forceps, sterile entomological needles, glass slides, disinfectant spray bottle (e.g. Virkon 1%) and a sharp waste container with disinfectant solution.

1. Enter cardboard cages with RVFV-exposed specimens to the insulation chamber.
2. Anesthetize specimens as previously described in **step 2** Subheading 3.5.1.
3. Open the cage and spread mosquitoes on the surface of the CO_2 horizontal-flow plate.
4. Grab one mosquito from its leg using entomological forceps and place it on a glass microscope slide for dissection.
5. Use sterile entomological needles to remove legs and wings (*see* **Note 25**). Discard limbs.
6. Repeat **steps 4–5** replacing needles and glass slides per mosquito.
7. Pipette 7 μL of FBS, remove the pipette tip from pipette and introduce the proboscis of each mosquito inside the tip for saliva recollection.
8. Label slide accordingly and record the time at which the proboscis was introduced inside the pipette tip. Wait 30 min.
9. Remove tip from proboscis and pipette its content in a 1.5 mL microcentrifuge tube containing 193 μL of DMEM, PNS 2%.
10. After salivation, decapitate mosquitoes using new sterile needles. Store head inside a 1.5 mL microcentrifuge tube containing 500 μL of DMEM, PNS 2% and 2 mm glass beads.
11. Store mosquito's body inside a 1.5 mL microcentrifuge tube containing 500 μL of DMEM, PNS 2% and 2 mm glass beads.
12. Label all tubes accordingly.
13. Decontaminate all the equipment and work benches, discard all disposable material following the approved biosafety protocols (19).
14. Store all samples at −75 °C until processing.

3.8 RVFV Detection and Quantification

In BSL3 laboratories, all samples should be handled on ice during bench work. After virus inactivation, RNA extractions can be performed without filtering facepiece (FFP) 3 masks or Purifying Air Powered Respirators (PAPR), upon previous risk assessment. All the equipment, BSC and work benches should be decontaminated, and disposable material should be discarded following the approved biosafety protocols [19].

3.8.1 Viral RNA Extraction

The following protocol is adapted for viral RNA extraction using NucleoSpin® RNA Virus kit (Macherey-Nagel, Düren, Germany). **Steps 1–5** are conducted in confined BSL3 laboratories. The use of filtering facepiece (FFP) 3 masks or Purifying Air Powered Respirators (PAPR) are mandatory, upon a previous risk assessment. From **step 6**, procedures can be performed in BSL3 laboratories without filtering facepiece (FFP) 3 masks or Purifying Air Powered Respirators (PAPR), as no infectious viruses are handled at that time.

1. Homogenize head and body samples on a mechanical homogenizer (30 Hz, 1 min). In head samples, verify that the head is homogenized, otherwise repeat step.
2. Centrifuge samples at 10,000 g, 4 °C for 5 min.
3. Add 600 μL of RAV1 buffer (with RNA carrier) to 1.5 mL microcentrifuge tubes according to the number of samples to be processed. Label accordingly.
4. Add 150 μL of the supernatant of each sample (medium with homogenized head/body) to individual tubes from **step 3** (*see* **Note 26**). Cover the tubes. Homogenize by inversion 3–5 times.
5. On a thermoblock, incubate all samples at 70 °C for 5 min. Give a centrifuge spin to lower condensation.
6. To continue with viral RNA extractions, follow manufacturer's instructions. Use sterile PBS as negative control. Thaw an aliquot of positive control (*see* **Note 26**). Elute with 50 μL nuclease-free water. Store at −80 °C.

3.8.2 RT-qPCR for RVFV Detection

The following protocol is adapted for RVFV G2 gene amplification using AgPath-ID™ One-Step RT-PCR Reagents (Applied Biosystems, Inc., Foster City, CA, U.S.A.) in a 20 μL reaction volume. RVFV primers, probe and RT-PCR conditions are given in Drosten et al. (20).

1. In a 1.5 mL microcentrifuge tube add (per sample including negative and positive extraction and amplification controls): 2× RT-PCR Buffer, 10 μL; 25× RT-PCR Enzyme mix, 0.8 μL; RVS (10 μM), 2 μL; RVAs (10 μM), 2 μL; RVP (10 μM), 0.4 μL; and nuclease-free water, 2.8 μL. Mix well performing up and down with the micropipette.
2. Dispense 18 μL of the previous mix in individual wells.
3. Add samples and controls for easier result interpretation. Seal plate.
4. For cycling profile, program thermocycler as follows: reverse transcription, 45 °C for 10 min; initial denaturation, 95 °C for 10 min; 45 cycles, 95 °C for 15 s and 57 °C for 35 s. Fluorescence detection at the latter.

5. For number of copies quantification, standard curves should be generated performing serial tenfold dilutions of in vitro RVFV RNA which was previously quantified (*see* **Note 27**).

3.9 Titration of RVFV-Infected Samples

Analyzing the mosquito's body and head, infection with the virus and dissemination of the virus are assessed respectively, while virus transmission is assessed by virus detection in the saliva.

3.9.1 Titration of Mosquito Head/Body

1. The day before titration, prepare a 300,000 cell/mL suspension (Vero cells in DMEM maintenance medium). Dispense 100 μL of the cell suspension in each well of a 96-well cell-culture plate (i.e. 30,000 cell/well). Incubate at 37 °C, 5% CO_2.
2. The day of titration, thaw body and head samples. Keep them on ice.
3. Centrifuge at 10,000 g for 5 min, 4 °C.
4. Take 200 μL of the supernatant; avoid touching the beads and dispense the volume in a 1.5 mL microcentrifuge tube. Label accordingly. Keep them on ice.
5. Prepare a 1/10 dilution using 20 μL of the supernatant and 180 μL of DMEM supplemented with FBS 2%, anti-anti 1× (*see* **Note 28**).
6. Prepare a 1/100 dilution using 20 μL of the previous dilution and 180 μL of DMEM supplemented with FBS 2%, anti-anti 1×.
7. Take the plate out of the incubator, place it inside the BSC and remove culture medium by aspiration.
8. Dispense 100 μL of the undiluted supernatant in the 1-A well of the plate.
9. Dispense 100 μL of the 1/10 dilution in the 2-A well.
10. Dispense 100 μL of the 1/100 dilution in the 3-A well.
11. Repeat **steps 8–10** with all the samples in the following wells.
12. Incubate at 37 °C, 5% CO_2 for 7 days.
13. On day 7, remove plate from incubator, place it inside the BSC and aspirate cell-culture medium from wells using a multichannel pipette.
14. Using a multichannel pipette, add 200 μL of crystal violet fixative per well.
15. Incubate at RT for at least 30 min.
16. Invert the plate to remove crystal violet fixative and gently rinse with distilled water until water runs clear.
17. Air-dry plates.

18. Determine presence (positive, clear areas on dark purple background) or absence (negative, dark purple background) of virus per well.
19. Decontaminate all the equipment, BSC, and work benches, discard all disposable material following the approved biosafety protocols [19].

3.9.2 Titration of Mosquito Saliva

1. The day before titration, seed a six-well cell-culture plate with an 800,000 cell/2 mL suspension per well using Vero cells in DMEM supplemented with FBS 10%. They must be confluent on the day of titration.
2. The day of titration, thaw saliva samples and keep them on ice.
3. Set the water bath at 42 °C and introduce the bottle containing solution B.
4. Melt the agarose solution in the microwave. Place it in the water bath.
5. Inside the BSC, aspirate medium from the wells, one plate at a time to avoid drying off the cells.
6. Add 265 μL of DMEM maintenance medium per well.
7. Add 35 μL of the saliva sample in each well, cover the plate with the lid and gently rock the plate.
8. Incubate for 1 h at 37 °C, 5% CO_2. Rock plates after 30 min.
9. By the end of the incubation period, mix the agarose solution with solution B and let it stand at room temperature to lower its temperature before dispensing it in the wells (*see* **Note 29**).
10. Slowly add 4 mL of the mix in each well using a sterile 5 mL disposable pipette and an electronic pipetting pump.
11. Wait until agarose solidifies and incubate for 7 days.
12. At day 7, place the plate in the BSC and remove the agarose from wells using a spatula. Be careful not to touch the cell monolayer.
13. Add 2 mL crystal violet fixative to each well.
14. Incubate at RT for at least 30 min.
15. Invert plate to eliminate crystal violet fixative and rinse with distilled water until water runs clear.
16. Air-dry plate.
17. Count the number of plaques (clear areas on dark purple background) per well. Report results as PFU/mL considering the volume of the inoculated saliva sample (0.035 mL).
18. Decontaminate all the equipment, BSC, and work benches, discard all disposable material following the approved biosafety protocols [19].

3.10 Calculation of Vector Competence Rates

To assess vector competence for RVFV, infection, disseminated infection and transmission rates (IR, DIR and TR, respectively), and transmission efficiency (TE) are estimated.

$$\mathrm{IR} = \frac{\#\mathrm{FEF\ RVFV\ positive\ bodies}}{\#\mathrm{total\ FEF}}$$

$$\mathrm{DIR} = \frac{\#\mathrm{FEF\ RVFV\ positive\ heads}}{\#\mathrm{FEF\ RVFV\ positive\ bodies}}$$

$$\mathrm{TR} = \frac{\#\mathrm{FEF\ RVFV\ positive\ saliva}}{\#\mathrm{FEF\ RVFV\ positive\ heads}}$$

$$\mathrm{TE} = \frac{\#\mathrm{FEF\ RVFV\ positive\ saliva}}{\#\mathrm{total\ FEF}}$$

4 Notes

1. Laboratory and field-caught mosquitoes present advantages and disadvantages. By using laboratory-reared mosquitoes there are no season or geographic restrictions, large sample sizes can be produced, and it is a characterized population; however, there is loss of microbiota and genetic variability. Whereas, by using field-caught populations, fully relevant microbiome and a high genetic variability are granted, though, with season and geographic restrictions, unknown individual history and, probably, lower sample sizes. When laboratory mosquitoes are used, it is recommended not to exceed the sixth laboratory generation (F6).
2. Most RVFV strains are Biosafety Level 3 enhanced (BSL-3E) pathogens and require government and institutional approvals, appropriate guidelines for handling and transportations, purpose-built facilities. Viral titer will depend on the scientific question/s. It is recommendable to mimic the viremia found in natural reservoirs (~7 $\log_{10}$ TCID_{50}/mL).
3. Climatic chambers could be used to simulate the environmental conditions that better respond the scientific question/s.
4. There are multiple ways to infect mosquitoes with RVFV. Mainly, by oral exposure directly from live viremic animals or using artificial blood feeding systems. If live animals are used, bioethics approvals and highly trained personnel with animal experimentation license are required. Artificial oral exposure to the virus facilitates approvals and wet-lab procedures.
5. Conventionally, three time points (besides T0, the day of virus exposure) are selected at 7, 14, and 21 days post-exposure (dpe) to assess infection, dissemination, and transmission of the virus. However early transmission of the virus could be evaluated at 3dpe.

6. RVFV can be detected in different mosquito body-parts (bodies, legs, head) and saliva by end-point RT-PCR or analyzed by RT-qPCR (*see* 3.8) to quantify the viral load and/or in mammalian cell culture to detect infectious virus (*see* 3.4.2).
7. Collecting late-instar larvae together with water from the field is recommendable to preserve most of the field-acquired microbiota in adult specimens and to increase survival if no extra food is planned.
8. The use dechlorinated and oxygenated tap water to rear laboratory colonies is the most common practice. It could be obtained letting tap water to rest in a dark container (or covered with a black plastic bag) for 48 h. However, if water from the original breeding site is logistically accessible, it could be used to preserve field-acquired microbiota in the adult population. If so, make sure to filtrate water to avoid introducing new larvae and/or predators. In the case of *Aedes albopictus* eggs, use autoclaved water to stimulate hatching.
9. Since *Aedes* mosquitoes lay eggs on moist surfaces, place a filter paper cone inside the cup so it contacts the water. In the case of *Culex* mosquitoes add 1 mL of hay infusion to the water as an attractant.
10. Filter paper cones containing *Aedes'* eggs could be preserved dry for long periods of time. To increase hatching rates, it is recommendable to let the eggs dry for 1 week but do not exceed 3 months. To stimulate egg hatching, use autoclaved water.
11. *Aedes* mosquitoes need a shorter starvation period since they are aggressive biters. *Culex* mosquitoes need longer starvation periods to increase appetite. In colonies reared between 22 °C—28 °C starvation periods of 24 h and 48 h are required for *Aedes* and *Culex pipiens* respectively.
12. There are many types of feeding membranes suitable for different species. Collagen membranes, pathogen-free chicken skins or Parafilm. For colony maintenance, Parafilm is a great option since it is less expensive, and no bioethical issues should be considered. To use Parafilm as a feeding membrane, cut a small piece and expand it from the ends using the thumbs to stretch. Parafilm loses its shine when the appropriate thickness is accomplished, stop when it first turns matte otherwise it is possible to have a leak with temperature exposure.
13. Optionally, adenosine 5′-triphosphate (ATP) disodium salt hydrate (0.5 mM) could be added as phagostimulant. For colony maintenance, fresh or frozen (previously thawed) could be used.

14. For experimental infections, a 1-hour feeding period is established to prevent virus titer loss. If feeding rates are not satisfactory, a second feeding could be done the next day preparing a new infectious meal and leaving females without sucrose to increase appetite. For mosquito colony maintenance, feeding periods could be longer and as frequent as needed.
15. South African virulent 56/74 RVFV strain viral stock kindly provided by Alejandro Brun, INIA. Ideally virus propagations protocols should be adapted according to virus strains and cell line used, since the multiplicity of infection (MOI) required may vary in different cell line clones, as well as the day of collection. A comprehensive description of RVFV titration, propagation and storage is provided by Smith et al., 2019 [21].
16. Each T75 may yield ~20 mL of virus suspension. Cytopathic effect is easily seen in mammalian cells. Since in C6/36 cells there is no cytopathic effect a growth curve must be required to establish the collection day. C6/36 can be used to increase virus titer and should be incubated at 28 °C.
17. Depending on the strain, MOI, and cell line, the time to harvest may vary between 1 and 7 dpi.
18. Prior harvest, hitting gently with the hand may help disrupt cells from the flask walls and liberate virus particles.
19. Make aliquots of smaller microtubes to prevent titer loss in virus stocks owing to thaw/refreeze cycles.
20. If RVFV stock is limited, smaller volumes can be used. For serial dilutions, 50 μL RVFV in 450 μL DMEM is enough to dispense 20 μL per well (A–H) which is enough to completely cover the cell monolayer.
21. Make sure not to prolong CO_2 exposure more than 3 min to prevent specimen death.
22. Negative control group could be smaller than virus-exposed groups since its purpose is to control mortality owing to the virus and check that there has been no contamination at the beginning (day 0) and at the end of the experiment.
23. Most of the times the gloves from the insulation chamber are larger than the technicians' hands, to facilitate manipulation, put latex gloves over the insulation chamber gloves for a better fit.
24. The number of specimens to be sacrificed per time point will depend on the total number of FEF.
25. Make sure to cut the limbs from its base, the presence of stumps difficult saliva extractions since mosquitoes are awake and move.
26. A positive control stock for several extractions can be prepared when samples are being inactivated. For this purpose, make a

dilution of RVFV in sterile PBS to get a weak positive for avoiding potential cross contamination. Then, add 150 μL of RVFV dilution in 600 μL of RAV1 buffer (with RNA carrier) and incubate at 70 °C for 5 min. Store at −75 °C.

27. The limit of sensitivity is 0.09 $TCID_{50}$ per reaction, which corresponded to a Ct value of 39.19 (**see** [22]).
28. Dilutions are used to identify false positives. Since contamination destroys cell monolayers as high virus titers do. By diluting samples, contamination is also diluted.
29. The temperature of the agarose solution should be ~37 °C to avoid cell damage and agarose solidification.

References

1. Turell MJ, Dohm DJ, Mores CN, Terracina L, Wallette DL Jr, Hribar LJ et al (2008) Potential for North American mosquitoes to transmit Rift Valley fever virus. J Am Mosq Control Assoc 24:502–507. https://doi.org/10.2987/08-5791.1
2. Gaudreault NN, Indran SV, Balaraman V, Wilson WC, Richt JA (2019) Molecular aspects of Rift Valley fever virus and the emergence of reassortants. Virus Genes 55(1):1–11. https://doi.org/10.1007/s11262-018-1611-y
3. Pepin M, Bouloy M, Bird BH, Kemp A, Paweska J (2010) Rift Valley fever virus (*Bunyaviridae: Phlebovirus*): an update on pathogenesis, molecular epidemiology, vectors, diagnostics and prevention. Vet Res 41(6):61. https://doi.org/10.1051/vetres/2010033
4. Bird BH, Albariño CG, Nichol ST (2007) Rift Valley fever virus lacking NSm proteins retains high virulence in vivo and may provide a model of human delayed onset neurologic disease. Virology 362(1):10–15. https://doi.org/10.1016/j.virol.2007.01.046
5. Crabtree MB, Kent Crockett RJ, Bird BH, Nichol ST, Erickson BR, Biggerstaff BJ et al (2012) Infection and transmission of Rift Valley fever viruses lacking the NSs and/or NSm genes in mosquitoes: potential role for NSm in mosquito infection. PLoS Negl Trop Dis 6(5): e1639. https://doi.org/10.1371/journal.pntd.0001639
6. Ikegami T, Makino S (2011) The pathogenesis of Rift Valley fever. Viruses 3(5):493–519. https://doi.org/10.3390/v3050493
7. Boshra H, Lorenzo G, Busquets N, Brun A (2011) Rift valley fever: recent insights into pathogenesis and prevention. J Virol 85(13): 6098–6105. https://doi.org/10.1128/JVI.02641-10
8. Chevalier V, Pepin M, Plee L, Lancelot R (2010) Rift Valley fever–a threat for Europe? Euro Surveill 15:19506
9. Rolin AI, Berrang-Ford L, Kulkarni MA (2013) The risk of Rift Valley fever virus introduction and establishment in the United States and European Union. Emerg Microbes Infect 2:e81. https://doi.org/10.1038/emi.2013.81
10. Arsevska E, Hellal J, Mejri S, Hammami S, Marianneau P et al (2016) Identifying areas suitable for the occurrence of Rift Valley fever in North Africa: implications for surveillance. Transbound Emerg Dis 63:658–674. https://doi.org/10.1111/tbed.12331
11. Gale P, Kelly L, Snary EL (2015) Pathways for entry of livestock arboviruses into Great Britain: assessing the strength of evidence. Transbound Emerg Dis 62:115–123. https://doi.org/10.1111/tbed.12317
12. Budasha NH, Gonzalez JP, Sebhatu TT, Arnold E (2018) Rift Valley fever seroprevalence and abortion frequency among livestock of Kisoro district, South Western Uganda (2016): a prerequisite for zoonotic infection. BMC Vet Res 14(1):271. https://doi.org/10.1186/s12917-018-1596-8
13. Nielsen SS, Alvarez J, Bicout DJ, Calistri P, Depner K, Drewe JA et al (2020) Rift Valley fever – epidemiological update and risk of introduction into Europe. EFSA J 18(3): 6041, 72 pp. https://doi.org/10.2903/j.efsa.2020.6041
14. Wright D, Kortekaas J, Bowden TA, Warimwe GM (2019) Rift Valley fever: biology and epidemiology. J Gen Virol 100(8):1187–1199. https://doi.org/10.1099/jgv.0.001296

15. Lumley S, Horton DL, Hernandez-Triana LLM, Johnson N, Fooks AR, Hewson R (2017) Rift Valley fever virus: strategies for maintenance, survival and vertical transmission in mosquitoes. J Gen Virol 98(5):875–887. https://doi.org/10.1099/jgv.0.000765
16. Rostal MK, Evans AL, Sang R, Gikundi S, Wakhule L, Munyua P (2010) Identification of potential vectors of and detection of antibodies against Rift Valley fever virus in livestock during interepizootic periods. Am J Vet Res. 71(5):522–526. https://doi.org/10.2460/ajvr.71.5.522
17. Hartman A (2017) Rift Valley fever. Clin Lab Med 37(2):285–301. https://doi.org/10.1016/j.cll.2017.01.004
18. Bohers C, Vazeille M, Bernaoui L, Pascalin L, Meignan K, Mousson L (2024) *Aedes albopictus* is a competent vector of five arboviruses affecting human health, greater Paris, France, 2023. Euro Surveill 29(20):2400271. https://doi.org/10.2807/1560-7917.ES.2024.29.20.2400271
19. Pondeville E, Failloux AB, Simard F, Volf P, Crisanti A, Haghighat-Khah RE et al (2023) Infravec2 guidelines for the design and operation of containment level 2 and 3 insectaries in Europe. Pathog Glob Health 117(3):293–307. https://doi.org/10.1080/20477724.2022.2108639
20. Drosten C, Göttig S, Schilling S, Asper M, Panning M, Schmitz H et al (2002) Rapid detection and quantification of RNA of Ebola and Marburg viruses, Lassa virus, Crimean-Congo hemorrhagic fever virus, Rift Valley fever virus, dengue virus, and yellow fever virus by real-time reverse transcription-PCR. J Clin Microbiol 40(7):2323–2330. https://doi.org/10.1128/JCM.40.7.2323-2330.2002
21. Smith MR, Schirtzinger EE, Wilson WC, Davis AS (2019) Rift Valley fever virus: propagation, quantification, and storage. Curr Protoc Microbiol 55(1):e92. https://doi.org/10.1002/cpmc.92
22. Busquets N, Xavier F, Martín-Folgar R, Lorenzo G, Galindo-Cardiel I, del Val BP et al (2010) Experimental infection of young adult European breed sheep with Rift Valley fever virus field isolates. Vector Borne Zoonotic Dis 10(7):689–696. https://doi.org/10.1089/vbz.2009.0205

Chapter 9

Experimental Infection of Ticks with SFTSV

Jiao Xu and Xue-Jie Yu

Abstract

Many members in the class *Bunyaviricetes* are tick-borne viruses. Experimental viral infection of ticks is greatly needed to better understand the tick–virus interactions. To date, multiple laboratory techniques for the viral infection of ticks have been documented. Here we describe two methods of infecting *Haemaphysalis longicornis* ticks with SFTSV: infecting ticks through microinjection and direct feeding on infected animal hosts. Both methods have been shown to be effective in infecting ticks.

Key words *Bunyaviricetes*, SFTSV, Ticks, Viral Infection, Microinjection, Tick feeding

1 Introduction

Ticks are obligate blood-sucking ectoparasites that can parasitize on animals and humans [1]. They are considered to be important pathogen vectors second only to mosquitoes as they could transmit over 170 pathogens worldwide, including viruses, bacteria, spirochetes, and protozoa [2]. Among the tick-borne viruses (TBVs), the members of the *Bunyaviricetes* like *Nairoviridae* and *Phenuiviridae* are of great importance to public health [3]. For example, Crimean-Congo Haemorrhagic Fever Virus (CCHFV) is considered as one of the most serious tick-borne bunyaviruses, causing viral hemorrhagic fever in humans as well as abortions in domestic animals. The main vectors of CCHFV are *Hyalomma* ticks. The severe fever with thrombocytopenia syndrome virus (SFTSV), transmitted by *Haemaphysalis longicornis* ticks, is responsible for an emerging hemorrhagic fever in humans [4, 5]. Heartland Bandavirus (HRTV) is closely related to *Amblyomma americanum* ticks [6].

To reveal the viral vector competency of ticks, the routes of transmission, and to understand their susceptibility to control measures, experimental viral infection of ticks is greatly needed. To date, multiple laboratory techniques for the viral infection of ticks

Hani Boshra (ed.), *Bunyaviruses: Methods and Protocols*, Methods in Molecular Biology, vol. 2893,
https://doi.org/10.1007/978-1-0716-4338-9_9,

have been documented [7], including infection through injection, immersion, co-feeding, capillary feeding, membrane-feeding, and direct feeding on infected hosts.

Here we describe two methods of infecting *Haemaphysalis longicornis* ticks with SFTSV: infecting ticks through injection and direct feeding on infected hosts. Both of them have been shown to be effective methods of viral infection.

2 Materials

2.1 SFTSV Culturing

1. 1× phosphate buffered saline (PBS, 0.01 M): 137 mM NaCl, 2.7 mM KCl, 10 mM Na_2HPO_4, 2 mM KH_2PO_4, PH 7.2–7.4.
2. Trypsin.
3. Dulbecco's modified Eagle's medium (DMEM) growth solution: 10% fetal bovine serum, 10 U/mL penicillin and streptomycin.
4. DMEM maintenance solution: 2% fetal bovine serum, 10 U/mL penicillin, and streptomycin.
5. Cell culture flask: T25, T75.
6. Vero E6 cells.

2.2 Tick Feeding

1. 25% chloral hydrate: dissolve 25 g chloral hydrate powders in 100 mL water.
2. Collar: neckband at 15–20 cm, depth at 9 cm.
3. Gauze: 300 mesh.
4. White foam board: 1 cm thick.
5. Adhesive: Karma adhesive, double-sided adhesive (DSA).
6. Tweezers.
7. New Zealand white rabbits.

2.3 Detection of SFTSV

1. RNA extraction kit (QIAGEN, Hilden, Germany).
2. RT First Strand cDNA Synthesis Kit (Servicebio, Wuhan, China).
3. One-step primer script RT-PCR kit (TaKaRa, Dalian, China).
4. 75% ethanol: diluted with DEPC water.
5. Sterilized magnetic beads.
6. Liquid nitrogen.
7. Primers and probe (Table 1).

Table 1
Gene sequences of primers and probe

Primers/Probe	Sequences (5′-3′)	References
Probe	Hex-TGCCTTGACGATCTT-MGB	[8]
SFTSV S F	ACCTCTTTGACCCTGAGTTWGACA	
SFTSV S R	CTGAAGGA GACAGGTGGAGATGA	
BNYS1-F	TCTTCTCCATCAAGAACAGC	
BNYS1-R	TTCGACAAAATTAGACCTCC	

2.4 Microinjection

1. 50 mL disposable centrifuge tubes.
2. Solution used for cleaning ticks: distilled water, 75% ethanol, 3% hydrogen peroxide, autoclaved pure water.
3. Paper towels.
4. Dumont fine forceps.
5. Double-sided tape.
6. 3″ × 6″ red dental wax paper
7. Monoject insulin syringe with a ½ inch, 29-gauge needle.
8. Custom Hamilton syringe with a 1 inch, 33-gauge needle with a 45° beveled tip.

3 Methods

3.1 SFTSV Inoculation and Culturing

1. Culture Vero E6 cells in T75 cm^2 cell flasks with DMEM growth solution.
2. When the cells enter the logarithmic growth phase and the cell aggregation reaches 70–80%, wash twice with sterile PBS.
3. Melt the SFTSV strains on ice away from light (*see* **Note 1**).
4. Inoculate 1 mL viral solution on the surface of the Vero E6 cells, then add 3 mL DMEM maintenance solution. The mock control directly uses 4 mL DMEM maintenance solution instead (*see* **Note 2**).
5. Incubate at 37 °C for 2 h.
6. Add 20 mL maintenance solution to the cell culture flasks.
7. Place the cell culture flasks into the cell culture incubator at 37 °C with 5% CO_2. Observe the morphology of cells every day for 3–7 days.
8. When 70% of the cells in the virus inoculation groups show lesions (Fig. 1), harvest the virus culture (*see* **Note 3**).
9. Store the viral solution at −80 °C.

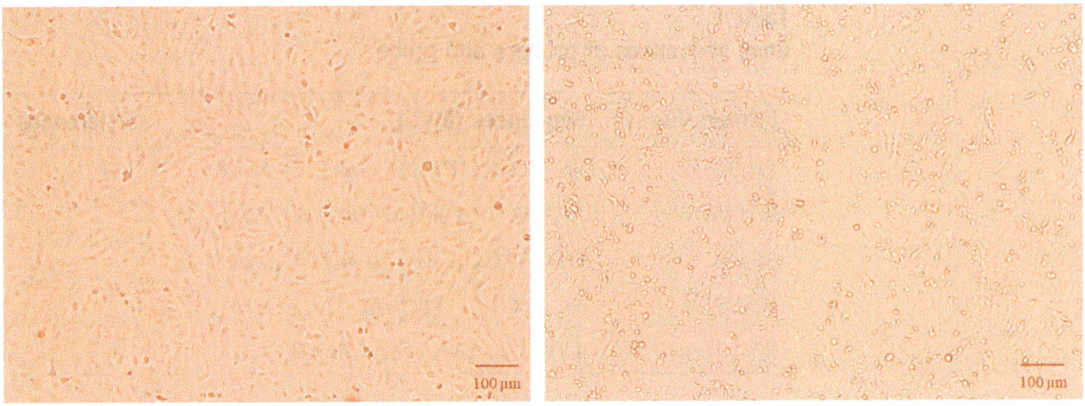

Fig. 1 Uninfected Vero E6 cells (**a**) and Vero E6 cells infected with SFTSV (**b**)

3.2 SFTSV-Free Tick Colonies

3.2.1 Tick Collection and Rearing

1. Collect ticks from the field using the manual flagging method (*see* **Note 4**). Select adult *H. longicornis* ticks for subsequent experiments (*see* **Note 5**).
2. Anesthetize a New Zealand white rabbit with 25% chloral hydrate by intraperitoneal injection at 1 mL/kg, wait for 5–10 min (*see* **Note 6**).
3. Put the rabbit into a large-sized rearing cage and shave the back of it with an electric razor. Put a collar on its neck (*see* **Note 7**).
4. Cut out a rectangular box from white foam board with an outer edge of 13 × 7 cm and an inner edge of 11 × 5 cm. Cut out a rectangle gauze with a size of 13 × 7 cm.
5. Apply Kamar adhesive to one side of the foam box, stick it to the back of the rabbit. Seal the perimeter with Kamar adhesive (*see* **Note 8**).
6. Apply double-sided adhesive (DSA) to the other side of the foam box and stick gauze on it (leave one side unsealed) (*see* **Note 8**).
7. Place the 5 mL centrifuge tube containing ticks in the foam box, seal the unsealed side with DSA, and open the centrifuge tube lid via gauze (*see* **Note 9**).
8. Open the gauze daily from day 3 to observe the blood-sucking status of ticks. Pick them up with tweezers when they are fully engorged and detached. Place each tick separately in a 5 mL collection tube (*see* **Note 10**).
9. A week after the engorged female ticks laying eggs, pick out each mother tick and some of its eggs (20–30 mg) separately and store them at −80 °C. Keep the remaining eggs in the tube until hatching.

10. After the eggs hatched, approximately 100 larvae from each female tick were placed into a 2 mL centrifuge tube individually.

3.2.2 Extraction of RNA

1. Rinse the samples in 75% ethanol for 5 min. Then immerse and wash the samples with pure water three times (5 min each time).
2. Transfer the samples into 2 mL EP tubes and immerse the tubes into liquid nitrogen (*see* **Note 11**).
3. Add 1 mL RNA extraction solution and one sterilized magnetic bead to the EP tubes. Grind for 10 min.
4. Centrifuge at 4°, 12,000 g for 10 min and transfer the supernatant to the new centrifuge tubes.
5. Add 200 μL chloroform, vortex to mix. Let it stand for 10 min at room temperature.
6. Centrifuge at 4° at 12,000 g for 15 min. Aspirate 500 μL of the upper colorless aqueous phase into the new centrifuge tubes.
7. Add 500 μL isopropanol to the solution, vortex and mix well.
8. Place on the ice for 15 min.
9. Centrifuge at 4° at 12,000 g for 10 min, discard the supernatant. The white precipitate at the bottom of the tube is the total RNA.
10. Add 1 mL 75% ethanol to clean the precipitate.
11. Centrifuge at 4° at 12,000 g for 5 min, discard the supernatant.
12. Open the tube and let it air-dry for 3–5 min. Add 20 μL DEPC water to fully dissolve the RNA.

3.2.3 Reverse Transcription

1. Configure the reaction mixture: 4 μL of 5 × Reaction Buffer, 1 μL SweScript RT I Enzyme Mix, 1 μL Oligo $(dT)_{18}$ Primer (100 μM), 2 μL RNA, 12 μL Nuclease-Free Water. 20 μL in total.
2. Set up procedures: 25 °C for 5 min, 50 °C for 30 min, 85 °C for 5 s.
3. Store the cDNA at −80 °C.

3.2.4 Real-Time PCR for SFTSV

1. Prepare the reaction mixtures on ice using the one-step Primer Script RT-PCR Kit. The reaction mixture contained 0.5 μL probe (10 μM), 1 μL PCR primer mix (20 μM of forward primer and reverse primer each) and 2 μL total RNA. The primers were targeted at the S segment of the SFTSV [8].
2. Set up the RT-PCR protocol: initial denaturation at 95 °C for 5 min followed by 40 cycles of denaturation at 95 °C for 10 s, annealing at 60 °C for 30 s, extension at 72 °C for 15 s, and an additionally final extension at 72 °C for 1 min.

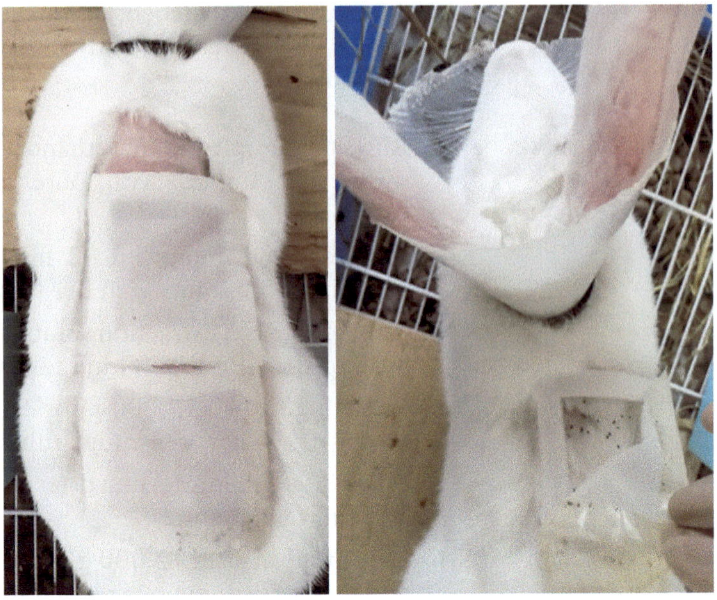

Fig. 2 New Zealand white rabbit blood feeding model

3.3 SFTSV Infection of Ticks through Feeding

1. Anesthetize a New Zealand white rabbit, intraperitoneally inject 10 mL of SFTSV stock solution (8×10^4 copies/mL) (*see* **Note 12**).
2. Feed the SFTSV free *H. longicornis* ticks on the infected rabbit. For detailed steps, see 3.2.1 (Fig. 2).
3. Select engorged ticks randomly for SFTSV detection to see whether they have been infected successfully.

3.4 Experimental Infection of Ticks with SFTSV by Microinjection

1. Wash the adult female *H. longicornis* ticks from SFTSV-free colony in the order of distilled water, 3% hydrogen peroxide, autoclaved pure water twice, 75% ethanol and autoclaved pure water twice. Blot ticks with paper towels to remove the residual water from the surface.
2. Capture ticks with Dumont fine tweezers and place them belly-up on double-sided tape attached to 3″ × 6″ red dental wax paper. Place ticks closely together in groups of 5 (*see* **Note 13**).
3. Tape a small strip of masking tape on the mouthparts of all five ticks (*see* **Note 14**).
4. Puncture a hole in the lower right quadrant of the exoskeleton using a Monoject insulin syringe fitted with an ½ inch, 29-gauge needle.
5. Inject 1 μL viral solution (6×10^5 viruses/mL) into each tick using a custom Hamilton syringe with a 1 inch, 33-gauge needle with a 45° beveled tip (*see* **Note 15**). Inject the same volume of PBS into each tick in the control group.

6. Pick up the injected ticks from the double-sided tape with fine tweezers and place them into the 2 mL preservation tubes. Blow carbon dioxide into the tubes to activate the ticks (*see* **Note 16**).
7. Maintain the ticks in an incubator with a relative humidity of 95 ± 5% (*see* **Note 17**).

4 Notes

1. Intense ultraviolet rays may damage nucleic acids, and high temperature inactivates proteins. Melting the viral solution away from light minimizes the damage to virus strains.
2. The purpose of setting up the control groups is to rule out that nonviral factors have an effect on the growth of cells.
3. After the SFTSV infection, Vero E6 cells will fall off and extend pseudopods when lesions appear.
4. Ticks should be collected on the sunny days between 9 am and 12 am in the peak time of tick activities. The collection sites should be bushes, grasslands, or green belts. When collecting ticks, hold one side of the cloth flag, spread it flat on the grasses, and walk forward slowly. Turn the cloth flag over every 10 m. Pick up ticks on the flags with tweezers and put them into 5 mL tubes. The tubes should be pre-engraved for vents on the lid to ensure oxygen circulation. Cloth flag should be pure white for better observation. All collectors need to wear white protective clothing to prevent tick bites.
5. It is extremely important to know the primary tick vector when performing virus infection. As the host and primary vector of SFTSV, the *Haemaphysalis longicornis* ticks are susceptible to SFTSV, which can greatly increase the success rate of virus infection. If the primary tick vector is not available, other secondary tick vectors can be used instead, but this may result in a lower success rate of virus infection. Furthermore, *H. longicornis* ticks have the reproductive strategy of obligate parthenogenesis [9]. They are able to reproduce without the need for male–female pairs, making them even better candidates for breeding and viral infection.
6. Select SFTSV free and healthy New Zealand white rabbit weighing 2–2.5 kg. Laboratory rearing for 1 week helps its adaptation to the environment.
7. The collar is used to prevent the rabbit from removing ticks with its tongue. Its depth should be appropriate to ensure normal feeding.

8. The Karma adhesive is sticky and could adhere tightly to the animals' skin. The double-sided adhesive is relatively less sticky, facilitating subsequent removal of the gauze to observe the ticks' blood-sucking status and to pick up the ticks.
9. The number of ticks feeding on one rabbit at one time is determined on the developmental stage of the ticks: 50 for adult ticks, 100 for nymphs, and 2000 for larvae.
10. When the tick is fully engorged, it will consciously leave the skin of the host. At this point, it could become ten or even hundreds times of its original size. The blood-saturated ticks should be picked up gently with tweezers to avoid death. It takes 2–5 days for larvae, 3–8 days for nymphs, and 6–15 days for adults to get engorged.
11. Ticks have hard carapaces and liquid nitrogen can make tick samples brittle and fragile.
12. Rabbits could maintain high-level SFTSV viremia for 6–20 days after inoculation with the SFTSV, providing a persistent infection source for ticks. Other laboratory animals such as mice and hedgehogs have also been shown to be good hosts for viral infections [10]. The procedure is similar to that of rabbits.
13. Each operator must wear disposable white gloves during operation.
14. The small strip of tape applied to the mouthparts of ticks is to further restrain them and leave most of their bodies exposed so that the injectors could observe the injection process clearly.
15. The needle should be fully inserted into the tick cavity to ensure placement and retention of the viruses. Some fluid may flow out from the injection site. Attention should be paid to not to over-inject the ticks, as this could result in hemolymph loss and tick death. After completing the injection, clean the Hamilton syringe before being used for the next time. Fill syringe from a beaker containing 3% hydrogen peroxide and then drained into a waste container and repeated 15 times. Fill the syringe from the beaker with sterile water and drain into the waste container, repeat 15 times. Be careful not to bend the piston of the Hamilton syringe because if it is bent, the piston will not move smoothly and meet the need of gentle touch required to inject the ticks.
16. After injection, the ticks will briefly stop moving and the carbon dioxide will help activate the ticks. Once the ticks begin to crawl and become active, the injection wound will quickly heal and they will likely survive. If the injection results in the death of the ticks, the spare ticks should be injected to ensure infection success.

17. As the *Haemaphysalis longicornis* ticks are reservoir hosts and vectors of SFTSV, once infected, they can remain infected until moulting or death.

References

1. Parola P (2004) Tick-borne rickettsial diseases: emerging risks in Europe. Comp Immunol Microbiol Infect Dis 27(5):297–304
2. Chen Z, Yang X, Bu F, Yang X, Yang X, Liu J (2010) Ticks (acari: ixodoidea: argasidae, ixodidae) of China. Exp Appl Acarol 51(4):393–404. https://doi.org/10.1007/s10493-010-9335-2
3. Fares M, Brennan B (2022) Virus-host interactions during tick-borne bunyavirus infection. Curr Opin Virol 57:101278. https://doi.org/10.1016/j.coviro.2022.101278
4. Yu X-J, Liang M-F, Zhang S-Y, Liu Y, Li J-D, Sun Y-L et al (2011) Fever with thrombocytopenia associated with a novel bunyavirus in China. N Engl J Med 364(16):1523–1532. https://doi.org/10.1056/NEJMoa1010095
5. Luo L-M, Zhao L, Wen H-L, Zhang Z-T, Liu J-W, Fang L-Z et al (2015) *Haemaphysalis longicornis* ticks as reservoir and vector of severe fever with thrombocytopenia syndrome virus in China. Emerg Infect Dis 21(10):1770–1776. https://doi.org/10.3201/eid2110.150126
6. Brault AC, Savage HM, Duggal NK, Eisen RJ, Staples JE (2018) Heartland virus epidemiology, vector association, and disease potential. Viruses 10(9). https://doi.org/10.3390/v10090498
7. Talactac MR, Hernandez EP, Fujisaki K, Tanaka T (2018) A continuing exploration of tick-virus interactions using various experimental viral infections of hard ticks. Front Physiol 9:1728. https://doi.org/10.3389/fphys.2018.01728
8. Zhuang L, Sun Y, Cui X-M, Tang F, Hu J-G, Wang L-Y et al (2018) Transmission of severe fever with thrombocytopenia syndrome virus by *Haemaphysalis longicornis* ticks, China. Emerg Infect Dis 24(5):868–871. https://doi.org/10.3201/eid2405.151435
9. Chen X, Xu S, Yu Z, Guo L, Yang S, Liu L et al (2014) Multiple lines of evidence on the genetic relatedness of the parthenogenetic and bisexual *Haemaphysalis longicornis* (Acari: Ixodidae). Infect Genet Evol 21:308–314. https://doi.org/10.1016/j.meegid.2013.12.002
10. Spengler JR, Estrada-Peña A, Garrison AR, Schmaljohn C, Spiropoulou CF, Bergeron É et al (2016) A chronological review of experimental infection studies of the role of wild animals and livestock in the maintenance and transmission of Crimean-Congo hemorrhagic fever virus. Antivir Res 135:31–47. https://doi.org/10.1016/j.antiviral.2016.09.013

Chapter 10

Methods for Thrips Transmission and Maintenance of Tomato Spotted Wilt Virus (TSWV)

Peyton LaBonte, Ryan Packer, Holly McInnes, Dorith Rotenberg, and Anna E. Whitfield

Abstract

Tomato spotted wilt virus (TSWV) is transmitted by insect pests from the Thripidae family, including *Frankliniella occidentalis*, commonly known as western flower thrips. For experimental purposes, researchers have developed methods for inoculating host plants with TSWV, allowing thrips to acquire TSWV, and verifying thrips acquisition. Plants can be inoculated with TSWV either mechanically or with thrips in the lab, but in nature, the virus is transmitted by thrips. For the study of tospovirus transmission biology and vector–virus interactions, efficient methods for mechanical and thrips transmission are essential. We have optimized these protocols for TSWV and western flower thrips, but they can also be adapted to other tospovirus-vector systems.

Key words Thrips vectors, *Orthotospovirus tomatomaculae*, Virus inoculation, Virus acquisition, Virus transmission

1 Introduction

Tomato spotted wilt virus (TSWV), a tospovirus transmitted by *Frankliniella occidentalis*, is one of the most economically significant plant viruses in agriculture [1]. Because of its importance as a major plant pathogen and a model tospovirid, TSWV is widely studied to better understand the interactions between tospoviruses and their arthropod vectors. *Frankliniella occidentalis*, commonly known as the western flower thrips, are small insect pests that cause significant agricultural losses through their feeding, oviposition, and transmission of tospoviruses. Western flower thrips are one of the most efficient vectors of tospoviruses and can transmit at least 7 [2] of the 26 described orthotospoviruses [3]. In nature, TSWV is transmitted in a circulative propagative manner by thrips; the virus is taken up through the midgut of the first instars, wherein it replicates and then disseminates to other organs including the

Hani Boshra (ed.), *Bunyaviruses: Methods and Protocols*, Methods in Molecular Biology, vol. 2893,
https://doi.org/10.1007/978-1-0716-4338-9_10,

tubular and principal salivary glands [4]. Dissemination of the virus to the salivary glands allows adult thrips to transmit TSWV through probing and feeding on various plant hosts and plant parts [5]. In the lab, TSWV can also be mechanically transmitted to leaf tissue. The use of both thrips-mediated and mechanical transmission is vital for maintaining TSWV isolates for experimentation.

Researchers have developed several protocols to maintain TSWV virus isolates in laboratory settings. Some fundamental TSWV maintenance techniques include mechanical (leaf rub) inoculation, thrips acquisition, measuring thrips acquisition efficiency, and thrips-mediated inoculation of the plant host [6, 7]. Mechanical inoculation involves rubbing infected plant sap protected in buffer onto a leaf dusted with an abrasive to lightly perforate the healthy leaf to allow the virus to enter the leaf cells and infect a new host. Mechanical inoculation is a quick and efficient technique to deliver TSWV to new plant hosts without having to maintain a thrips colony [8]. However, if TSWV is maintained only through mechanical transmission, it will lose the ability to be transmitted by thrips. For example, repeated mechanical inoculation of a TSWV isolate in de Oliveira Resende et al. created a morphologically defective isolate without a viral envelope [8] which resulted in the virus losing its ability to be transmitted by thrips [9]. To avoid this phenomenon, it is typical to maintain TSWV strains in a laboratory or greenhouse setting by cycling between mechanical inoculation and thrips transmission. The thrips acquisition protocol requires the first instar thrips to ingest TSWV virions through feeding from infected plant tissue. As the thrips mature, the virus disseminates from the midgut to tissues vital for transmission [4]. The efficiency of thrips acquisition, i.e. the number of individuals testing positive for the virus in a subsample obtained from a cohort of exposed larvae, can be determined by endpoint reverse transcription (RT)—PCR or real-time RT-quantitative PCR (RT-qPCR), a detection method that can also be used as a tool to better understand the interaction between TSWV and thrips [6]. Thrips-mediated transmission is useful as it mimics the natural route of transmission in nature and allows researchers to investigate virus–vector interactions. Thrips-mediated transmission has several advantages over mechanical transmission. For example, disruption of TSWV transmission by targeting aspects specific to thrips-mediated transmission cannot be observed via mechanical inoculation [10]. Here we describe our optimized protocol, using the instruments in our laboratory, for thrips transmission of TSWV that enables long-term maintenance of virus isolates that are vector-transmissible and highly, efficient transmission assays.

2 Materials

All reagents should be kept at room temperature unless specified otherwise. To avoid contamination or unwanted degradation of RNA or proteins, purchase plastic consumables packaged as certified nuclease and protease-free. Dispose of infected material properly according to local, state, and federal laws.

2.1 Mechanical Inoculation

1. Sodium sulfite buffer (10 mM): Add 0.063 g to a 50 mL centrifuge tube and fill it up to the 50 mL line with ultrapure water and cap tightly. Invert to mix and keep on ice until used. Make fresh for each inoculation.
2. Silicon carbide.
3. 50 mL sterile conical centrifuge tube.
4. Cheesecloth or other loosely woven material.
5. TSWV-infected plant material: keep chilled until inoculation.
6. Healthy host plant(s).
7. Autoclaved mortar and pestle or Fisherbrand polyethylene bag (Ca: 19-075-388G).
8. Sterile cotton swabs: 2.
9. Disposable gloves.

2.2 Thrips TSWV Acquisition Access Period (AAP)

1. Age synchronized first instar western flower thrips (*see* **Note 1**).
2. TSWV-infected plant material.
3. 16 oz. deli cup with thrips-proof modified lid (*see* Fig. 1).
4. Whatman® paper with 2 opposing hole punches (*see* Fig. 1).
5. Deionized water.
6. 2 mL microcentrifuge tube without lid.
7. Parafilm.
8. 25 °C Incubator with lights.

2.3 Thrips TSWV Acquisition Efficiency

2.3.1 Single Thrips RNA Extraction

1. 1.7 mL microcentrifuge tubes.
2. Adult thrips: noninfected and thrips from the TSWV AAP.
3. Two fine-tipped paint brushes.
4. Sterile micro pestle: Axygen® (Ca: PES-15-B-SI).
5. 50% Chelex® 100 slurry (*see* **Note 2**).
6. Ice.
7. Heat block (96 °C).
8. Cold centrifuge (4 °C).

Fig. 1 A sealed deli cup used for thrips TSWV acquisition access periods (AAP)

2.3.2 Verso cDNA Synthesis

1. ThermoFisher Verso cDNA synthesis kit (Ca: AB1453B).
2. 0.2 mL PCR strip tubes.
3. Thrips RNA from single thrips RNA extraction protocol.
4. Ice.

2.3.3 RT-qPCR

1. 1.7 mL microcentrifuge tubes.
2. Bio-Rad SYBR® Green (Ca: 1725270).
3. Actin primers: forward primer—5′ GGTATCGTCCTG GACTCTGGTG 3′ and reverse primer—5′ GGGAAGGGCGTAACCTTCA3′ [6].
4. TSWV nucleocapsid (N) qPCR primers: forward primer—5′ GCTTCCCACCCTTTGATTC3′ and reverse primer—5′ ATAGCCAAGACAACACTGATC3′ [6].
5. cDNA from Verso cDNA synthesis.
6. Nuclease-free water.
7. Ice.
8. qPCR 96-well plates (thin-walled).
9. qPCR plate seals.
10. qPCR thermocycler (Bio-Rad, CFX Connect™).

2.4 Thrips Inoculation Access Period (IAP)

1. Viruliferous adult thrips.
2. Healthy host plant(s).
3. Thrips proof chamber.
4. Hot Shot® No-Pest Strip$_2$®.
5. Plastic container with a lid (~102 L).

3 Methods

3.1 Mechanical Inoculation

1. Chill the mortar, pestle, and sodium sulfite solution on ice until cold to the touch (approximately 20 min for the mortar and pestle, and 5 min for the sodium sulfite solution).
2. Collect leaf tissue from a TSWV-infected source plant. For the best results, collect symptomatic, living tissue from multiple locations across the infected plant. This will optimize the chances of collecting high-quality infectious material (*see* **Note 3**).
3. Select the plants to be inoculated (*see* **Note 4**).
4. Label plants with a tag that identifies date, initials, passage number, plant name, and virus strain.
5. Transfer silicon carbide to a 50 mL falcon tube and secure the opening with tightly stretched cheesecloth. Dust the plants with a light layer of silicon carbide. All the leaves that will be inoculated should be dusted (*see* **Note 5**).
6. Leaving the mortar and pestle on ice, add ~200–500 g of symptomatic source tissue (about 3–7 tomato leaflets) to the mortar. Then add ~5 mL of chilled sodium sulfite solution and immediately begin to grind until the tissue is fine (*see* **Note 6**).
7. Add an additional 7–12 mL of the sodium sulfite solution to the mortar and grind until homogenous (*see* **Note** 7).
8. Place a gloved hand behind the leaf to be inoculated so that the leaf rests lightly against fingers (be careful not to brush off the silicon carbide). Take two cotton swabs in the other gloved hand and dip them in the inoculum slurry. Bring the wet swabs to the captured leaf and gently rub the entire leaf surface. Leave plants to sit for 10 min (*see* **Note 8**) (*see* Fig. 2).
9. After 10 min, use a gentle stream of tap or deionized water to rinse off the inoculum mixture and silicon carbide from the leaves. Plants can be left to grow in their respective environments (*see* **Note 9**).

   Decontamination:

10. Clean the mortar and pestle with 10% bleach. Autoclave mortar and pestle between uses.

Fig. 2 Mechanical inoculation of plant hosts

11. Wipe down the inoculation area and discard waste as appropriate for the biosafety measures needed for the virus you are working with in the lab.

    Quicker method: (*see* **Note 10**).

3.2 Thrips TSWV Acquisition Access Period (AAP)

1. Construct a "leaf bouquet": Select two visibly TSWV-infected leaves from a plant host that is 10–13 days post mechanical inoculation, acquisition efficiency decreases on day 14, and remove selected leaves so that a long length of the petiole is connected to the base of each leaf. Place the petioles into the 2 mL microcentrifuge tube. Fill the 2 mL tube with the diH2O and stretch the parafilm over the top of the 2 mL tube to keep the petioles in place and prevent diH2O from escaping. Flip the "leaf bouquet" with the parafilm facing downward to ensure no leaks; if leaks are present, add more parafilm to seal the top or re-wrap (*see* **Note 11**) (*see* Figs. 3 and 4).
2. Acquire a clean deli cup with lid and add Whatman® paper to the bottom of the cup. Label the deli cup with the strain, date, and name. Place the "leaf bouquet" onto the Whatman® paper and adjust bouquet placement to maximize the leaf surface touching the Whatman® paper (*see* Fig. 1).

Fig. 3 Symptomatic leaves chosen for a leaf bouquet for thrips acquisition access period

3. Collect first instar thrips (L1) that have emerged from their eggs within the last 24 h, in a deli cup (*see* **Note 12**) (*see* Fig. 5).
4. Turn the deli cup, containing L1 thrips, over, so that its opening aligns with the opening of the leaf bouquet cup. Tap the bottom and sides of the upside-down cup to knock the L1 thrips into the cup with the infected leaves, then add the lids back to both deli cups (*see* Fig. 6).
5. Cut a small strip of parafilm and seal the top of the deli cup to prevent escapees.
6. Place the deli cup into the incubator at 25 °C for 24 h.
7. After the 24 h are up, add 1–2 green beans to the leaf bouquet cup to prevent thrips from desiccating and to act as a food source. Add more green beans as necessary until the thrips reach the adult stage (*see* **Note 13**).
8. Once thrips have reached the adult stage, they are ready to be tested for TSWV and/or used in bioassays (*see* Fig. 7).

Fig. 4 "Leaf bouquet" of infected *Emilia sonchifolia* leaves wrapped in parafilm in a 2 mL tube full of water

3.3 Thrips TSWV Acquisition Efficiency

Single Thrips RNA Extraction

1. Set the heat block to 96 °C and a refrigerated centrifuge to 4 °C.
2. Collect 2–4 recently eclosed male and female adults (*see* Fig. 8) from the uninfected colony with a fine-tipped brush and place them individually into 1.7 mL microcentrifuge tubes so that each tube contains only a single thrips. These "uninfected" thrips will serve as the TSWV-negative, thrips-positive control for the acquisition efficiency test. Place tubes on ice to restrict thrips movement (*see* **Note 14**).
3. Move the TSWV exposed thrips individually into their corresponding tubes using a different brush for viruliferous thrips. There should be one thrips per tube. Place tubes on ice (*see* **Note 15**). We typically test ten exposed thrips using

Fig. 5 First instar (larval) thrips to be used for thrips acquisition of TSWV

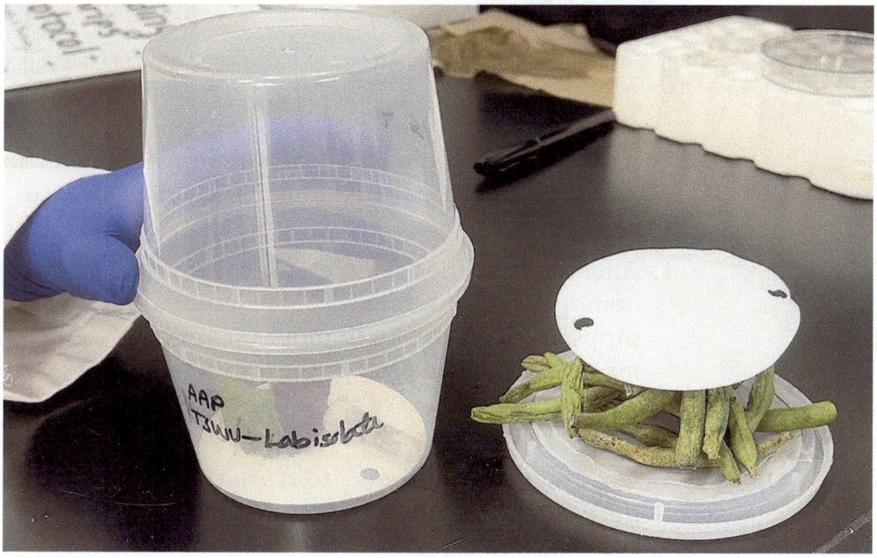

Fig. 6 Transferring first instar thrips from beans to the leaf bouquet

Fig. 7 Viruliferous adult western flower thrips used for inoculation access period (IAP)

qRT-qPCR and obtain a good representation of the overall acquisition efficiency and this correlates with virus titer and inoculation.

4. After all the thrips samples are collected, the samples can be homogenized. Starting with the "uninfected" negative control thrips, use a pipette tip to move the thrips to around the 55 uL mark near the bottom of the tube. Add 50 uL of ultrapure water to the bottom of the tube so that the thrips sit at the top of the water line (*see* **Note 16**).
5. Use a sterile plastic pestle to crush the individual thrips against the wall of the tube. Confirm that the thrips is visibly breaking apart against the side of the tube (*see* **Note 17**).
6. Return the tube to the ice and use the pestle to further homogenize the thrips against the side of the tube in the water solution.

Fig. 8 Female (left) and male (right) adult western flower thrips. Modified from Han et al. [11], permission of John Wiley and Sons

7. To make 50% Chelex® 100 slurry: mix 0.5 g Chelex® 100 in 1 mL of ultrapure water and vortex. Add 50 uL of Chelex® 100 slurry and continue to grind thrips over ice for an additional 30–60 s (*see* **Note 18**).
8. Repeat **steps 4–7** until all thrips are homogenized.
9. Place the tubes onto the 96 °C heat block and let incubate for 5 min.
10. Transfer the tubes to the 4 °C cold centrifuge and spin down at 13,000rcf for 5 min.
11. Move the centrifuged tubes back onto ice, taking care not to disrupt the Chelex® 100 pellet. Keep them on ice until RNA is used for cDNA synthesis (*see* **Note 19**).

Table 1
The standard cDNA synthesis reaction mix for the RT step using individual thrips for RT-qPCR

10 uL Verso enzyme reaction	
5× cDNA synthesis buffer	2 uL
dNTP mix	1 uL
RT enhancer	0.5 uL
Verso enzyme mix	0.5 uL
Random hexamer	0.5 uL
Anchored oligo dT	0.5 uL
RNA prep from Chelex® 100	5 uL

3.4 Verso cDNA Synthesis

Conduct all steps over ice. Follow kit instructions unless otherwise specified.

1. Thaw components of the cDNA kit and spin down tubes.
2. Calculate the volumes of reagents needed to make a Verso cDNA master mix or your preferred cDNA synthesis method. For a 10 uL Verso Enzyme Reaction, follow the guidelines below and multiply the volumes by the number of reactions necessary (Table 1) (*see* **Note 20**).
3. Aliquot 5 uL of master mix into each 0.2 mL PCR tube.
4. Use 5 uL of supernatant from the single thrips RNA extraction as the reaction template (for a 10 uL Verso cDNA reaction) (*see* **Note 21**). Be sure to include a negative control of 5 uL water in one reaction tube.
5. Mix tubes and spin down. Repeat twice more.
6. Incubate tubes at 42 °C for 30 min, 95 °C for 2 min, and hold at 6 °C. Store samples at −20 °C until ready for qPCR.

3.5 RT-qPCR

Conduct all steps on ice.

1. Using the Verso cDNA synthesis product, dilute 10 uL of the cDNA reaction in 28 uL ultrapure water (*see* **Note 22**).
2. Following the Bio-Rad SYBR® Green protocol (Bio-Rad, Hercules, CA, USA), create two master mixes for amplification-one using the actin primers (internal reference gene [7]) and the other with the TSWV primers (nucleocapsid gene). (*see* **Note 23**).
3. Vortex master mixes for 3 s, pause for 3 s, and repeat twice more.

4. Aliquot 16 uL of master mix into each reaction mixing tube (1.7 mL tube) (*see* **Note 24**).
5. Add 8 uL of diluted cDNA to each tube of aliquoted master mix (*see* **Note 25**).
6. Vortex well and spin down the tubes to mix. Keep on ice.
7. Obtain a RT-qPCR plate and plastic plate seal. Keep the plate on ice (*see* **Note 26**).
8. Prior to transferring each sample to the RT-qPCR plate, vortex for 3 s, pause for 3 s, and repeat twice more (*see* **Note 27**).
9. For each reaction mix, load two 11 uL technical replicates to two individual microplate wells (there will be four wells for each thrips sample- two are for TSWV amplification, two are for actin amplification).
10. Seal the 96-well plate using the qPCR plate seal. Use caution not to smudge the plastic seal parts that cover the plate and use the paper-backing of the seal to ensure that it is tightly adhered to the plate, to avoid leaks (*see* **Note 28**).
11. Spin down the plate in a plate centrifuge for about 10 s.
12. Place the plate in the qPCR machine.
13. According to the qPCR machine specifications, select the relevant wells to scan and select the SYBR family channel.
14. Start a two-step amplification plus melt program with the following thermocycle protocol (based on Bio-Rad specifications):
    (a) Polymerase Activation & DNA Denaturation: 95 °C for 3 min.
    (b) Denaturation: 95 °C for 10 s.
    (c) Annealing/Extension and plate read: 60C for 30 s.
        (i) 39 cycles
    (d) Melt Curve: 65–95 °C at 0.5 °C increments every 5 s.
15. After the run has finished, export and save the results.
16. To determine which samples are positive, look at the slope and the cycle threshold (Ct) values. An exponential reaction progression slope and a Ct value of less than 32 are indicative of TSWV acquisition. If a sample is positive for both TSWV and actin, that thrips successfully acquired TSWV. If the sample is positive for only actin, that thrips was successfully processed but did not acquire TSWV. Actin values are generally reproducible and consistent between technical replicates and individual thrips, using the above methods. (*see* **Note 29**).

3.6 Thrips Inoculation Access Period (IAP)

1. Thrips from the AAP can be used for the IAP once they have reached adulthood (2 weeks in a 25 °C incubator) (*see* **Note 30**) (*see* Fig. 7).
2. Well-watered, healthy plants to be inoculated should be placed in a thrips-proof cage prior to thrips exposure (*see* **Note 31**).
3. The deli cup containing the adult viruliferous thrips should be opened and any green beans, remaining plant tissue, and Whatman® paper should be tapped with a stiff instrument within the cup to return any clinging thrips into the deli cup for transfer. Carefully turn the deli cup upside down inside the chamber directly over the plants. The bottom of the cup should be tapped to release all the thrips directly onto the plant leaves. Close the chamber and seal the door with tape (*see* **Note 32**) (*see* Fig. 9).
4. Leave thrips to feed for about 72 h (*see* **Note 33**).
5. After the thrips feeding period, transfer the plants and their tray to a 102-liter plastic container with pest strips (insecticide

Fig. 9 Transferring viruliferous adult western flower thrips on tomato plants for thrips inoculation of TSWV

chamber) for at least an hour. After insecticide exposure, remove plants with gloved hands from the insecticide chamber and let the plants grow in a conducive environment (*see* **Note 34**). In our lab we always place plants in cages with thrips-proof mesh to prevent other insects from feeding on plants and transmitting other unwanted pathogens. Monitor plants for symptoms such as chlorosis, mosaic patterns, and/or necrotic spots; once symptoms appear, plant tissue can be used as desired (*see* **Note 35**).

These protocols can also be adapted for other thrips species and tospoviruses.

4 Notes

1. Thrips were age synchronized using the method developed by Han & Rotenberg [10].
2. Chelex® 100 slurry must be made fresh and should be used within an hour. This metal-chelating agent and single thrips RNA extraction technique was first developed and described by Boonham et al. [7].
3. Material can be used immediately, stored at 4 °C for 1–3 days prior to inoculation, or kept at −80 °C for up to a year prior to inoculation. If frozen, care must be taken to not allow tissue to thaw before grinding.
4. In general, younger plants are easier to rub inoculate than older plants. For tomatoes, the best age to inoculate is the 2–3 true leaf stage. You may need to optimize your inoculation age for other plant hosts.
5. For even dispersal of silicon carbide, transfer powder to the 50 mL Falcon tube and cover the tube opening with cheesecloth (or another wide toothed fabric). With the cheesecloth secured, this will act as a shaker that allows for a light dusting of the leaf surface.
6. Grinding mixture should never be at a paste consistency—it should be more of a thin soup. Add more sodium sulfite solution as necessary.
7. Volumes can be adjusted depending on how many plants need to be inoculated—keep in mind the size and number of leaves that will be inoculated. Volumes above can sufficiently inoculate 7–12 plants.
8. The pressure used for mechanical inoculation depends on the tenderness of the leaf (consider leaf age, plant species, and cultivar). In general, the goal is to lightly scratch the surface without tearing the leaf. Inoculate as many leaves as desired but

avoid newly emerging apical leaves. If infecting multiple plants, inoculate 1–2 leaves per plant first and then return to each plant to complete inoculations. This ensures that even when inoculating a large group of plants, each plant is exposed to fresh inoculum.

9. Visible symptoms can become apparent as early as 5 days, but symptom development in young tomatoes usually takes 15–20 days.
10. Tissue can also be ground in a transparent plastic bag over ice (or on an ice pack), using a sharpie to pulverize tissue. We recommend the Fisherbrand polyethylene bag (Ca: 19075388G).
11. Leave enough petiole on the inoculum source so the leaves can be securely placed in the 2 mL microcentrifuge tube. Source leaves should be symptomatic but still viable. Remove sections of the leaf that are primarily healthy to ensure thrips feed from viruliferous material. Be sure not to overtighten the parafilm around the petioles as it may damage the vascular system.
12. Adult thrips are given at least 24 h to lay eggs in fresh green beans and then removed to leave green beans to incubate. After 72 h, an initial flush of first instars eclose from the green beans and is ready for collection. After an additional 16 h, a second flush is ready. The age synchronized protocol described in Badillo et al. [12] and Han and Rotenberg [10] produces larval thrips that have emerged between 0 and 16 h and are highly competent for thrips acquisition of TSWV.
13. The thrips should reach adulthood in about 11 days.
14. Collecting older adults may result in poor Ct values.
15. Keeping the deli cup on ice slows down the thrips and can make collection easier; however, if condensation forms on the cup, the water buildup can kill them—so swiftness is key.
16. Make sure to always know where the thrips is in the tube as thrips are difficult to see and can be easily lost in tubes.
17. It is easiest to capture thrips on the side of the tube, above the water line, while the thrips is dry.
18. Try to remove excess liquid from the pestle before removing it from the tube. Remove the pestle from the tube and close the lid. Keep the ground thrips tube on ice.
19. Use RNA as soon as possible as this extraction method may leave some contaminants behind that may degrade RNA.
20. This reaction volume can be halved if desired, but we advise starting out with a 10 uL reaction.
21. Use filter tips for RNA.

22. Can add water directly to cDNA. May need to increase dilution if downstream qRT-PCR issues occur.
23. Make enough master mix for the number of reactions multiplied by about one and a fifth to account for any pipetting errors.
24. Use filter tips from this step forward. Pipet to mix in between aliquots.
25. Vortex tube to suspend cDNA before adding it to the reaction tube.
26. We use the Bio-Rad Microseal® 'B' seal (Ca: MSB1001). Only touch the breakaway sides of the seal and set the seal paper side down while not in use.
27. Load negative controls on the plate first.
28. Use the matte side of the paper backing. Tear off removable plastic sides.
29. The following formula should be used to determine acquisition efficiency: *(#viruliferous thrips/# total AAP thrips) x 100 = acquisition efficiency.*
30. Thrips will reach adulthood in about approximately 2 weeks in a 25 °C incubator. At lower temperatures, development to adulthood requires more time.
31. The age of plants to use for the IAP is dependent on the host, but tomato plants at the 1–3 true leaf stage work best for tomato inoculation.
32. A paint brush handle, metal forceps, or scissors can be used to tap the bottom of the deli cup firmly and ensure all the thrips are released into the chamber.
33. Do not let thrips feed long enough that new thrips can eclose from their eggs.
34. Consult pest strip instructions to determine the quantity of and amount of time to leave strips in the chamber. It is dependent on the age and strength of the pest strips used. Also look to confirm there is no thrips activity before removing pest strips.
35. The time it takes for visible symptoms to appear is dependent on the host plant. Tomatoes usually take about 5–10 dpi for symptoms to appear.

Acknowledgments

We thank Kirsten Lahre for her work optimizing and training lab members on these protocols. This research was supported by USDA-FNRI 58-6034-2-008 and BARD IS-5270-20-R, and the North Carolina Agricultural Foundation award number 23–46.

References

1. Scholthof K-BG, Adkins S, Czosnek H, Palukaitis P, Jacquot E, Hohn T, Hohn B, Saunders K, Candresse T, Ahlquist P, Hemenway C, Foster GD (2011) Top 10 plant viruses in molecular plant pathology. Mol Plant Pathol 12(9):938–954. https://doi.org/10.1111/j.1364-3703.2011.00752.x
2. Oliver JE, Whitfield AE (2016) The genus Tospovirus: emerging Bunyaviruses that threaten food security. Annu Rev Virol 3:101–124. https://doi.org/10.1146/annurev-virology-100114-055036
3. Lefkowitz EJ, Dempsey DM, Hendrickson RC, Orton RJ, Siddell SG, Smith DB (2018) Virus taxonomy: the database of the international committee on taxonomy of viruses (ICTV). Nucleic Acids Res 46(D1):708–717. https://doi.org/10.1093/nar/gkx932
4. Montero-Astúa M, Ullman DE, Whitfield AE (2016) Salivary gland morphology, tissue tropism and the progression of tospovirus infection in Frankliniella occidentalis. Virology 493:39–51. https://doi.org/10.1016/j.virol.2016.03.003
5. Schneweis DJ, Whitfield AE, Rotenberg D (2017) Thrips developmental stage-specific transcriptome response to tomato spotted wilt virus during the virus infection cycle in *Frankliniella occidentalis*, the primary vector. Virology 500:226–237. https://doi.org/10.1016/j.virol.2016.10.009
6. Rotenberg D, Krishna Kumar NK, Ullman DE, Montero-Astúa M, Willis DK, German TL, Whitfield AE (2009) Variation in *tomato spotted wilt virus* Titer in *Frankliniella occidentalis* and its association with frequency of transmission. Phytopathology 99(4):404–410. https://doi.org/10.1094/PHYTO-99-4-0404
7. Boonham N, Smith P, Walsh K, Tame J, Morris J, Spence N, Bennison J, Barker I (2002) The detection of tomato spotted wilt virus (TSWV) in individual thrips using real time fluorescent RT-PCR (TaqMan). J Virol Methods 101(1–2):37–48. https://doi.org/10.1016/s0166-0934(01)00418-9
8. de Oliveira Resende R, de Haan P, de Avila AC, Kitajima EW, Kormelink R, Goldbach R, Peters D (1991) Generation of envelope and defective interfering RNA mutants of tomato spotted wilt virus by mechanical passage. J Gen Virol 72(10):2375–2383. https://doi.org/10.1099/0022-1317-72-10-2375
9. Nagata T, Nagata-Inoue AK, Prins M, Goldbach R, Peters D (2000) Impeded thrips transmission of defective tomato spotted wild virus isolates. Phytopathology 90:454–459. https://doi.org/10.1094/PHYTO.2000.90.5.454
10. Han J, Rotenberg D (2021) Integration of transcriptomics and network analysis reveals co-expressed genes in *Frankliniella occidentalis* larval guts that respond to tomato spotted wilt virus infection. BMC Genomics 22:810. https://doi.org/10.1186/s12864-021-08100-4
11. Han J, Klobasa W, de Oliveira L, Rotenberg D, Whitfield AE, Lorenzen MD (2024) CRISPR/Cas9-mediated genome editing of *Frankliniella occidentalis*, the western flower thrips, via embryonic microinjection. Insect Mol Biol. https://doi.org/10.1111/imb.12913
12. Badillo-Vargas IE, Rotenberg D, Schneweis DJ, Hiromasa Y, Tomich JM, Whitfield AE (2012) Proteomic analysis of Frankliniella occidentalis and differentially expressed proteins in response to tomato spotted wilt virus infection. J Virol 86(16):8793–8809. https://doi.org/10.1128/JVI.00285-12

Chapter 11

In Vitro Evaluation of Bunyavirus T Cell Immunity

Anita K. McElroy

Abstract

Characterization and quantitation of T cell responses following infection and/or vaccination can provide insight into mechanisms of host cell immunity that provide resolution of acute infection or protection from future infection or disease. While these types of studies are very advanced for viruses such as HIV, influenza, and SARS-CoV-2, they are less well developed for most of the Bunyaviruses. Cytotoxic CD8T cells are especially relevant in the context of viral infections since they recognize virus-infected cells via interaction of the T cell receptor with virally derived peptides presented in the context of MHCI. CD4T follicular cells are especially important for augmenting the antiviral antibody response. This chapter provides methods for characterizing T cell responses post infection/vaccination in both mice and humans as well as several methods for quantifying virus-specific T cells with the goal of arming bunyavirus researchers with the tools needed to move the field forward.

Key words T cells, Bunyavirus, ELISPOT, Flow cytometry

1 Introduction

Adaptive immunity, humoral and cellular, is generated following exposure to a pathogen and contribute to both pathogen clearance and protection from disease following a subsequent exposure. While much attention has been focused on humoral (antibody-mediated) immune responses in the context of Bunyaviral infection, measurement of T cell-mediated responses is relevant for characterizing the natural immune response to infection and following vaccination. An especially challenging aspect of the evaluation of T cell responses is the identification and characterization of virus-specific T cells. This is due in part to the host-specific genetic heterogeneity of T cell function since the T cell receptor (TCR) must engage viral peptides in the context of host major histocompatibility complex (MHC) proteins: MHCI for CD8 T cells and MHCII for CD4 T cells.

The *Bunyaviricetes* class is large and encompasses many medically relevant viruses, including, but not limited to, old- and

Hani Boshra (ed.), *Bunyaviruses: Methods and Protocols*, Methods in Molecular Biology, vol. 2893, https://doi.org/10.1007/978-1-0716-4338-9_11,

new-world Hantaviruses (family *Hantaviridae*), Rift Valley fever virus (RVFV, family *Phenuviridae*), Crimean-Congo hemorrhagic fever virus (CCHFV, family *Nairoviridae*), Lassa virus (family *Arenaviridae*), and La Crosse virus (LACV, family *Peribunyaviridae*). Ex vivo characterization of T cells from infected individuals has provided insight into the pathogenesis of Hantaviruses [1–7], Lassa virus [8], and severe fever with thrombocytopenia syndrome virus (SFTSV) [9–11]. In the case of other viruses, such as CCHFV, RVFV, or LACV, accessing acute phase human samples for T cell research has proved challenging; therefore, investigators have turned to animal models to understand the roles of T cells in protection from or progression of disease [12–14]. Studies of the role of T cells in vaccine-mediated immunity for bunyaviruses have also relied heavily on animal models and in some cases have revealed a unique role for T cells in providing protection [15, 16]. Despite sharing a phylogenetic order, these viruses have unique modes of transmission, pathogenesis, and immunity; it is important that investigators continue to dissect the immunologic mechanisms at work during both natural infection and vaccine-mediated immunity.

This chapter uses examples from our studies of Rift Valley fever virus (RVFV), a prototypic Bunyavirus in the *Phenuviridae* family, but could be easily adapted for use with any Bunyavirus (or any other pathogen). I highly recommend the weblog of Holden Maecker at Stanford University for protocols, references, and tips and tricks for flow cytometry and immune monitoring (https://maeckerlab.typepad.com/); this resource has been very helpful for our group's work. Flow cytometry-based T cell characterization assays of blood or immune cells derived from organs following infection or vaccination are discussed first. These permit quantification of both the timing and magnitude of acute responses and can be combined with studies of other immune cells (e.g. B cells, NK cells, or myeloid cells) to get an overall picture of the host immune response. However, these acute characterization studies do not measure the *virus-specific* response. The allelic diversity present in human MHC proteins limits the direct comparison of antigen-specific T cell assay results between individuals, but antigen-specific T cells can be identified, quantitated, and characterized in humans [17]. The use of inbred mouse models for the study of T cells that target epitopes contained within viral antigens is much more tractable, and the lack of genetic diversity provides an opportunity to identify all immunodominant viral epitopes, and, in some models, enables the development of tetramers to directly stain virus-specific T cells [9]. Finally, inbred mouse models also permit adoptive transfer studies to define mechanistic cell-based immune correlates of protection from disease [18].

To quantitate virus-specific T cells, viral proteins must be selected (usually proteins that are produced in high quantity during an infection, often, but not always, structural proteins), and then overlapping peptides are generated [19]. These peptides can be used individually or as pools in ELISPOT or flow cytometry-based assays. In either case, the readout is production of a cytokine (typically IFN-γ, but IL-2, TNF-α, and others have been used) upon recognition of the peptide by the T cell receptor. Since bunayviruses are relatively small, it would be feasible to do this for all the viral proteins, but the highly expressed nucleocapsid or glycoproteins are typically chosen. ELISPOT assays have the advantage of not requiring a flow cytometer and are less expensive and more sensitive than flow cytometry-based assays. However, flow cytometry-based assays can identify which cell is producing the cytokine being measured and are more easily multiplexed, such that an investigator can get a sense of the functionality of a virus-specific T cell based upon magnitude and diversity of elaborated cytokines following peptide stimulation. These features can be quite relevant in the context of heterologous protection against related viruses or for assessment of immune correlates of vaccine-mediated protection. Another benefit of using flow cytometry is that activation-induced markers (such as CD25, CD69, CD154, OX-40), which are upregulated on the surface of virus-specific T cells following TCR engagement, are often more readily detectable than intracellular cytokines [20]. In summary, there are numerous ways to assess T cells in the context of bunyaviral infection or vaccination, and these various methods provide unique but complementary information. The investigator should choose the methodology that best addresses their specific research questions, and these decisions can be informed by the costs and logistical challenges inherent to each method.

2 Materials

2.1 Preparation of Host Cells

1. Ficoll 1077 if not using cell preparation tubes (CPT, BD).
2. Phosphate buffered saline (PBS) without Mg and Ca.
3. RPMI10: RPMI media with 10% FBS.
4. Red blood cell (RBC) lysis buffer.
5. Tubes for collecting anticoagulated whole blood (EDTA, Li Heparin or Na Citrate).
6. 15 and/or 50 mL conicals.
7. 5 mL syringes.
8. 70 uM filter units that fit into a 50 mL conical.
9. Tabletop centrifuge.
10. Hemacytometer.

2.2 Characterization of T Cell Responses Following Infection or Vaccination Using Flow Cytometry

1. Host cells: These could be derived either from blood or organs.
2. Fluorophore-conjugated antibodies. Table 1 contains common markers for T cells. There are many additional options; this is not an exhaustive list, just a starting point. Specific fluorophores are not noted as this would be specific to the instrumentation available at your institution. Consult with your flow cytometry core for guidance on panel design and see published examples that were used for previous studies [21–24] (*see* **Note 1**).
3. PBS without Mg and Ca.
4. RPMI10.
5. Flow wash buffer: PBS with 2% FBS.
6. Live/dead dye.
7. BD Fix/Perm and Perm/Wash.
8. Compensation beads.
9. 5 mL snap cap tubes.
10. Flow cytometer capable of measuring the fluorophores in the selected panel.
11. Vortex.
12. Tabletop centrifuge.

2.3 Ex Vivo Stimulation with Antigen-Specific Peptides and Quantitation by ELISPOT

1. Host cells: These could be derived from either blood or organs.
2. PBS without Mg and Ca.
3. RPMI10.
4. Peptides specific to the viral protein of interest (*see* **Note 2**).
5. Capture and detection antibodies for ELISPOT (e.g. ELISpot Flex: Mouse IFN-γ (HRP), Mabtech).

Table 1
Common T cell markers for immune characterization

Marker	Surface or intracellular	Purpose	Mice, human, or both
CD3	Surface	Identifies all CD3 T cells	Both
CD4	Surface	Identifies CD4 T cells	Both
CD8	Surface	Identified CD8 T cells	Both
CD44	Surface	Activation marker	Mice
Ki-67	Intracellular	Proliferation marker	Both
Granzyme B	Intracellular	Cytotoxicity marker	Both
CD38	Surface	Activation marker	Human
HLADR	Surface	Activation marker	Human

6. Positive control stimulation, often PMI/ionomycin mix (e.g. 500× stimulation mix without Brefeldin A, BioLegend).
7. ELISPOT plates (e.g. MSIPS4W10, Millipore).
8. 96-well format titer tubes.
9. Tissue culture incubator 37 °C and 5% CO_2.
10. Vortex.
11. ELISPOT reader (these could theoretically be visualized manually with a stereoscope or magnifying glass, but a reader is highly recommended).

2.4 Ex Vivo Stimulation with Antigen-Specific Peptides and Quantitation by Intracellular Flow Cytometry

1. Host cells: These could be derived from either blood or organs.
2. Fluorophore-conjugated antibodies: Table 2 contains commonly used markers for T cells. There are many additional options; this is not an exhaustive list, just a starting point. Specific fluorophores are not noted as this would be specific to the instrumentation available at your institution. Consult with your flow cytometry core for guidance on panel design and see published examples that were used for previous studies [22, 24].
3. PBS without Mg and Ca.
4. RPMI10.
5. Peptides specific to viral protein of interest (*see* **Note 2**).
6. Positive control stimulation, often PMI/ionomycin mix (e.g. 500× stimulation mix without Brefeldin A, BioLegend).
7. Brefeldin A stock solution in DMSO.
8. Live/dead stain.
9. 96-well V-bottom plate.
10. Cell culture incubator capable of 37 °C and 5% CO_2.

Table 2
Common T cell markers for ex vivo stimulation assays

Marker	Surface or intracellular	Purpose	Mice, human, or both
CD3	Surface	Identifies all CD3 T cells	Both
CD4	Surface	Identifies CD4 T cells	Both
CD8	Surface	Identified CD8 T cells	Both
IFN-γ	Intracellular	Cytokine marker	Both
IL-2	Intracellular	Cytokine marker	Both
TNF-α	Intracellular	Cytokine marker	Both
CD44	Surface	Acute activation marker	Mice

11. Vortex.
12. Tabletop centrifuge.
13. Flow cytometer capable of measuring the fluorophores in the selected panel.

3 Methods

3.1 Preparation of Host Cells

1. Collect whole anticoagulated blood (EDTA, Li Heparin or Na Citrate) or harvest organs on various days post infection (dpi). For measurement of acute phase responses consider 7, 10, and 14 dpi or vaccination; for measurement of memory responses consider 21–28 dpi or vaccination.
2. Peripheral blood mononuclear cells (PBMCs): This only applies to larger animals/humans since you can expect approximately 1×10^6 cells per mL and in smaller animal models this is often limiting. To purify PMBCs, collect blood directly into cell preparation tubes (CPT) and follow manufacturer's instructions. If the cost of CPTs is prohibitive, then mix anticoagulated whole blood 1:1 with PBS and layer over top an equal volume of Ficoll 1077. Centrifuge $500 \times g$ for 30 min with NO brake. Transfer the interphase layer to new tube, wash in PBS, and spin $500 \times g$ for 5 min. Suspend in PBS and count cells. Cells are now ready to use.
3. Organs: In mice, a readily available organ for immune studies is the spleen (*see* **Note 3**). Immediately after harvest, the spleen should be placed in RPMI10 until ready to process. Processing should be done within the hour. Place a 70 um filter unit into the top of a 50 mL conical tube and pour the spleen into the filter. Using the blunt end of a 5 cc syringe plunger, smash the spleen through the filter. Rinse with 5 mL of RMPI 10. Repeat these steps until the organ is completely dissociated, usually requires at least 20 mL of RMPI10. Only fibrinous material should remain in the filter. Centrifuge the conical tube $500 \times g$ for 5 min to pellet the splenocytes. Remove the supernatant and suspend the pellet in 5 mL of RBC lysis solution; incubate for 5 min at ambient temperature. Bring up the volume to 25 mL with RPMI10 and count the cells. They are ready to use. An entire mouse spleen can yield up to 1×10^8 cells, so many stains can be performed from one spleen. The spleen is mostly B cells, but about 30% of the cells will be T cells (*see* **Note 4**).

3.2 Characterization of Immune Responses Following Infection or Vaccination Using Flow Cytometry

1. Whole blood: As little as 100 uL of fresh whole blood can be stained. Using quantities of fluorophore-conjugated antibody/cell number suggested by the manufacturer, add surface stain antibodies directly to whole blood (*see* **Note 5**). Incubate for 30 min at ambient temperature. Wash twice in 2 mL of flow wash buffer, centrifuging at 500 × g for 5 min each time. To the pellet after the washes, add 500 uL of BD Fix/Perm and incubate for 20 min at ambient temperature. Wash twice in 2 mL of Perm/Wash, centrifuging at 600 × g for 5 min each time. Vortex the pellet after the washes, then add intracellular stain antibodies and incubate for 45 min at ambient temperature. Wash twice in 2 mL of Perm/Wash, centrifuging at 600 × g for 5 min each time. The final pellet can be suspended in PBS for acquisition on the flow cytometer.
2. For PBMCs or splenocytes: aliquot 1×10^6 cells to 5 mL snap cap tube(s). Wash in PBS then perform a live/dead dye stain per manufacturer's instructions. Wash in flow buffer once, and then stain as above for whole blood.
3. Once stains are complete, samples can be acquired immediately or stored at 4 °C overnight. Prolonged storage before acquisition is not recommended.
4. Prepare compensation beads for each of the fluorophores in the panel, including any "dump" channel you might have (e.g. live/dead dye or cells you plan to gate out, e.g. CD19 or CD14 cells).
5. After acquisition of all compensation and sample tubes and application of a compensation matrix to your data, set up a gating strategy that will permit you to quantitate your activated, proliferating and/or cytotoxic CD4 and CD8 T cells as show in Fig. 1. Be sure to include a singlet gate (FSC-A × FSC-H) to eliminate clumped cells and a time gate (SSC × time) to eliminate fluctuations that occur at the start and stop of acquisition.

3.3 Ex Vivo Stimulation with Antigen-Specific Peptides and Quantitation by ELISPOT

1. The day prior to the assay, follow manufacturer's instructions to coat the ELISPOT plate with the capture antibody and incubate overnight at 4 °C. These are species-specific assays and are available for many different cytokines, but IFN-γ is the most common.
2. The day of the assay, wash the ELISPOT plate 4× with PBS and block with 200ul of RPMI10 per well at 37 °C for at least 1 h.
3. Prepare host cells as noted in Subheading 3.1. Since these assays are often used to quantitate memory responses, later timepoints such as 21 or 28 dpi or vaccination should be included.

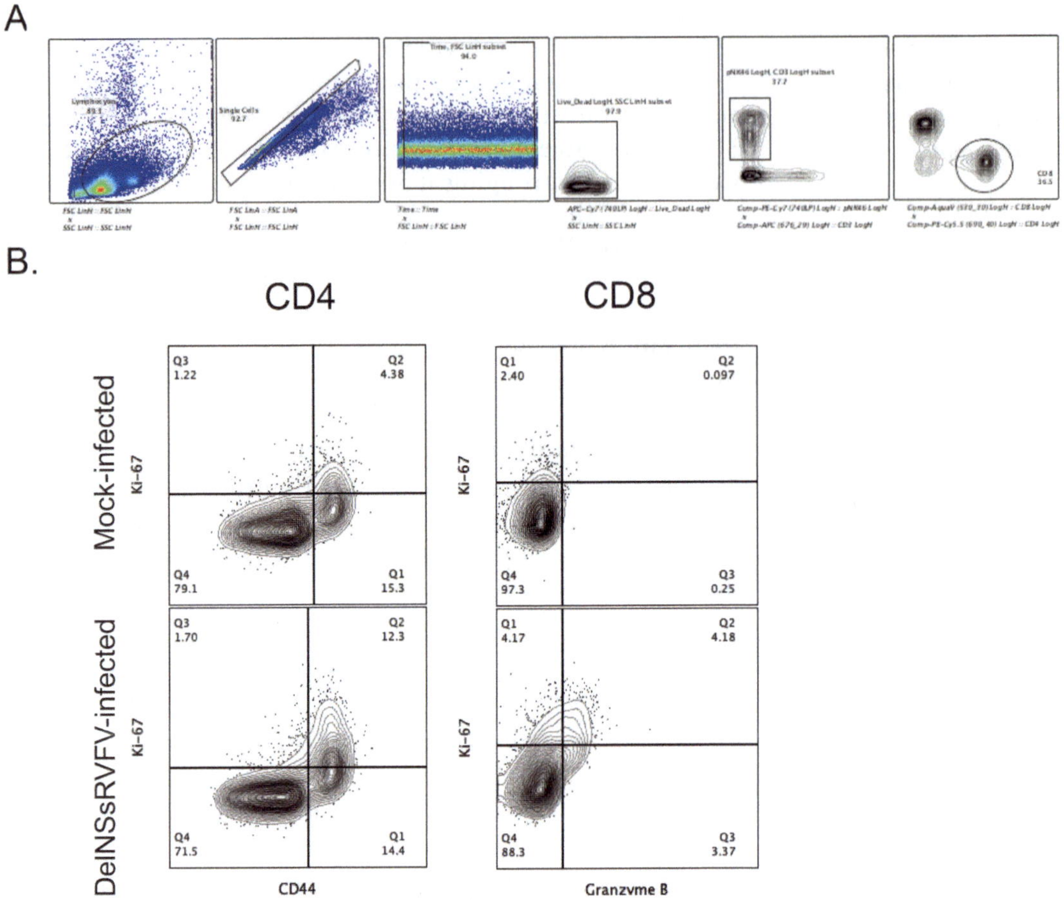

Fig. 1 Acute characterization of T cells. Total splenocytes obtained from mock- or DelNSsRVFV-infected C57BL/6 mice on day 6 post infection were stained for markers of T cell activation, proliferation, and cytotoxicity. Suggested gating strategy; lymphocytes→singlets→homogenous events on time gate→live cells→CD3+ cells→CD4 or CD8+ cells (**a**). An increase in activated (CD44+) and proliferating (Ki-67+) CD4T cells and an increase in proliferating (Ki-67+) and cytotoxic (GranzymeB+) CD8T cells was noted (**b**)

4. Suspend cells in RPMI10 to 6×10^6 cells/mL. Aliquot 150 uL of this suspension to four tubes in the top row of a titer tube rack. These four tubes will be a negative control, a positive control and duplicates of the experimental peptide. Repeat with an additional 4 tubes for each specimen to be tested. Add 150 uL of RPMI10 to tubes in rows 2–4 of the titer tube rack.
5. Make up stimulation mixes.
   (a) Positive control: RPMI10 + PMI/ionomycin mix (usually 500× stock, diluted to 2×). Make up sufficient volume to have 150 uL × # of specimens to be tested +1.
   (b) Peptide mix: Peptides are typically shipped lyophilized. Suspend these in DMSO to make 10 or 20 mg/mL stocks

and store them at −20 °C or −80 °C. Dilute the peptide (s) in RPMI10 to 4 ug/mL. (the final concentration once mixed with cells will be 2 ug/mL) (*see* **Note 6**). Make up sufficient volume to have 150 uL × 2 × # of specimens to be tested +1.

(c) Negative mix: This assesses for spontaneous background activity. Mix the same volume of DMSO used to make the peptide mix with RPMI10. Make up sufficient volume to have 150 uL × # of samples to be tested +1.

6. Add 150 uL of the respective stimulation mix to the cells in the appropriate well in the top row of the titer tube rack. Mix by pipetting. Make twofold serial dilutions by mixing 150 uL of row 1 with the 150 uL of media in row 2; repeat down to row 4. One titer tube rack/plate can therefore quantitate virus-specific T cells with controls from six specimens.
7. Dump off the RPMI10 block from the coated ELISPOT plate. Transfer 100 uL of each cell + stimulation mix from the titer tube rack to the ELISPOT plate and incubate at 37 °C in a 5% CO_2 incubator for 24 h. The top row will have 3×10^5 cells/well, the second row 1.5×10^5 cells/well, the third row 7.5×10^4 cells/well, and the fourth 3.75×10^4 cells/well.
8. After the 24 h have elapsed, dump off the cells + stim from the ELISPOT plate and wash 4× in 200 uL of PBS per well. Follow the remainder of the manufacturer's protocol for detection of the spots.
9. Use an ELISPOT reader to count spots in each well. Convert data to spots per 1×10^5 cells. Average the experimental duplicates, and subtract the background. The readout is virus-specific spot forming units per 1×10^5 cells. For example, 28 days following vaccination of C57BL/6 mice with a live attenuated strain of RVFV (DelNSsRVFV), splenocytes routinely have 50–100 virus-specific SFU/1×10^5 cells following stimulation with a nucleocapsid pool in an IFN-γ ELISPOT assay [22].

3.4 Ex Vivo Stimulation with Antigen-Specific Peptides and Quantitation by Intracellular Flow Cytometry

1. Prepare host cells as noted in Subheading 3.1. Since these assays are often used to quantitate memory responses, later timepoints such as 21 or 28 dpi or vaccination should be included.
2. Suspend cells to 1×10^7 cells/mL in PBS. Aliquot 100 uL per well of a 96-well V-bottom plate. Centrifuge 500 × g for 5 min, and dump supernatant.
3. Suspend pellets in stimulation mix as appropriate. Make up sufficient mix for 200 uL per well.

(a) Negative stimulation: DMSO (same volume as used for peptides) + RPMI10 + Brefeldin A (10 ug/mL final concentration).

(b) Positive stimulation: PMA/ionomycin (1× final concentration) + RPMI10 + Brefeldin A (10ug/ml final concentration).

(c) Experimental stimulation: Peptide (individual or pool, 2 ug/mL of *each peptide* final concentration) in RPMI10 + Brefeldin A (10 ug/mL final concentration).

4. Incubate at 37 °C in 5% CO_2 incubator for 6 h.
5. Centrifuge 500 × g for 5 min, dump supernatant, and wash cells in 200 uL of PBS. Centrifuge 500 × g for 5 min, dump supernatant, and suspend pellet in live/dead stain (dilute and incubate according to manufacturer). Centrifuge 500 × g for 5 min and dump supernatant.
6. Wash cells in 200 uL of flow buffer, centrifuge plates as before, and then suspend pellets in surface stains (diluted in flow buffer); 100 uL per well. Incubate for 30 min at ambient temperature.
7. Add 100 uL of flow buffer per well. Centrifuge 500 × g for 5 min and dump supernatant. Wash cells in 200 uL of flow buffer, centrifuge plates as before, and suspend pellets in BD Fix/Perm. Incubate 20 min at ambient temperature. Centrifuge 600 × g for 5 min and dump supernatant.
8. Wash pellets in 200 uL of Perm/Wash. Centrifuge 600 × g for 5 min and dump supernatant. Repeat.
9. Suspend pellets in intracellular stains diluted in Perm/Wash. Incubate 45 min at ambient temperature (*see* **Note** 7).
10. Centrifuge 600 × g for 5 min and dump supernatant. Wash pellets in 200 uL of Perm/Wash. Centrifuge 600 × g for 5 min and dump supernatant. Repeat.
11. Final cell pellets can be suspended in PBS and immediately acquired or stored overnight at 4 °C and acquired the following day.
12. Prepare compensation beads for each of the fluorophores in the panel, including any "dump" channel you might have (e.g. live/dead dye or cells you plan to gate out, e.g. CD19 or CD14 cells).
13. After acquisition of all compensation and sample tubes and application of a compensation matrix to your data, the cytokine producing CD4 and CD8 T cells can be quantitated as shown in Fig. 2.

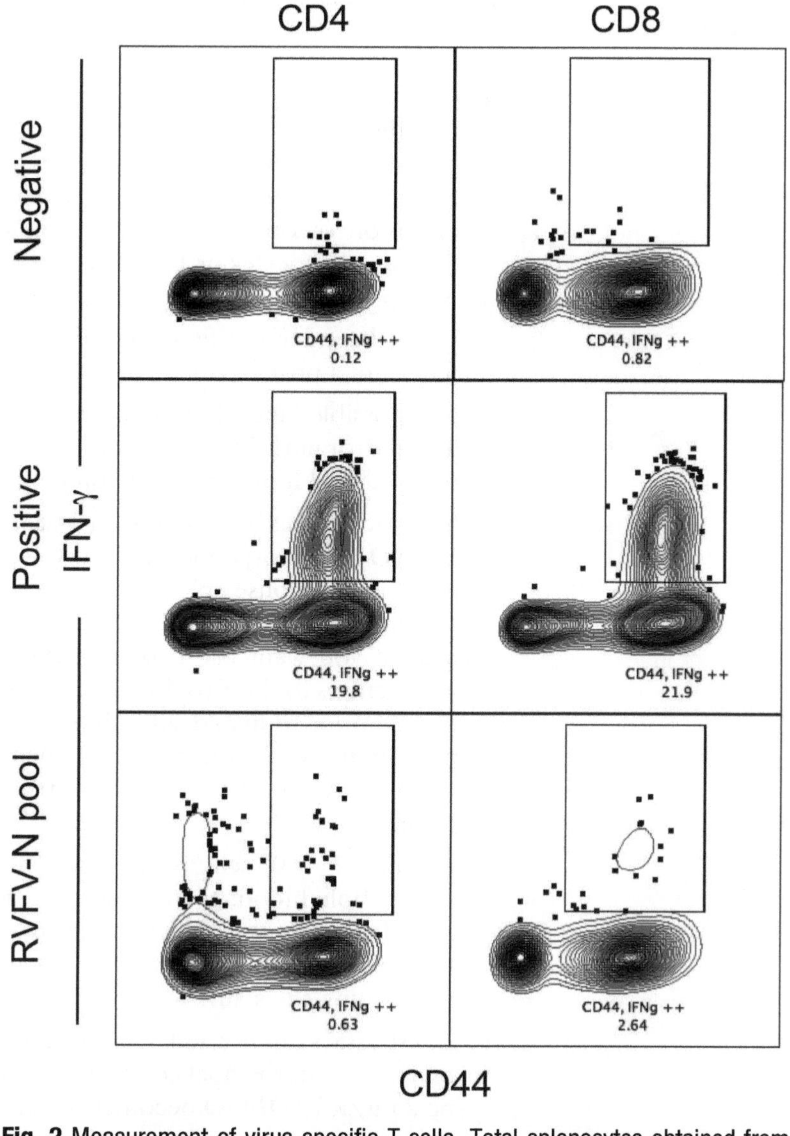

Fig. 2 Measurement of virus-specific T cells. Total splenocytes obtained from DelNSsRVFV vaccinated CC057/Unc mice at 28 days post vaccination were incubated with DMSO (negative), PMI/ionomycin (positive), or RVFV N peptide pool ($n = 59$ peptides spanning the entire open reading frame) for 6 h and then flow cytometry was performed to quantitate IFN-γ producing T cells

14. Data can be shown as percent of CD4 or CD8 T cells that produce each cytokine following antigen stimulation or you can use Boolean gating to define the percentage of cells that produce one, two, or all three cytokines thought to represent more functional T cells. Use https://niaid.github.io/spice/ to help you visualize this data.

4 Notes

1. Choose the brightest fluorophores for intracellular and activation markers, and use dimmer fluorophores for surface markers that are present in high quantities (e.g. CD3, CD4, CD8). Sometimes makers are included in a "dump" channel to permit exclusion of cells when gating, such as CD14 or CD20 to exclude monocytes or B cells, respectively.
2. Peptides are custom-made to >70% purity. For screening and pools, have 15mers generated that overlap by 11 aa along the length of a protein [19].
3. It is possible (and often of interest) to measure T cells in other affected organs. Detailed methods for processing liver, brain, footpad, and lung for flow cytometry are available [22, 25, 26].
4. Splenocytes can be suspended in freezing media (90% FBS 10% DMSO) and stored in liquid nitrogen for later use; but in our experience, mouse splenocytes die quickly during storage, so you must incorporate a live/dead dye stain prior to the surface stain if you want use frozen splenocytes. See manufacturer's instructions for how to do this staining but be sure to wash cells in PBS prior to live/dead staining, as these dyes stain proteins present in cells with disrupted membranes and any FBS or other protein in the sample can prevent the dye from working properly.
5. Estimate 1×10^5 cells in 100 uL of whole blood. When staining fresh whole blood, there is no need for a live/dead stain. If a live/dead stain was desired for some reason, do an RBC lysis step and wash the cells in PBS to remove any protein prior to performing the stain.
6. These assays can be used with a single peptide or up to 300 peptides [27] but the final concentration of EACH peptide should be 2 ug/mL. DMSO becomes the limiting factor in how many peptides can be used in a pool of peptides. Keep final DMSO concentration below 1%.
7. If you decide to use activation-induced cell markers (AIM) instead of intracellular cytokine markers for identification of antigen-specific T cells, the incubation times are longer for stimulation and the stains are included in the surface stains; no need to perform intracellular staining. See [20] for details.

Acknowledgements

I would like to thank Dr. Rama Akondy, Ashoka University, India for her kindness and patience in teaching me many technical aspects of immunology; she has a unique love for T cells, and it is contagious. The McElroy lab is supported by the NIH (R01AI171200, R56AI171920, R01AI150792 and UC7AI180311), the RK Mellon Institute, Burroughs Wellcome (CAMS 1013362.02), and the Coalition for Epidemic Preparedness Innovations (CEPI).

References

1. Xie M, Dong Y, Zhou Y, Ren H, Ji Y, Lv S (2013) Levels of HTNV-specific CD8+ T lymphocytes in PBMC from the patients with hemorrhagic fever with renal syndrome. Intern Emerg Med 8(6):503–508. https://doi.org/10.1007/s11739-011-0633-4
2. Terajima M, Ennis FA (2011) T cells and pathogenesis of hantavirus cardiopulmonary syndrome and hemorrhagic fever with renal syndrome. Viruses 3(7):1059–1073. https://doi.org/10.3390/v3071059
3. Ma Y, Wang J, Yuan B, Wang M, Zhang Y, Xu Z, Zhang C, Zhang Y, Liu B, Yi J, Yang K, Yang A, Zhuang R, Jin B (2013) HLA-A2 and B35 restricted hantaan virus nucleoprotein CD8+ T-cell epitope-specific immune response correlates with milder disease in hemorrhagic fever with renal syndrome. PLoS Negl Trop Dis 7(2):e2076. https://doi.org/10.1371/journal.pntd.0002076
4. Koivula TT, Tuulasvaara A, Hetemaki I, Makela SM, Mustonen J, Sironen T, Vaheri A, Arstila TP (2014) Regulatory T cell response correlates with the severity of human hantavirus infection. J Infect 68(4):387–394. https://doi.org/10.1016/j.jinf.2013.11.007
5. Ma Y, Yuan B, Zhuang R, Zhang Y, Liu B, Zhang C, Zhang Y, Yu H, Yi J, Yang A, Jin B (2015) Hantaan virus infection induces both Th1 and ThGranzyme B+ cell immune responses that associated with viral control and clinical outcome in humans. PLoS Pathog 11(4):e1004788. https://doi.org/10.1371/journal.ppat.1004788
6. Liu B, Ma Y, Zhang Y, Zhang C, Yi J, Zhuang R, Yu H, Yang A, Zhang Y, Jin B (2015) CD8low CD100- T cells identify a novel CD8 T cell subset associated with viral control during human Hantaan virus infection. J Virol 89(23):11834–11844. https://doi.org/10.1128/JVI.01610-15
7. Rasmuson J, Pourazar J, Mohamed N, Lejon K, Evander M, Blomberg A, Ahlm C (2016) Cytotoxic immune responses in the lungs correlate to disease severity in patients with hantavirus infection. Eur J Clin Microbiol Infect Dis 35(4):713–721. https://doi.org/10.1007/s10096-016-2592-1
8. Port JR, Wozniak DM, Oestereich L, Pallasch E, Becker-Ziaja B, Muller J, Rottstegge M, Olal C, Gomez-Medina S, Oyakhliome J, Ighodalo Y, Omomoh E, Olokor T, Adomeh DI, Asogun D, Ogbani-Emovon E, Hartmann K, Krasemann S, Nelson EV, Escudero-Perez B, McElroy AK, Gunther S, Munoz-Fontela C (2020) Severe human Lassa fever is characterized by nonspecific T-cell activation and lymphocyte homing to inflamed tissues. J Virol 94(21). https://doi.org/10.1128/JVI.01367-20
9. Barbeau DJ, Cartwright HN, Harmon JR, Spengler JR, Spiropoulou CF, Sidney J, Sette A, McElroy AK (2021) Identification and characterization of Rift Valley fever virus-specific T cells reveals a dependence on CD40/CD40L interactions for prevention of encephalitis. J Virol 95(23):e0150621. https://doi.org/10.1128/JVI.01506-21
10. Zong L, Yang F, Liu S, Gao Y, Xia F, Zheng M, Xu Y (2023) $CD8^+$ T cells mediate antiviral response in severe fever with thrombocytopenia syndrome. FASEB J 37(1):e22722. https://doi.org/10.1096/fj.202201343RR.
11. Li J, Han Y, Xing Y, Li S, Kong L, Zhang Y, Zhang L, Liu N, Wang Q, Wang S, Lu S, Huang Z (2014) Concurrent measurement of dynamic changes in viral load, serum enzymes, T cell subsets, and cytokines in patients with severe fever with thrombocytopenia syndrome. PLoS One 9(3):e91679. https://doi.org/10.1371/journal.pone.0091679
12. Rao D, Meade-White K, Leventhal S, Mihalakakos E, Carmody A, Feldmann H, Hawman DW (2023) CD8(+) T-cells target the Crimean-Congo haemorrhagic fever virus Gc protein to control the infection in wild-type mice. EBioMedicine 97:104839. https://doi.org/10.1016/j.ebiom.2023.104839
13. Wonderlich ER, Caroline AL, McMillen CM, Walters AW, Reed DS, Barratt-Boyes SM, Hartman AL (2018) Peripheral blood biomarkers of disease outcome in a monkey model of Rift Valley fever encephalitis. J Virol 92:3. https://doi.org/10.1128/JVI.01662-17
14. Winkler CW, Myers LM, Woods TA, Carmody AB, Taylor KG, Peterson KE (2017)

Lymphocytes have a role in protection, but not in pathogenesis, during La Crosse virus infection in mice. J Neuroinflammation 14(1):62. https://doi.org/10.1186/s12974-017-0836-3

15. Hinkula J, Devignot S, Akerstrom S, Karlberg H, Wattrang E, Bereczky S, Mousavi-Jazi M, Risinger C, Lindegren G, Vernersson C, Paweska J, van Vuren PJ, Blixt O, Brun A, Weber F, Mirazimi A (2017) Immunization with DNA plasmids coding for crimean-Congo hemorrhagic fever virus capsid and envelope proteins and/or virus-like particles induces protection and survival in challenged mice. J Virol 91(10). https://doi.org/10.1128/JVI.02076-16
16. Lopez-Gil E, Lorenzo G, Hevia E, Borrego B, Eiden M, Groschup M, Gilbert SC, Brun A (2013) A single immunization with MVA expressing GnGc glycoproteins promotes epitope-specific CD8+-T cell activation and protects immune-competent mice against a lethal RVFV infection. PLoS Negl Trop Dis 7(7):e2309. https://doi.org/10.1371/journal.pntd.0002309
17. Harmon JR, Barbeau DJ, Nichol ST, Spiropoulou CF, McElroy AK (2020) Rift Valley fever virus vaccination induces long-lived, antigen-specific human T cell responses. NPJ Vaccines 5(1):17. https://doi.org/10.1038/s41541-020-0166-9
18. Doyle JD, Barbeau DJ, Cartwright HN, McElroy AK (2022) Immune correlates of protection following Rift Valley fever virus vaccination. NPJ Vaccines 7(1):129. https://doi.org/10.1038/s41541-022-00551-4
19. Maecker HT, Dunn HS, Suni MA, Khatamzas E, Pitcher CJ, Bunde T, Persaud N, Trigona W, Fu TM, Sinclair E, Bredt BM, McCune JM, Maino VC, Kern F, Picker LJ (2001) Use of overlapping peptide mixtures as antigens for cytokine flow cytometry. J Immunol Methods 255(1–2):27–40. https://doi.org/10.1016/s0022-1759(01)00416-1
20. Poloni C, Schonhofer C, Ivison S, Levings MK, Steiner TS, Cook L (2023) T-cell activation-induced marker assays in health and disease. Immunol Cell Biol 101(6):491–503. https://doi.org/10.1111/imcb.12636
21. Arunkumar G, Devadiga S, McElroy AK, Prabhu S, Sheik S, Abdulmajeed J, Robin S, Sushama A, Jayaram A, Nittur S, Shakir M, Kumar KGS, Radhakrishnan C, Sakeena K, Vasudevan J, Reena KJ, Sarita RL, Klena JD, Spiropoulou CF, Laserson KF, Nichol ST (2019) Adaptive immune responses in humans during nipah virus acute and convalescent phases of infection. Clin Infect Dis 69(10):1752–1756. https://doi.org/10.1093/cid/ciz010
22. Harmon JR, Spengler JR, Coleman-McCray JD, Nichol ST, Spiropoulou CF, McElroy AK (2018) CD4 T cells, CD8 T cells, and monocytes coordinate to prevent rift valley fever virus encephalitis. J Virol 92(24). https://doi.org/10.1128/JVI.01270-18
23. McElroy AK, Akondy RS, Davis CW, Ellebedy AH, Mehta AK, Kraft CS, Lyon GM, Ribner BS, Varkey J, Sidney J, Sette A, Campbell S, Stroher U, Damon I, Nichol ST, Spiropoulou CF, Ahmed R (2015) Human Ebola virus infection results in substantial immune activation. Proc Natl Acad Sci USA 112(15):4719–4724. https://doi.org/10.1073/pnas.1502619112
24. McElroy AK, Akondy RS, Harmon JR, Ellebedy AH, Cannon D, Klena JD, Sidney J, Sette A, Mehta AK, Kraft CS, Lyon MG, Varkey JB, Ribner BS, Nichol ST, Spiropoulou CF (2017) A case of human Lassa virus infection with robust acute T-cell activation and long-term virus-specific T-cell responses. J Infect Dis 215(12):1862–1872. https://doi.org/10.1093/infdis/jix201
25. Singer BD, Mock JR, D'Alessio FR, Aggarwal NR, Mandke P, Johnston L, Damarla M (2016) Flow-cytometric method for simultaneous analysis of mouse lung epithelial, endothelial, and hematopoietic lineage cells. Am J Physiol Lung Cell Mol Physiol 310(9):L796–L801. https://doi.org/10.1152/ajplung.00334.2015
26. Hameed M, Rai P, Makris M, Weger-Lucarelli J (2023) Optimized protocol for mouse footpad immune cell isolation for single-cell RNA sequencing and flow cytometry. STAR Protoc 4(3):102409. https://doi.org/10.1016/j.xpro.2023.102409
27. Chevalier MF, Bobisse S, Costa-Nunes C, Cesson V, Jichlinski P, Speiser DE, Harari A, Coukos G, Romero P, Nardelli-Haefliger D, Jandus C, Derre L (2015) High-throughput monitoring of human tumor-specific T-cell responses with large peptide pools. Onco Targets Ther 4(10):e1029702. https://doi.org/10.1080/2162402X.2015.1029702

Chapter 12

Analysis of *Leishbuviridae* from Trypanosomatids

Danyil Grybchuk, Alexei Yu Kostygov, and Vyacheslav Yurchenko

Abstract

Over the last decade, considerable progress has been made in unraveling RNA virus diversity. This has contributed to our understanding of the evolution of these viruses, which include emerging zoonotic human pathogens. Current success has been greatly facilitated by the development of next-generation sequencing platforms instrumental for meta-transcriptomic studies. However, due to the rapid evolution of RNA viruses, there are numerous "blind spots" waiting to be explored; one of those is the RNA virome of unicellular eukaryotes. Here, we present the pipeline, which has been successfully used to characterize various types of RNA viruses, including *Leishbuviridae* (*Bunyaviricetes, Hareavirales*) in the parasitic flagellates of the family Trypanosomatidae. The pipeline relies on axenic in vitro cell culture and double-stranded RNA enrichment, followed by direct RNA-sequencing. A detailed procedure description starting from the initial total RNA preparation to the final assembly of the viral segments is provided.

Key words Trypanosomatidae, *Leishbuviridae*, Protists, Virus Discovery, RNA isolation, dsRNA, NGS

1 Introduction

Transcription and replication of RNA viruses, regardless of genome strandedness and polarity, requires production of both "+" and "−" RNA strands in the cell cytoplasm by viral RNA-dependent RNA polymerase (RdRp) [1–4]. This trait of RNA viruses was intensely studied half-a-century ago as a curious deviation from the Central dogma of molecular biology. At the core of this research were chemical and enzymatic methods allowing to distinguish single-stranded (ss) and double-stranded (ds) RNA species. Differential precipitation of ssRNA and dsRNA with 2 M and 5 M lithium chloride (LiCl), respectively, was first reported by David Baltimore as part of his Nobel Prize winning studies on the

The original version of the chapter has been revised. A correction to this chapter can be found at https://doi.org/10.1007/978-1-0716-4338-9_22

Hani Boshra (ed.), *Bunyaviruses: Methods and Protocols*, Methods in Molecular Biology, vol. 2893,
https://doi.org/10.1007/978-1-0716-4338-9_12,

replication of poliovirus [5]. Soon after that, the ssRNA/ssDNA-specific S1 nuclease was first used to reveal the role of terminal secondary structures in replication of *Mammalian orthoreovirus 3*, MRV-3 [6] and LiCl fractionation of dsRNA was optimized for bulk detection of plant viruses [7]. Both LiCl- and S1 nuclease-based approaches were actively used till the early 2000s for the search of fungal dsRNA viruses fueled, in part, by the interest in potential biocontrol agents for phytopathogenic fungi [8, 9]. For example, LiCl was used to detect a partitivirus and a mitovirus in *Ophiostoma* spp., a causative agent of the Dutch elm disease [10, 11], while S1 nuclease-based approach yielded the discovery of an endornavirus in hypovirulent strain of *Helicobasidium mompa* (violet root rot) [12] and numerous divergent RNA mycoviruses in various fungi [13–16].

Even though negative-sense (−)ssRNA viruses do not accumulate dsRNA in the cell to avoid destruction by cellular RNA interference systems [17], their "+" and "−" RNAs recombine in vitro upon nucleic acid extraction. This phenomenon allows all ssRNA viruses to be detected by gel electrophoresis of S1 nuclease- and/or LiCl-treated total RNA samples, albeit with lower sensitivity as compared to dsRNA viruses. Combined with the observation that protist and fungal RNA viruses are often cryptic and persist in low copy numbers [18–21], the approaches introduced above require high amount of total RNA for analysis.

Individual viral bands can be cut out from the agarose gel and amplified by reverse-transcription (RT)-PCR. There are three methods for the viral RNA amplification, which differ in RNA pretreatment and the type of oligonucleotides used for the first strand cDNA synthesis: (i) no pretreatment and RT-PCR using specific primer with random six nucleotides at the 3′ end; (ii) *Escherichia coli* poly(A) polymerase treatment and RT-PCR using specific primer with 6–8 Thymine nucleotides at the 3′ end [22]; and (iii) attachment of an oligonucleotide using the T4 RNA ligase followed by RT-PCR with a complementary specific primer [23] or ligation of a self-priming hairpin oligonucleotide [24]. Methods (ii) and (iii) are suitable for the recovery of the complete genomic segment(s), which can be then sequenced conventionally by "primer walking" [25]. However, they rely on the free hydroxyl group at the 3′ end and, thus, are not suitable for viruses, which have modifications at this position. Conversely, method (i) does not depend on the 3′ end chemistry but yields only partial sequences.

Nowadays, the wide availability of next-generation sequencing (NGS) methods makes such RNA-amplification methods redundant. Moreover, modern bioinformatic software, such as Trinity [26] and SPAdes [27], allow de novo assembly of viral genomes with no prior knowledge about them. Therefore, these tools are applicable not only for the detection of known pathogens but also for the discovery of novel RNA viruses. Over the last decade, there was a burst of studies reporting thousands of sequences of RNA viruses in meta-transcriptomes across various biotopes and ecosystems [28–31]. These data significantly expanded our understanding of diversity and evolution of RNA viruses but suffered from one considerable drawback: they provided no information on the viral hosts.

A typical RNA-seq data analysis pipeline includes the following steps: raw read quality control and trimming, mapping and filtering of trimmed reads followed by their assembly (reference-based or de novo) into contigs (transcripts and/or viral genomic segments), identification and evaluation of their relative abundance within and between datasets [32, 33]. Quality control software performs a series of analyses to identify known problems in the read data, while trimming programs, such as Trimmomatic [34], solve these problems by discarding reads with overall high sequencing error rate, deleting low-quality bases from the 3′ termini, and removing adaptor sequences introduced at the 5′ ends of each read as a part of the sequencing procedure. Next, the "clean" reads are mapped onto the reference genome if one is available. Routinely, this is done with Bowtie2 and SAMtools bundle [35, 36] or BBtools [37]. For eukaryotic organisms, this step is crucial to determine exon–intron boundaries and untranscribed regions [38, 39]. The advantage of working with RNA viruses is that their genomes are usually simple and their genes are distinct from those of the hosts [40]. As such, the initial mapping can be useful to set apart viral-from host reads to speed up further assembly and minimize the probability of chimeric sequence appearance. The assembly of viral sequences is usually done in the reference-free (de novo) mode [26, 27], since appropriate reference is absent in most cases. The resulting assembled sequences (contigs) usually contain the complete genomic segment(s) of a virus. Some gaps and fragmentation may occur if viruses are scarce and/or contain secondary structures.

The identification of viral contigs in the databases is done via amino acid sequence homology search. Since the input data are in nucleotides, this requires translating contigs in all six frames and comparing each translation product against the database using BLASTx first realized in BLAST+ software [41]. This procedure is computationally intense and requires both an efficient software specifically designed to work with genomic/transcriptomic data, such as DIAMOND [42], and use of the clustered databases, where searches are performed only against a subset of representative sequences with a given all-to-all amino acid identity threshold, for example, UniRef [43]. In addition, a more sensitive search for viral proteins can be done using the hidden Markov model approach that compares query proteins against a profile build from an amino acid sequence alignment of a specific set of homologous proteins [44, 45]. For RNA viruses, the best candidate is RdRp [46, 47].

Another piece of information retrieved from the assembled data is the coverage of viral contigs (i.e., how many reads went into constructing a given contig). This value may help to estimate abundance of viral RNAs relative to each other or even to the host transcriptome, if total RNA was sequenced. In order to calculate coverage, the reads are first aligned back to the assembled contigs using the mapping software mentioned above. Then, the number of reads aligned to each contig is counted using a custom script. To facilitate comparison within a single sequencing experiment or between different ones, the read count for each contig is normalized per length of the latter in kilobases and per million reads. For single-end reads, the coverage value is called RPKM (reads per kilobase per million). For paired-end reads, a complementary pair of read mates aligned to the same sequence is counted as one read and the resultant value is called FPKM (fragments per kilobase per million). This allows comparison between single- and paired-end sequencing experiments [32].

In this chapter, we describe a pipeline for molecular characterization of novel viruses of parasitic flagellates belonging to the family Trypanosomatidae. It has been used to document numerous groups of RNA viruses including -ssRNA viruses currently classified as the family *Leishbuviridae* (class Bunyaviricetes) [48–52]. The pipeline starts with an axenic (ideally, clonal) culture of a flagellate, which removes any ambiguity in respect to the host of studied viruses. It proceeds with the viral dsRNA detection by enzymatic (DNase I/ S1 nuclease) enrichment. For this purpose, total RNA is isolated from the late log-phase culture to ensure that (i) viral genome is actively transcribing and replicating to produce comparable amounts of + and – RNA strands and (ii) the host cell density is high enough to maximize the total RNA yield. Finally, we provide a full bioinformatic workflow to assemble and identify RNA viruses of trypanosomatids.

2 Materials

2.1 Total RNA Isolation

1. Phosphate buffer saline, PBS.
2. Insulin syringe (optional).
3. TRIzol (or TRI Reagent) for RNA extraction.
4. Chloroform.
5. iPrOH.
6. 70% EtOH.
7. Deproteination solution (optional, *see* **Note 2**): 50 mM Tris-HCl pH 9.3, 1% 2-mercaptoethanol.
8. Nuclease-free water.
9. Multipurpose refrigerated centrifuge.
10. Refrigerated microcentrifuge.

2.2 dsRNA Enrichment for Screening and NGS

1. DNase I at 2 units/μL.
2. S1 nuclease from *Aspergillus oryzae* at 100 u/μL.
3. 200 mM EDTA.
4. ssRNA precipitating solution (optional, *see* **Note 4**): 3.3 M LiCl, 250 mM NaCl, 25 mM Tris-HCl pH 8.
5. 3 M NaOAc (optional, *see* **Note 4**).
6. 96% EtOH (optional, *see* **Note 4**).
7. Heating block for 1.5-mL Eppendorf tubes.
8. RNA purification kit allowing elution in a small volume.

2.3 Gel Electrophoresis

1. 15–25 cm long 0.8% agarose gel prepared with 1× TAE buffer.
2. Horizontal electrophoresis chamber, power supply.
3. 1× TAE buffer (40 mM Tris, 20 mM acetic acid, 1 mM EDTA)
4. Ethidium bromide 0.5 μg/mL solution in distilled water (or an alternative fluorescent dye).
5. Gel documentation system.

2.4 NGS Sequencing and Bioinformatic Analysis

1. Hardware: Workstation with 16 CPU cores and 64 GB RAM, Linux OS.
2. Software: FastQC, Trimmomatic (0.40), Bowtie2 (2.4.4), SAMtools (1.13), Trinity (2.4.0), DIAMOND (2.0.14), BLAST+ (2.13.0).

3 Methods

3.1 Total RNA Isolation

1. Harvest at least 10^8 trypanosomatid cells from a late log-phase axenic culture by centrifugation at 2500 × g for 15 min at 4 °C.
2. Resuspend the pellet in sterile 1 × PBS and centrifuge again as above.
3. Resuspend the pellet in 1 mL of TRIzol, transfer the suspension into a 2-mL Eppendorf tube, and break any clumps by pipetting (*see* **Note 1**).
4. Place samples at −80 °C for later processing or proceed to the next step.
5. Add 0.5 mL (1:2 of the added TRIzol volume) of 100% chloroform and shake vigorously.
6. Centrifuge at 20,000 × g for 15 min at 4 °C in a microcentrifuge.
7. Transfer about 700 μL of the upper phase into a new 1.5-mL Eppendorf tube (*see* **Note 2**).
8. Add 1 volume (about 700 μL) of prechilled at −20 °C iPrOH and mix gently by inverting the tube.

9. Incubate at −20 °C for at least 2 h, better overnight.
10. Centrifuge at 20,000 × g for 15 min at 4 °C in a microcentrifuge.
11. Discard the supernatant, add 500 μL of 70% ethanol to the pellet and centrifuge again.
12. Repeat the **step 11** with 150 μL of 70% ethanol.
13. Remove the liquid, dry the pellet, dissolve it in 25 μL of nuclease-free water, and measure the concentration (*see* **Note 3**).
14. Store at −80 °C for up to 1 month or proceed to the next step (Subheading 3.2).

3.2 DsRNA Enrichment for Screening and NGS

1. Take about 50 μg of total RNA in 21.5 μL. If the concentration is substantially lower or more RNA is needed for the analysis to increase sensitivity, split the sample and process several 21.5 μL-aliquots in parallel.
2. Add 1 μL of DNase I and 2.5 μL of 10× DNAse Reaction Buffer.
3. Incubate at 37 °C for 60 min and proceed to the next steps without heat inactivation.
4. Add 0.5 μL of S1 nuclease directly into the DNase mix.
5. Incubate at 37 °C for 60 min (*see* **Note 4**).
6. Add 0.5 μL 200 mM EDTA and incubate at 70 °C for 10 min for the enzyme inactivation.
7. At this stage, the dsRNA can be visualized in agarose gel (Subheading 3.3) or purified for NGS.
8. For NGS, purify dsRNA by the RNA purification kit, elute in 15 μL of nuclease-free water preheated to 50 °C (*see* **Note 5**).
9. Check 1–5 μL of resultant dsRNA prep by agarose gel electrophoresis.
10. Use purified dsRNA for Illumina paired-end RNA sequencing with TruSeq Stranded library prepared using random hexamers.

3.3 DsRNA Visualization

1. Add 0.2–1 μL of 6× loading dye to 1–5 μL of the kit-purified dsRNA prep (Subheading 3.2, **step 8**) or 5 μL of DNase I/ S1 nuclease-treated total RNA (Subheading 3.2, **step** 7), mix, and load onto a 0.8% agarose gel.
2. Run the gel for at least 1.5 h at maximum voltage of 5 V/cm for proper band separation (*see* **Note 6**).
3. Stain the gel in 0.5 μg/mL ethidium bromide solution for 15 min (*see* **Note** 7).

4. Destain the gel in pure distilled water for 15 min (*see* **Note 8**).
5. Analyze the band patterns using a gel documentation system.

3.4 NGS Sequencing Data Processing and Bioinformatic Analyses

1. Sequence dsRNA to the minimal depth of 3 Gb. Check the quality of the reads using FastQC.
2. Trim the reads and sort them into left and right paired/unpaired with Trimmomatic. The software requires the file with adaptor sequences, which are available from the sequencing company (for example, for the Illumina paired-end reads the file is called TruSeq3-PE-2.fa). Trimmomatic can read .gz files, thus, it is not necessary to unpack them beforehand. The typical command line for trimming of paired-end RNA-seq Illumina is:

```
trimmomatic PE -threads 16 -phred33 <yourdata>_1.fastq.gz
<yourdata>_2.fastq.gz -baseout <yourdata>_trimmed ILLUMINA-
CLIP:TruSeq3-PE-2.fa:2:20:10 LEADING:3 TRAILING:3 SLIDINGWIN-
DOW:4:15 MINLEN:50
```

where **PE** is a specification of paired-end reads, -**threads** is the number of CPU cores, -**phred33** is type of quality score used by the sequencing company, ***_fastq.gz** are left and right reads provided by the sequencing company, -**baseout** is the basename of the output files, of which there will be four: left paired (1P), right paired (2P), left unpaired (1U), where the right mate did not pass the quality control, and right unpaired (2U), where the left mate was dropped (the exact description of the other parameters can be found in the Trimmomatic manual).

3. If the reference genome of the host organism is available, it is highly recommended to decontaminate the dataset from eukaryotic reads before the assembly. This not only speeds up the whole procedure but also often results in cleaner virus assemblies. Otherwise, direct de novo assembly of the entire dataset is also possible. For direct assembly skip to **step 6**.
4. Map the trimmed reads (**step 2**) onto the host genome using Bowtie2. This step requires a reference genome sequence in fasta format, which can be downloaded from the NCBI website by browsing available genomes. First, the reference genome is indexed with a command:

```
bowtie2-build <yourgenome>.fasta <yourgenome>
```

Then mapping is performed using:

```
bowtie2 -x <yourgenome> -1 <yourdata>_trimmed_1P -2 <yourda-
ta>_trimmed_2P -U <yourdata>_trimmed_1U,<yourdata>_trimmed_2U
```

```
--end-to-end --very-sensitive -p 16 -S reads_to_<yourgenome>.
sam 2> reads_to_<yourgenome>_bowtie2log.txt
```

where option **-x** specifies the base-name of index files created by the previous command in this step, **-1**, **-2**, and **-U** specifies trimmed reads generated in **step 2** (note that unpaired reads are provided under a single parameter and their filenames are separated with a comma without a whitespace), **--end-to-end** is the alignment mode, in which the entire read length is considered as opposed to local alignment, where read terminals can be disregarded, **--very-sensitive** is a preset of alignment parameters, **-S** specifies the name of the output SAM file with read mapping, and ***_bowtie2log.txt** is a run-log with alignment statistics and errors.

5. Extract the unmapped reads with samtools fastq and grep programs using the SAM file generated in **step 4** as an input.

   For the right paired reads run:

```
samtools fastq -@ 16 -f 68 reads_to_<yourgenome>.sam > <your-
data>_decontam_1P;
```

For the left paired reads run:

```
samtools fastq -@ 16 -f 132 reads_to_<yourgenome>.sam >
<yourdata>_decontam_2P;
```

where **-@** is the number of CPU cores and **-f** is a SAM flag (full interactive explanation can be found at the following link https://broadinstitute.github.io/picard/explain-flags.html). Getting the unpaired reads is more complicated and requires first outputting all unmapped mates 1 and 2 separately and then eliminating paired mates already present in 1P and 2P files. For this purpose, the following commands can be used:

```
samtools fastq -@ 16 -f 72 reads_to_<yourgenome>.sam >
tmp_reads72;
grep -vf <(grep '@' <yourdata>_decontam_1P) <(grep '@'
tmp_reads72) | grep -A 3 -f - tmp_reads72 | sed 's/^--$//g' |
sed '/^\s*$/d' > <yourdata>_decontam_1U;
```

and for 2 U:

```
samtools fastq -@ 16 -f 136 <yourdata>_decontam_2P >
tmp_reads136;
grep -vf <(grep '@' <yourdata>_decontam_2P) <(grep '@'
tmp_reads136) | grep -A 3 -f - tmp_reads136 | sed 's/^--$//g'
| sed '/^\s*$/d' > <yourdata>_decontam_2U; rm tmp_reads*
```

6. Assemble reads de novo with Trinity software using paired-end protocol (program options are self-explanatory).

```
Trinity --seqType fq --min_contig_length 200 --max_memory 64G
--CPU 16 --left <yourdata>_decontam_1P,<yourdata>_decontam_1U
--right <yourdata>_decontam_2P,<yourdata>_decontam_2U --out-
put trinity_out_<yourdata>_decontam > trinity_<yourdata>_de-
contam.log 2> trinity_<yourdata>_decontam.errlog;
```

If decontamination against reference eukaryotic genome was not performed, substitute "decontam" with "trimmed" in the command above. The resulting output will be the folder with specified name (note that its name must begin with "trinity_out"), which will contain assembled contigs in the "Trinity.fasta" file. Move this file to the working directory and proceed.

7. The following steps will describe the procedure of mapping reads onto assembled contigs. This is done similarly to **step 4** but with Trinity.fasta and original trimmed reads generated in **step 2**. Run.

```
bowtie2-build Trinity.fasta Trinity
```

then

```
bowtie2 -x Trinity -1 <yourdata>_trimmed_1P -2 <yourdata>_-
trimmed_2P -U <yourdata>_trimmed_1U,<yourdata>_trimmed_2U --
end-to-end --very-sensitive -p 16 -S Trinity.sam 2> Trinity_-
bowtie2log.txt
```

Make sure to save the error-log from Bowtie2 (Trinity_-bowtie2log.txt) containing the information on the number of mapped reads, which is important to calculate per-million factor.

8. Use SAMtools to process the output SAM-file (indexing, sorting and converting to binary). This step ensures efficient search in the mapping database. In addition, the resultant *.bam and *.bai files can be loaded together with the contig file Trinity.fasta into a genome browser if visual inspection and/or graphical representation of contig coverage is desired.

Run the series of commands:

```
samtools faidx Trinity.fasta;
samtools view -S -F 0x4 -b -u -t Trinity.fasta Trinity.sam |
samtools sort –o Trinity_sort;
mv Trinity_sort Trinity_sort.bam;
samtools index Trinity_sort.bam Trinity_sort.bam.bai;
```

9. Now, the sorted *.bam file can be used to extract the read count for an individual contig. For this purpose, a loop over Trinity.fasta.fai can be used. The loop reads in contig name and length and passes them to samtools view for read extraction. Then, samtools fastq is used to convert reads back to the text file, from which the read headers are extracted. If both left and right mates are present among contig's reads, only one is kept in accordance with FPKM calculation strategy. The surviving headers are counted. The loop outputs contig name, length, and read count. The output is passed to awk program, which performs per-million and per-kilobase normalization, and the resultant file with contigs' coverage is recorded as Trinity_FPKMs.txt. First, record the per-million variable from the Bowtie2 log file generated in **step 7** with the following command:

```
num_of_reads=`head -n1 Trinity_bowtie2log.txt | awk '{print $1/1000000}'`
```

then run the loop (note the "pipe" symbol at the end of the loop before awk):

```
while read c d e; do
 samtools view -S -F 4 Trinity_sort.bam $c -o tmp.bam;
 samtools fastq tmp.bam > tmp.fastq;
 echo $c $d `grep '@' tmp.fastq | sed 's/\/.*$//' | sort -u | wc -l`;
 rm tmp*;
done < Trinity.fasta.fai |
awk -v b="$num_of_reads" '{print$1"\t"$2"\t"$3*1000/$2/b}' > Trinity_FPKMs.txt
```

(*see* **Note 9**).

10. The identification of virus contigs is best performed in two steps. Firstly, the contigs are searched against nucleotide database with BLASTn. This search is usually fast and very specific. It helps to sift out noncoding sequences, such as ribosomal RNA, which are rather abundant even if the read decontamination (**step 4**) was performed. Furthermore, it reduces the number of potential target contigs that need to be checked via computationally greedy BLASTx-search. The BLASTn is best performed against a custom database built from the host genome or genomes of closely related organisms, although NCBI "nt" database can be used as well. To build a custom database run the following command.

```
makeblastdb -in <yourgenome(s)> -out <yourgenome(s)>_blastndb -dbtype nucl
```

then, perform the search

```
blastn -query Trinity.fasta -db <yourgenome(s)>_blastndb -out
Trinity_blastn.txt -evalue 1e-5 -outfmt 6 -num_threads 16
```

where -**query** is the file with assembled contigs, -**db** is the search database (custom or "nt"), -**out** is the output file, -**evalue** is the minimal e-value to record a hit (the lower it is—the higher confidence hits are retrieved, default "10" is too high and can result in false-positive hits) and -**outfmt** is type of the output, for further processing the tabulated output (**6**) is the best.

11. Secondly, the contig sequences that were not identified in the previous step (absent from Trinity_blastn.txt file) are extracted from the Trinity.fasta file and BLASTx search against UniProt50 database is performed using DIAMOND software. For sequence extraction fasta file must be linearized, that is, the nucleotide sequence must occupy exactly one line. This can be done with the following Unix shell commands:

```
sed '/^>/ s/.$/&,,,/' Trinity.fasta | tr -d '\n' | sed $'s/>/\\
\n&/g' | sed $'s/,,,/\\\n/g' | tail -n+2 > Trinity_lin.fasta
```

Then, for sequence extraction, run:

```
grep -vf <(awk -F '\t' '{print$1}' Trinity_blastn.txt | sed 's/
.*$//' | sort -u) <(grep '>' Trinity_lin.fasta | sed 's/ .*$//
') | grep -A 1 -f - Trinity_lin.fasta | sed 's/^--$//g' | sed
'/^\s*$/d' > Trinity_for_blastx.fasta
```

Now, prepare a search database for DIAMOND. Download UniRef50 from https://ftp.uniprot.org/pub/databases/uniprot/uniref/uniref50/uniref50.fasta.gz and unpack it with gunzip command in the terminal. Fasta-headers of UniRef50 file contain useful information, such as name and accession of the protein, scientific name, and NCBI taxonomy ID of the organism/group. However, both DIAMOND and BLAST+ disregard this information as it is separated with whitespaces in the fasta-header. To avoid the information loss and other bugs, all fasta headers in uniref50.fasta file can be fixed with the following command:

```
sed 's/ /__/;s/ /_/g;s/:/_/;s/\//_/;s/,/_/;s/;/_/;s/\[/_/;s/
\]//' uniref50.fasta | tr -d '(' | tr -d ')' > uniref50_h.fasta
```

Prepare DIAMOND database from a file with proper fasta-headers:

```
diamond makedb -in uniref50_h.fasta -d uniref50DMND
```

Finally, run DIAMOND BLASTx search:

```
diamond blastx -d uniref50DMND -q Trinity_for_blastx.fasta -o
Trinity_blastx.txt --outfmt 6 qseqid stitle pident bitscore
evalue length qstart qend sstart send --ultra-sensitive -t
DMND_TMP -k 1 -p 16
```

where -**d** is DIAMOND database prepared from UniRef50, -**q** is file with contigs selected for BLASTx, -**o** is the output table with hits, --**outfmt 6** is the specification of tabular output format (the same as in BLASTn, **step 11**), -**t** is path to the temporary file, which must be stored locally if the computation is run on a remote server, -**k** is the number of hits per query, and -**p** is the number of CPUs.

12. The information on contigs' coverages and its BLASTn/BLASTx hits can be integrated for better overview using the first column (i.e., contig name) as a key. At this point, more conserved virus proteins, such as RdRp, can be readily identified based on "Tax=" label of the BLASTx hit. A simple text search with "vir" through integrated results file might already return virus hits. Finding more divergent viral proteins, such as, for example, a leishbuviral glycoprotein, is more complicated and requires multiple lines of evidence. These include: (i) correspondence between the contig length and size of the genomic segment observed in the gel (this condition may be violated in case of read assembly errors); (ii) elevated coverage of putative viral contig compared to host mRNA (*see* **Note 10**). Genomic segments originating from the same virus should have comparable coverage. However, the level of expression of genes from these segments may vary, which will be reflected in FPKM values; (iii) since RNA viruses evolve rapidly and encode their genetic information very efficiently, it is expected that the putative viral contig will contain a large ORF (or several overlapping ORFs) that span 70–95% of contig length but will return null result from BLASTx searches (so-called ORFans). If a contig satisfies all three criteria above, it is highly likely to be of viral origin.
13. All *Bunyaviricetes* have complementary sequences at their genomic segments' termini. These can be used as additional markers of viral contigs and facilitate assessing completeness of the latter. For this purpose, numerous RNA secondary structure prediction web-servers (RNAfold, RNAstructure, IPknot, to name just a few) can be employed. The completeness of the genomic segment can be assessed by the position of the ORF in the contig and presence/absence of start and stop-codons. For

short segments up to 1500 nt, the entire contig sequence can be used for prediction although results may be difficult to interpret. Therefore, it is better to extract the first and the last ~200 nt of the contig and connect them with a stretch of 100 uridines. Alternatively, the two fragments can be input directly to RNAcofold server (http://rna.tbi.univie.ac.at/cgi-bin/RNAWebSuite/RNAcofold.cgi). If terminal complementary nucleotides are present, they will pair to form a typical "panhandle" structure with 5′- and 3′-ends brought together. The unpaired nucleotides before and after the panhandle can be removed.

4 Notes

1. If any clumps of cells are visible, use an insulin syringe to break them.
2. If the middle (white) phase containing proteins is too thick, it is advisable to shake the sample again, separate it into multiple tubes, top each tube to 1 mL with TRIzol, and go back to the **step 5**. The resultant RNA can be combined at later steps, if necessary. When it is known in advance that the studied organisms contain abundant surface proteins, this problem can be mitigated by incubating the cells for 2–3 min in 1 mL of freshly prepared deproteination solution (Subheading 2.1, **step 7** (Total RNA Isolation, Deproteination solution)) and subsequent washing with PBS.
3. At this step, RNA concentration is usually too high and the solution is viscous. Therefore, it is advisable to dilute a small aliquot ten-fold before measuring the concentration.
4. LiCl precipitation of ssRNA is an alternative to the S1 nuclease removal of ssRNA. For that, after **step 5** of protocol 3.2 (dsRNA Enrichment for Screening and NGS), inactivate DNase I by adding 0.5 μL 200 mM EDTA and incubating for 10 min at 70 °C. Then, add 37.5 μL of ssRNA precipitating solution (Subheading 2.2, **step 4** (dsRNA Enrichment for Screening and NGS)) to 25 μL of the DNase I-treated sample and keep the mix at 4 °C for at least 16 h. Next, centrifuge the sample at 20,000 × g for 30 min at 4 °C in a microcentrifuge, take supernatant, and proceed with standard nucleic acid precipitation protocol: add 1/10 volume (6.3 μL) of 3 M NaOAc, 2.5 volumes (172 μL) of prechilled 96% EtOH, and place the tube at −20 °C for at least 2 h, followed by **steps 10** and **13** of the protocol 3.1 (Total RNA Isolation). In our experience, the LiCl approach results in sharper bands in gel when compared to the nuclease-treated sample. This is likely due to the absence of salts and other chemicals present in the buffers for enzymatic

digestion. As such, this might enhance weaker bands and improve sensitivity. In addition, the smear of undigested low molecular weight RNA, which sometimes appear with S1 nuclease method, is never observed after precipitation with LiCl. On the other hand, the LiCl treatment does not completely eliminate ribosomal RNA, which can mask viral bands. Of note, LiCl precipitation can be performed after the S1 nuclease treatment. In this case the advantages of both methods are combined.

5. We routinely use Zymoclean Gel RNA Recovery Kit for this purpose.
6. A higher voltage can cause overheating and distortion of the gel. It is advisable to use a large electrophoresis chamber and wide loading pockets for best band visualization.
7. Although ethidium bromide is cheap, in many labs it is not used due to safety concerns. In such a case, an alternative fluorescent dye with high sensitivity to RNA (e. g. EliDNA™ PS Green Plus) should be used according to the manufacturer's instructions.
8. Post-staining ensures even distribution of a signal throughout the whole gel area, which is crucial to achieve maximum sensitivity of the method. Destaining procedure removes the unwanted background.
9. It is also possible to get the read count without looping through a contig list using the following command:

```
samtools idxstats *.bam
```

   (index file *.bai should be present in the folder). However, it counts multiple alignments of the same read as well as both left and right mates, thus, overestimating the read count.

10. In our experience, if dsRNA-enriched preparations are used for both gel electrophoresis and RNA-seq, contigs with FPKM of at least 30 are visible on the gel as faint bands. Viral dsRNAs usually have FPKM values in the range of 10^2–10^4, similar coverages are observed for trypanosomatid ribosomal RNA. Host protein-coding genes have FPKM values between 1 and 50.

Acknowledgments

The work on *Leishbuviridae* in Yurchenko laboratory is primarily supported by the Grant Agency of the Czech Republic (24-328 10009S) and the European Union's Operational Program "Just Transition" (LERCO CZ.10.03.01/00/22_003/0000003).

References

1. Rampersad S, Tennant P (2018) Replication and expression strategies of viruses. In: Tennant P, Fermin G, Foster JE (eds) Viruses: molecular biology, host interactions, and applications to biotechnology. Academic Press/Elsevier, London, pp 55–82
2. Baltimore D (1971) Expression of animal virus genomes. Bacteriol Rev 35(3):235–241
3. Field AK, Lampson GP, Tytell AA, Hilleman MR (1972) Demonstration of double-stranded ribonucleic acid in concentrates of RNA viruses. Proc Soc Exp Biol Med 141(2):440–444
4. Zinzula L, Tramontano E (2013) Strategies of highly pathogenic RNA viruses to block dsRNA detection by RIG-I-like receptors: hide, mask, hit. Antivir Res 100(3):615–635
5. Baltimore D (1966) Purification and properties of poliovirus double-stranded ribonucleic acid. J Mol Biol 18(3):421–428
6. Muthukrishnan S, Shatkin AJ (1975) Reovirus genome RNA segments: resistance to S_1 nuclease. Virology 64(1):96–105
7. Diaz-Ruiz JR, Kaper JM (1978) Isolation of viral double-stranded RNAs using a LiCl fractionation procedure. Prep Biochem 8(1):1–17
8. Ghabrial SA, Castón JR, Jiang D, Nibert ML, Suzuki N (2015) 50-plus years of fungal viruses. Virology 479–480:356–368
9. García-Pedrajas MD, Cañizares MC, Sarmiento-Villamil JL, Jacquat AG, Dambolena JS (2019) Mycoviruses in biological control: from basic research to field implementation. Phytopathology 109(11):1828–1839
10. Hong Y, Cole TE, Brasier CM, Buck KW (1998) Evolutionary relationships among putative RNA-dependent RNA polymerases encoded by a mitochondrial virus-like RNA in the Dutch elm disease fungus, *Ophiostoma novo-ulmi,* by other viruses and virus-like RNAs and by the *Arabidopsis* mitochondrial genome. Virology 246(1):158–169
11. Crawford LJ, Osman TA, Booy FP, Coutts RH, Brasier CM, Buck KW (2006) Molecular characterization of a partitivirus from *Ophiostoma himal-ulmi.* Virus Genes 33(1):33–39
12. Osaki H, Nakamura H, Sasaki A, Matsumoto N, Yoshida K (2006) An endornavirus from a hypovirulent strain of the violet root rot fungus, *Helicobasidium mompa.* Virus Res 118(1–2):143–149
13. Kozlakidis Z, Hacker CV, Bradley D, Jamal A, Phoon X, Webber J, Brasier CM, Buck KW, Coutts RH (2009) Molecular characterisation of two novel double-stranded RNA elements from *Phlebiopsis gigantea.* Virus Genes 39(1):132–136
14. Magae Y (2012) Molecular characterization of a novel mycovirus in the cultivated mushroom, *Lentinula edodes.* Virol J 9:60
15. Kanhayuwa L, Kotta-Loizou I, Ozkan S, Gunning AP, Coutts RH (2015) A novel mycovirus from *Aspergillus fumigatus* contains four unique dsRNAs as its genome and is infectious as dsRNA. Proc Natl Acad Sci USA 112(29):9100–9105
16. Niu Y, Yuan Y, Mao J, Yang Z, Cao Q, Zhang T, Wang S, Liu D (2018) Characterization of two novel mycoviruses from *Penicillium digitatum* and the related fungicide resistance analysis. Sci Rep 8(1):5513
17. Weber F, Wagner V, Rasmussen SB, Hartmann R, Paludan SR (2006) Double-stranded RNA is produced by positive-strand RNA viruses and DNA viruses but not in detectable amounts by negative-strand RNA viruses. J Virol 80(10):5059–5064
18. Pearson MN, Beever RE, Boine B, Arthur K (2009) Mycoviruses of filamentous fungi and their relevance to plant pathology. Mol Plant Pathol 10(1):115–128
19. Khan HA, Nerva L, Bhatti MF (2023) The good, the bad and the cryptic: the multifaceted roles of mycoviruses and their potential applications for a sustainable agriculture. Virology 585:259–269
20. Robinson JI, Beverley SM (2018) Concentration of 2'C-methyladenosine triphosphate by *Leishmania guyanensis* enables specific inhibition of *Leishmania RNA virus 1 via* its RNA polymerase. J Biol Chem 293(17):6460–6469
21. Kuhlmann FM, Robinson JI, Bluemling GR, Ronet C, Fasel N, Beverley SM (2017) Antiviral screening identifies adenosine analogs targeting the endogenous dsRNA *Leishmania RNA virus 1* (LRV1) pathogenicity factor. Proc Natl Acad Sci USA 114(5):E811–E819
22. Cashdollar LW, Esparza J, Hudson GR, Chmelo R, Lee PW, Joklik WK (1982) Cloning the double-stranded RNA genes of reovirus: sequence of the cloned *S2* gene. Proc Natl Acad Sci USA 79(24):7644–7648
23. Imai M, Richardson MA, Ikegami N, Shatkin AJ, Furuichi Y (1983) Molecular cloning of double-stranded RNA virus genomes. Proc Natl Acad Sci USA 80(2):373–377
24. Niu Y, Zhang T, Zhu Y, Yuan Y, Wang S, Liu J, Liu D (2016) Isolation and characterization of

a novel mycovirus from *Penicillium digitatum*. Virology 494:15–22

25. Sterky F, Lundeberg J (2000) Sequence analysis of genes and genomes. J Biotechnol 76(1):1–31
26. Haas BJ, Papanicolaou A, Yassour M, Grabherr M, Blood PD, Bowden J, Couger MB, Eccles D, Li B, Lieber M, Macmanes MD, Ott M, Orvis J, Pochet N, Strozzi F, Weeks N, Westerman R, William T, Dewey CN, Henschel R, Leduc RD, Friedman N, Regev A (2013) *De novo* transcript sequence reconstruction from RNA-seq using the trinity platform for reference generation and analysis. Nat Protoc 8(8):1494–1512
27. Bankevich A, Nurk S, Antipov D, Gurevich AA, Dvorkin M, Kulikov AS, Lesin VM, Nikolenko SI, Pham S, Prjibelski AD, Pyshkin AV, Sirotkin AV, Vyahhi N, Tesler G, Alekseyev MA, Pevzner PA (2012) SPAdes: a new genome assembly algorithm and its applications to single-cell sequencing. J Comput Biol 19(5):455–477
28. Li CX, Shi M, Tian JH, Lin XD, Kang YJ, Chen LJ, Qin XC, Xu J, Holmes EC, Zhang YZ (2015) Unprecedented genomic diversity of RNA viruses in arthropods reveals the ancestry of negative-sense RNA viruses. elife 4
29. Shi M, Lin XD, Tian JH, Chen LJ, Chen X, Li CX, Qin XC, Li J, Cao JP, Eden JS, Buchmann J, Wang W, Xu J, Holmes EC, Zhang YZ (2016) Redefining the invertebrate RNA virosphere. Nature 540(7634):539–543
30. Shi M, Lin XD, Chen X, Tian JH, Chen LJ, Li K, Wang W, Eden JS, Shen JJ, Liu L, Holmes EC, Zhang YZ (2018) The evolutionary history of vertebrate RNA viruses. Nature 556(7700):197–202
31. Zayed AA, Wainaina JM, Dominguez-Huerta-G, Pelletier E, Guo J, Mohssen M, Tian F, Pratama AA, Bolduc B, Zablocki O, Cronin D, Solden L, Delage E, Alberti A, Aury JM, Carradec Q, da Silva C, Labadie K, Poulain J, Ruscheweyh HJ, Salazar G, Shatoff E, Tara Oceans Coordinatorsdouble, d, Bundschuh R, Fredrick K, Kubatko LS, Chaffron S, Culley AI, Sunagawa S, Kuhn JH, Wincker P, Sullivan MB, Acinas SG, Babin M, Bork P, Boss E, Bowler C, Cochrane G, de Vargas C, Gorsky G, Guidi L, Grimsley N, Hingamp P, Iudicone D, Jaillon O, Kandels S, Karp-Boss L, Karsenti E, Not F, Ogata H, Poulton N, Pesant S, Sardet C, Speich S, Stemmann L, Sullivan MB, Sungawa S, Wincker P (2022) Cryptic and abundant marine viruses at the evolutionary origins of Earth's RNA virome. Science 376(6589):156–162
32. Kukurba KR, Montgomery SB (2015) RNA sequencing and analysis. Cold Spring Harb Protoc 2015(11):951–969
33. Yang IS, Kim S (2015) Analysis of whole transcriptome sequencing data: workflow and software. Genomics Inform 13(4):119–125
34. Bolger AM, Lohse M, Usadel B (2014) Trimmomatic: a flexible trimmer for Illumina sequence data. Bioinformatics 30(15):2114–2120
35. Langmead B, Salzberg SL (2012) Fast gapped-read alignment with Bowtie 2. Nat Methods 9(4):357–359
36. Ramirez-Gonzalez RH, Bonnal R, Caccamo M, Maclean D (2012) Bio-SAMtools: ruby bindings for SAMtools, a library for accessing BAM files containing high-throughput sequence alignments. Source Code Biol Med 7(1):6
37. Bushnell B, Rood J, Singer E (2017) BBMerge – accurate paired shotgun read merging *via* overlap. PLoS One 12(10):e0185056
38. Trapnell C, Roberts A, Goff L, Pertea G, Kim D, Kelley DR, Pimentel H, Salzberg SL, Rinn JL, Pachter L (2012) Differential gene and transcript expression analysis of RNA-seq experiments with TopHat and cufflinks. Nat Protoc 7(3):562–578
39. Au KF, Jiang H, Lin L, Xing Y, Wong WH (2010) Detection of splice junctions from paired-end RNA-seq data by SpliceMap. Nucleic Acids Res 38(14):4570–4578
40. Cross ST, Michalski D, Miller MR, Wilusz J (2019) RNA regulatory processes in RNA virus biology. Wiley Interdiscip Rev RNA 10(5):e1536
41. Camacho C, Coulouris G, Avagyan V, Ma N, Papadopoulos J, Bealer K, Madden TL (2009) BLAST+: architecture and applications. BMC Bioinformatics 10:421
42. Buchfink B, Reuter K, Drost HG (2021) Sensitive protein alignments at tree-of-life scale using DIAMOND. Nat Methods 18(4):366–368
43. Suzek BE, Wang Y, Huang H, McGarvey PB, Wu CH, UniProt C (2015) UniRef clusters: a comprehensive and scalable alternative for improving sequence similarity searches. Bioinformatics 31(6):926–932
44. Prakash A, Jeffryes M, Bateman A, Finn RD (2017) The HMMER web server for protein sequence similarity search. Curr Protoc Bioinformatics 60: 3.15.11–13.15.23
45. Steinegger M, Meier M, Mirdita M, Vöhringer H, Haunsberger SJ, Söding J (2019) HH-suite3 for fast remote homology

detection and deep protein annotation. BMC Bioinformatics 20(1):473

46. Wolf YI, Silas S, Wang Y, Wu S, Bocek M, Kazlauskas D, Krupovic M, Fire A, Dolja VV, Koonin EV (2020) Doubling of the known set of RNA viruses by metagenomic analysis of an aquatic virome. Nat Microbiol 5(10):1262–1270
47. Neri U, Wolf YI, Roux S, Camargo AP, Lee B, Kazlauskas D, Chen IM, Ivanova N, Zeigler Allen L, Paez-Espino D, Bryant DA, Bhaya D, Consortium RNAVD, Krupovic M, Dolja VV, Kyrpides NC, Koonin EV, Gophna U (2022) Expansion of the global RNA virome reveals diverse clades of bacteriophages. Cell 185(21):4023–4037
48. Grybchuk D, Akopyants NS, Kostygov AY, Konovalovas A, Lye LF, Dobson DE, Zangger H, Fasel N, Butenko A, Frolov AO, Votýpka J, d'Avila-Levy CM, Kulich P, Moravcová J, Plevka P, Rogozin IB, Serva S, Lukeš J, Beverley SM, Yurchenko V (2018) Viral discovery and diversity in trypanosomatid protozoa with a focus on relatives of the human parasite *Leishmania*. Proc Natl Acad Sci USA 115(3):E506–E515
49. Grybchuk D, Kostygov AY, Macedo DH, Votýpka J, Lukeš J, Yurchenko V (2018) RNA viruses in *Blechomonas* (Trypanosomatidae) and evolution of *Leishmaniavirus*. MBio 9(5): e01932–e01918
50. Grybchuk D, Macedo DH, Kleschenko Y, Kraeva N, Lukashev AN, Bates PA, Kulich P, Leštinová T, Volf P, Kostygov AY, Yurchenko V (2020) The first non-LRV RNA virus in *Leishmania*. Viruses 12(2):168
51. Macedo DH, Grybchuk D, Režnarová J, Votýpka J, Klocek D, Yurchenko T, Ševčík J, Magri A, Urda Dolinská M, Záhonová K, Lukeš J, Servienė E, Jászayová A, Serva S, Malysheva MN, Frolov AO, Yurchenko V, Kostygov AY (2023) Diversity of RNA viruses in the cosmopolitan monoxenous trypanosomatid *Leptomonas pyrrhocoris*. BMC Biol 21(1):191
52. Klocek D, Grybchuk D, Macedo DH, Galan A, Votýpka J, Schmid-Hempel R, Schmid-Hempel P, Yurchenko V, Kostygov AY (2023) RNA viruses of *Crithidia bombi*, a parasite of bumblebees. J Invertebr Pathol 201:107991

Chapter 13

Recombinant Protein Production and Purification of Rift Valley Fever Virus Nucleoprotein from *Escherichia coli* Expression Systems

Daniel Desmecht and Hani Boshra

Abstract

The recombinant expression and purification of viral proteins are a key component in the study of the immune response of viruses, as well as the creation of diagnostic techniques for the detection of viruses. For structurally simple proteins, one commonly used technique is the production of recombinant proteins in bacterial expression systems, which enable the large-scale synthesis and purification of recombinant viral proteins. In this technique, the cDNA encoding for a viral protein is cloned into a bacterial expression vector (with an appropriate purification tag), produced in a modified bacterial culture, and optimized for maximum protein production in a minimal amount of time. In this chapter, a protocol for the production of Rift Valley fever virus nucleoprotein is described. This protein has been previously shown to highly antigenic (and thus, is used in diagnostic tests), and has also been shown to be a potent inducer of the T-cell response following Rift Valley fever virus infection. The protocol outlined in this chapter describes the cloning of the cDNA of RVFV nucleoprotein into a bacterial expression vector, which also contains a fusion protein for optimal protein expression and solubilization, as well as a poly-histidine tag for efficient purification This chapter also describes the steps required for bacterial transformation, culture, lysis, purification and dialysis of RVFV nucleoprotein, resulting in a recombinant protein preparation, which can be upscaled to produce milligram quantities of protein product, which in turn can be used for downstream immunological and diagnostic applications.

Key words Rift Valley fever virus, Recombinant protein, *E. coli*, Protein production

1 Introduction

Rift Valley fever virus (also known as *Phlebovirus riftense*) is a bunyavirus of the *Phenuviridae* family (genus *Phlebovirus*) [1, 2]. A zoonotic virus that has been shown to be lethal in both humans and ruminant animals, Rift Valley fever virus (RVFV) is transmitted via mosquitoes (primarily of the Aedes and Culex species), with outbreaks usually occurring following periods of heavy rainfall [3, 4]. Since its discovery in the early twentieth century, RVFV outbreaks have been localized to Africa (both

Hani Boshra (ed.), *Bunyaviruses: Methods and Protocols*, Methods in Molecular Biology, vol. 2893,
https://doi.org/10.1007/978-1-0716-4338-9_13,

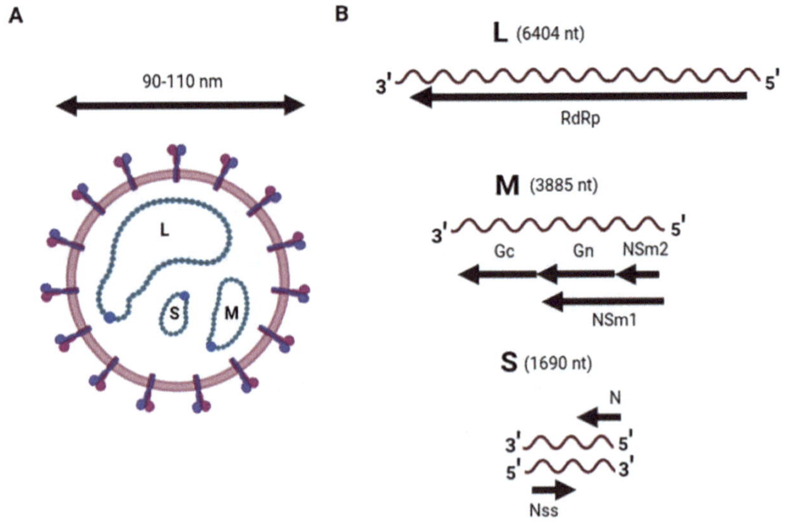

Fig. 1 An overview of the strategy to purify bacterially recombinant RVFV nucleoprotein fused to thioredoxin (trx-N). (**a**) Upon cloning the cDNA of RVFV nucleoprotein into a sequencing vector, the fragment is then digested and recloned into the pET-32a expression vector containing the cDNA for thioredoxin containing a poly-histidine tag. Following transformation of *E. coli* BL21 cells, the bacteria are induced by IPTG, and the cells are lysed. The lysate is then incubated with Ni-NTA resin, with the recombinant protein ultimately being eluted using imidazole. (**b**) A Coomassie blue SDS-PAGE indicating the presence of trx-N and its degradation product

Northern and Sub-Saharan) as well as the Arabian Peninsula, where outbreaks still periodically occur [5]. In humans, the symptoms associated with RVFV infection vary from mild headaches, retinitis, and encephalitis, with fatalities occurring in about 1% of all reported cases. In livestock, RVFV can induce abortion storms in pregnant ewes, with fatality rates of up to 80% in young lambs [6].

The constant threat of natural outbreaks, as well as its possible use as a bioterrorist agent has made RVFV one of the most studied species of bunyavirus; and as such, significant progress has been made in elucidating its mode of transmission, replication cycle, tropism (from both a species and organ/tissue level), as well as the host immune response to infection (in both humans and livestock). As a result, several vaccine candidates have been developed, using a variety of platforms such as: attenuated viruses (i.e. MP12, Clone 13), inactivated viruses, recombinant viruses (i.e. vaccinia/capripox, adenovirus, lentiviral vectors), and DNA vaccines [7].

As seen in Fig. 1, RVFV is composed of a lipid-bilayered, enveloped virion, with the surface of each particle measuring 90–110 nM in diameter, with a tri-segmented, RNA genome, composed of an L-, M- and S-segments, measuring 6404 bp, 3885 bp, and 1690 bp, respectively. While the L-and

M-segments of the genome are in the negative sense, the S-segment is ambisense, with one ORF (NSs) being in the positive sense, while the nucleoprotein is in the negative sense. The L-segment encodes for the viral RNA-dependent RNA polymerase (RdRp, also known as the L-protein), which is responsible for both the replication of the viral genome, as well as its transcription into positive-sense mRNA [1, 6].

The M-segment encodes for two glycoproteins (Gn and Gc) as well as the nonstructural proteins NSm1 and NSm2, having molecular weights of 78 kDa and 14 kDa, respectively. The precise roles of both NSm1 and NSm2 in RVFV pathogenesis have not been clearly defined.

The M-segment also encodes for two glycoproteins, Gn and Gc. These glycoproteins form heterodimers, which ultimately form an ordered icosahedral shell of 122 capsomers at the exterior of the virus particle [8]. From an immunological standpoint, it should also be noted that neutralizing antibodies generated against RVFV are also localized to the glycoproteins. This suggests that, apart from forming the structural exterior of the virus, RVFV glycoproteins are also responsible for adhesion and cellular entry of the virus into the host cell [9].

As previously mentioned, the S-segment encodes for two open reading frames (ORFs), NSs and the nucleoprotein. NSs has been shown to localize within the nucleus of an infected host cell, and has been implicated in the inhibition of the host immune response; specifically, NSs has been found to have an antagonistic activity against the IFN antiviral response [10]. Recombinant expression of NSs has shown that these proteins can multimerize, forming filaments that can cause physical damage to the infected host nucleus [11, 12].

The S-segment also encodes for the nucleoprotein, which associates with viral genomic RNA to form ribonucleoprotein (RNP) complexes, which in turn, associate with glycoproteins to localize to the interior of the virion [13].

From an immunological standpoint, RVFV nucleoprotein has been shown to be the most immunogenic component of the virus, with infected animals and individuals generating a potent non-neutralizing antibody response [14–16]. Currently, diagnostic kits used to detect RVFV primarily detect this protein. Current antigen test kits are based on either direct or competitive Enzyme-Linked Immunosorbent Assays (ELISAs), the most immunodominant epitope (and thus, the basis of all animal and human RVFV antigen kits) and are based on the nucleoprotein of the virus.

Furthermore, RVFV nucleoprotein has been shown to promote partial protection when expressed as a DNA vaccine [14–16], while providing nearly complete protection against virulent RVFV challenged when expressed as a fusion protein with ubiquitin

[17]. These results suggested that RVFV nucleoprotein possesses at least one T-cell epitope; and that when expressed with ubiquitin (previously characterized to function as a self-adjuvant, increasing CD8+ T-cell responses), can induce non-neutralizing antibodies. Whether or not these non-neutralizing antibodies play a role in the host immune response is, yet, unclear.

As RVFV nucleoprotein does not have any known glycosylation sites, nor any cysteines (and thus, no disulfide bridges), it can be readily expressed in bacterial expression systems. Here, we describe a protocol for the expression of RVFV nucleoprotein in *Escherichia coli*, where the cDNA encoding for the nucleoprotein is expressed as a fusion protein with thioredoxin, a thermally stable and highly soluble bacterial protein. When expressed together with RVFV nucleoprotein, higher amounts of soluble protein can be produced and extracted from bacterial culture [18]. The thioredoxin-RVFV nucleoprotein fusion protein also contains: (1) an N-terminal histidine tag, to enable its purification through Nickel-resin purification columns, as well as; (2) an enterokinase cleavage site, located between the thioredoxin and RVFV nucleoprotein open reading frames (ORF). Upon purification of the fusion protein, subsequent cleavage with enterokinase will permit the cleavage of the fusion protein into a 1:1 mixture of both proteins. As RVFV nucleoprotein readily multimerizes, the protein mixture can then be separated by size-exclusion chromatography, resulting in highly pure preparations of recombinant protein. The recombinant RVFV nucleoprotein can then be used for a variety of uses, including: (1) a T-cell antigen for stimulating T-cells from animals infected/vaccinated with RVFV virus/antigens; (2) structural studies involving the use of nucleoprotein and/or their interactions with RNA, or; (3) a substrate for ELISAs or other assays to detect previous exposure to RVFV.

2 Materials

As the technique involves the transformation, culturing, and induction of bacteria (specifically *E. coli*), steps should be performed using appropriate laboratory protective material; this includes protective eyewear, lab coat, and lab gloves. Also, all bacterial manipulation should be performed in a sterile space, such as within 6 inches of a Bunsen burner or a biosafety cabinet designed for use with bacteria. All bacterial media should be prepared using an autoclave, and all buffers should be sterilized (either through autoclaving or filtration using a 0.22 μm syringe/bottle top filter).

2.1 Reagents for Cloning RVFV Nucleoprotein into a Sequencing Vector

1. pET-32a bacterial expression vector (Novagen #69015).
2. primers encoding for RVFV nucleoprotein. An example of a primer pair that can be used in pET-32a is the following, with restrictions sites NcoI and XhoI already added to the forward and reverse primers (underlined), respectively (*see* **Note 1**):

   Forward primer: 5′GCCATG GACAACTATCAA GAGCTTGCGATCCA′3.

   Reverse primer: 5′CTCGAG TTAGGCTGCTGTCTTG TAAGCCTGAGCGGCTGCCA'3.
3. Template cDNA encoding for RVFV nucleoprotein (based on RVFV reference strain ZH501 or ZH548).
4. DH5-alpha (multiple manufacturers) or Top10 (Invitrogen) cloning grade bacterial cells.
5. 10 mL bacterial cell culture tubes (sterile).
6. Bacterial mini-prep kit.
7. DNA ligase (*see* **Note 2**).
8. Super Optimal broth with Catabolite repression (also known as SOC) medium (available through various commercial manufacturers).
9. 1000× ampicillin solution: 1 g ampicillin powder in 10 mL water to give a concentration of 100 mg/mL, syringe filtered through a 0.22 μm filter.
10. Luria-Broth (LB) powder (*see* **Note 3**).
11. LB-Agar powder.
12. 150 mM bacterial culture plates.

2.2 Reagents for Cloning RVFV Nucleoprotein into the Thioredoxin Expression Vector

1. *Escherichia coli* BL21 chemically competent bacterial cells (*see* **Note 4**).
2. Isopropyl ß-D-1-thiogalactopyranoside (IPTG) powder.

2.3 Reagents for the Production and Purification of Recombinant Thioredoxin-RVFV Nucleoprotein

1. 1 M imidazole solution (in PBS, pH 7.5).
2. 4 L Glass conical flasks (autoclavable).
3. 250 mL or 500 mL high-speed centrifuge tubes.
4. Lysozyme powder..
5. Enterokinase enzyme.
6. Enterkinase buffer: 50 mM Tris-HCl 100 mM NaCl 2 mM $CaCl_2$ pH 8.0 (*see* **Note 5**).
7. Nickel-nitrilotriacetic acid (Ni-NTA)-conjugated agarose resin (*see* **Note 6**).
8. Dialysis tubing (minimum 10,000 MW size cutoff).

9. Sterile PBS (calcium-free and magnesium-free).
10. 4–20% SDS-PAGE gels and accessories (including running buffer and gel stains).
11. 1.5 mL centrifuge tubes.
12. Superdex 200 size exclusion chromatography columns—(*see* **Note** 7).

3 Methods

3.1 Preparation of Media and Cloning Products

1. Prepare 1000× ampicillin stock solution by measuring 10 g of ampicillin powder in distilled/deionized H_2O (ddH_2O) to a final volume of 10 mL. Vortex in a Falcon tube and filter using a 0.22 um syringe filter. Aliquot in 1 mL tubes, and freeze at − 20 °C.
2. Prepare 4 L of LB Broth, as well as at least 100 mL LB-Agar plates, according to the manufacturer's instructions, and autoclave at 121 °C for at least 30 min to sterilize. In general, LB Broth should be made at 25 g/L in ddH_2O. LB-Agar plates, unless otherwise indicated by the manufacturer, should be prepared by adding 40 g of LB-Agar (commonly composed of 10 g sodium chloride, 10 g tryptone, 5 g yeast extract, and 15 g agar) to 1 liter of ddH_2O.
3. After autoclaving, place the LB-Agar solution in a 50 °C water bath, and wait for the flask to cool down to the temperature of the water bath. When the flask reaches approximately 50 °C, add the 1000× ampicillin stock solution, diluted to 1× (i.e. for 100 mL LB-Agar solution, add 1 mL of stock solution).
4. Swirl for several seconds, and pour approximately 10–15 mL of LB-stock solution into the 150 mM bacterial culture plates in a sterile environment (such as a biosafety cabinet or laminar flow hood). Wait at least 2 h for the LB-Agar to harden, then keep all the prepared plates at 4 °C until further use.

3.2 Cloning

1. Begin PCR amplification of RVFV nucleoprotein cDNA by using the primers described in Subheading 2, using a high-fidelity DNA polymerase (*see* **Note** 7). Run the PCR product on a 1% agarose gel, and cut out the band migrating to approximately 750 bp. Purify the agarose band using a commercial DNA extraction kit.
2. Ligate the DNA fragment encoding for RVFV nucleoprotein into a DNA sequencing vector (such as pGEM T (Promega) or a TA cloning kit (Invitrogen)). If a blunt-ended polymerase is used, use an appropriate blunt-ended cloning vector (i.e. Zero Blunt or Zero Blunt-Topo (Invitrogen)). Follow the

manufacturer's instructions to ensure the proper cloning, transformation and selection of DH5-alpha (multiple manufacturers) or Top10 cells (Invitrogen).

3. Pick several bacterial clones, and amplify them using a small-scale bacterial mini-prep (5–10 mL) kit. Store the purified DNA samples at −20 °C.
4. Using sequencing primers specific to the DNA sequencing vector used, ensure that the sequence of RVFV nucleoprotein does not have any errors (*see* **Note 8**).
5. Perform a restriction digest using 2–5 μg of cloned sequencing vector using both NcoI and XhoI restriction enzymes (*see* **Note 9**) overnight (according to the restriction digestion conditions described by the manufacturer). Also, digest 1–2 μg of the pET-32a vector using NcoI and XhoI restriction enzymes.
6. The following day, run both the digested vector, as well as the cloned sequencing vector (containing the RVFV nucleoprotein) in 2 separate lanes on a 1% agarose gel. Cut out the bands corresponding to the digested pET-32a vector (5.9 kb), as well as the approximately 750 bp fragment in the cloning vector (corresponding to RVFV nucleoprotein). Keep in bands in separate 1.5 mL tubes.
7. Ligate approximately 0.1 ug of pET-32a vector with between 1 and 2 μg of RVFV nucleoprotein cDNA, according to the manufacturer's instructions (*see* **Note 10**).
8. Take 1–2 μL of the ligase reaction and chemically transform 50–100 μL of either DH5-alpha or Top10 cells, according to the manufacturer's instructions. In both cases, an aliquot of frozen cells is thawed on ice. Once thawed, 1–2 μL of the ligation mixture is added to the tubes and left to incubate on ice for about 20–30 min. A "heat shock" is then performed, where the tube is incubated in a 42 °C water bath for 30–45 s, followed by another incubation on ice for another 2 min. 100 μL of SOC medium (pre-warmed to 37 °C) is then added to the tube, which is then placed in a 37 °C shaking incubator (agitating 200 rpm or more) for 1 h. Finally, using two LB-Ampicillin agar plates (LB plates), spread 10 μL of the tube contents into one plate, while spreading the rest of the contents of the tube onto another LB plate. Incubate overnight at 37 °C.
9. The following day, pick five individual bacterial colonies, and inoculate each colony in a 10 mL bacterial culture tube, with 2 mL of LB-ampicillin broth (LB-Amp broth) at 100 μg/mL using a 1000× dilution of ampicillin stock), and incubate overnight 37 °C shaking incubator (agitating 200 rpm or more) for at least 8 h (or overnight, depending on convenience).

10. The following day, purify the plasmid DNA of each tube using a DNA miniprep purification kit. Verify the insert via PCR, restriction digestion (using NcoI or XhoI) or DNA sequencing (using either sequence-specific primers, or T7 promoter/termination primers.
11. Freeze all insert-containing plasmids at −20 °C for transformation into BL21 cells.

3.3 Induction of RVFV Nucleoprotein in BL21 E. coli

1. Transform 50–100 uL of BL21 cells using 1–2 μL using the "heat shock" method described in **Step 12**. Also, as in **Step 12**, prepare 2 LB-Amp plates—one with 10 μL of the transformation reaction, spreading out the rest of the transformation reaction on the other. Incubate overnight at 37 °C.
2. The following day, look for clear, isolated colonies on the LB-Amp plates. Pick 5 of them, and incubate each colony in 5 mL of LB-Amp broth, leaving incubating in a 37 °C shaker for 6–8 h.
3. In a 4 L glass conical flask prepare 1 L of LB-Amp and prewarm at 37 °C.
4. Add the contents of all five culture tubes from **Step 17** to the 4 L glass conical flask, and incubate at 37 °C, agitating at 200 rpm for 4 hours.
5. Add 1 mM of IPTG, diluting the 1000× stock solution to 1× (i.e. for 1 L of culture solution, add 1 mL of 1000× IPTG stock solution). Continue to incubate for 4 h at 37 °C (*see* **Note 11**).
6. Centrifuge the bacteria for 10 min at 10,000 × g. Decant the medium.

3.4 Purification of RVFV-Thioredoxin Nucleoproteins (trx-N)

1. Resuspend the pellet in 25 or 50 mL of PBS-Lysozme (1 mg/mL), and leave overnight at −20 °C.
2. The following day, thaw the pellet for 1–2 h in a 37 °C water bath.
3. Centrifuge the lysate at 15,000–20,000 × g for 1 h.
4. Collect the supernatant. Add 20 mM Imidazole (from the stock solution of 1 M imidazole powder in PBS). The trx-N fusion protein can now be purified with Ni-NTA resin (*see* **Note 12**).
5. Incubate the fusion protein mixture with at least 2 mL of Ni-NTA resin for at least 30 min at 4 °C.
6. Wash the resin at least five times (i.e. five column volumes) with PBS-20 mM imidazole. If you are doing a batch purification (i.e. the incubation is in a 50 mL centrifuge tube, use at least 25 mL imidazole with each wash, and centrifuge at 2000–3000 × g for 10 min.

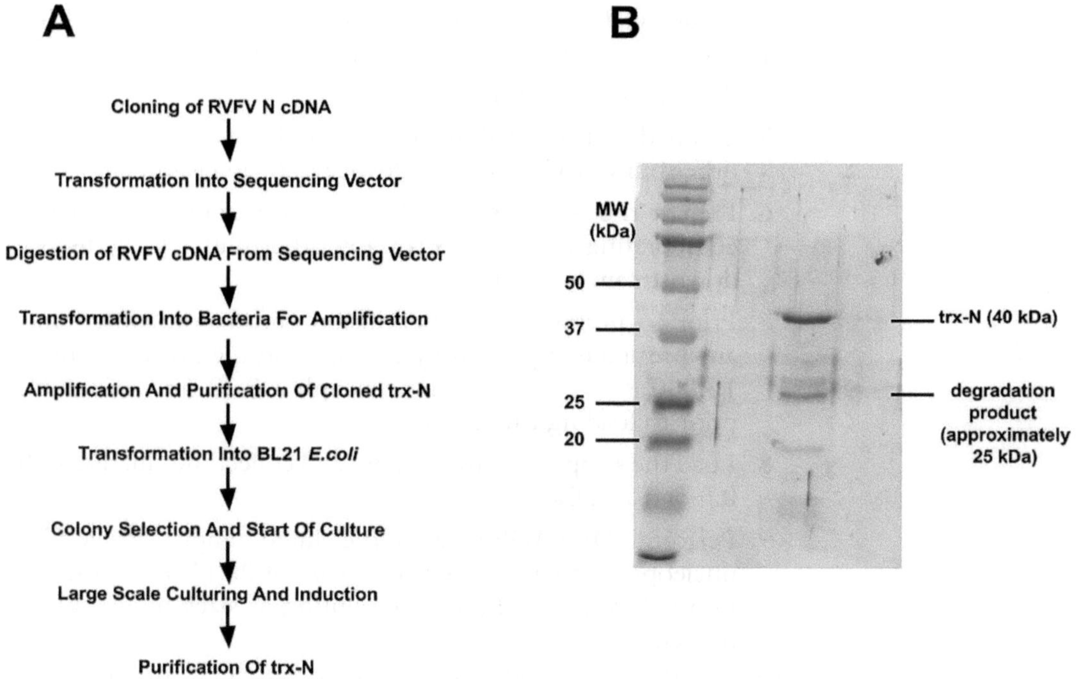

Fig. 2 Schematic of Rift Valley fever virus. (**a**) The virion is composed of a lipid-bilayered, enveloped virion, with the surface of each particle measuring 90–110 nM in diameter, with a tri-segmented, RNA genome (**b**). While the L-segment encodes for the RNA-dependent RNA-polymerase (RdRp), the M-segments encode for both glycoproteins (Gn and Gc), as well as the nonstructural proteins NSm1 and NSm2. The S-segment encodes for another nonstructural protein (NSs) and the nucleoprotein (N). All open reading frames are encoded in the negative sense, with the exception of NSs

7. Perform a final wish with PBS-50 mM imidazole (*see* **Note 13**).
8. To elute incubate the resin with PBS-500 mM imidazole for 5–10 min.
9. Centrifuge the resin and collect the supernatant. Run 10 μL on an SDS-PAGE. As seen in Fig. 2, you should see two bands, a 40 kDa band (trx-N) and a 25 kDa band (degradation fragment).
10. Using the dialysis tubing (pre-equilibrated in PBS), dialyze the eluate in 1 L PBS, changing the dialysis buffer twice.
11. Collect the dialyzed sample. The supernatant can be used directly for ELISA diagnostics. However, if the user wishes to remove the thioredoxin fusion protein, proceed with **Step 32**.

3.5 (Optional) Digestion and Removal of Thioredoxin

1. Dialyze the protein sample in enterokinase digestion buffer (see Materials) and leave overnight at 4 °C with stirring.
2. Add enterokinase enzyme to the trx-N mixture according to the manufacturer's instructions. Typically, an enzyme-substrate ratio of 1:200 will ensure nearly complete digestion in 24 h.

3. Incubate at room temperature, while rocking, for 16–24 h.
4. (Optional). Verify that the protein has been completely digested via SDS-PAGE.
5. Upon digestion, dialyze the digested trx-N in PBS (or any other buffer to be used for downstream applications.
6. Prepare the GF200 size-exclusion chromatography column by equilibrating it with PBS (or any other buffer to be used for downstream applications).
7. Prior to performing size-exclusion chromatography, remove any precipitated protein by centrifuging the protein sample at 15,000 × g, or by filtering through a 0.22 μm syringe filter prior to loading the sample.
8. Load the sample into the column and collect the eluate at either 0.5 or 1 mL fractions.
9. Perform SDS-PAGE on all of the eluted samples. While RVFV nucleoprotein has a theoretical mass of 25 kDa, multimerization will result in the protein eluting at sizes of 150 kDa or higher.
10. Identify the fractionated proteins via SDS-PAGE. Under reduced, denatured conditions, RVFV nucleoprotein should migrate at a size of approximately 25 kDa.
11. Depending on subsequent use, aliquot and store at −20 °C.

4 Notes

1. The restriction sites used in the primer can be replaced with other sites corresponding to those found in pET-32a. These include: EcoRV, BamHI, SacI, SalI, HindIII, EagI, and AvaI. Should any of these sites be used, ensure that the cDNA encoding for RVFV nucleoprotein does not have them.
2. Should another cloning strategy (i.e. In-Fusion™ (Clontech)) be used, please modify the primers according to the manufacturer's instructions.
3. While different types of LB-broth are commercially available, ensure that the LB to be used does not contain glucose, as this inhibits the induction of the IPTG induction.
4. A broad variety of BL21 *E. coli* are commercially available, offering bacteria that are phenotypically deficient for genes encoding for various proteases. Based on the author's own experience, these phenotypes do not significantly alter the degree of degradation of the RVFV-nucleoprotein fusion product.

5. The composition of the enterokinase buffer may differ, according to the type of commercial enzyme used. Therefore, follow the manufacturer's instructions should their composition of enterokinase buffer differ from that described in this protocol.
6. Other types of protein binding resins are available for the purification of His-tagged proteins, such as Cobalt-based protein binding resins. Based on the author's own experience, the protein yield, as well as the purity of the sample does not differ significantly between Nickel- or Cobalt-based resins.
7. While high-fidelity Taq polymerases leave T-A overhangs, many "proofreading" polymerases may result in blunt-ended PCR products. Keep this in mind when choosing the appropriate sequencing vectors.
8. Restriction digestion may be performed directly on the PCR product without the use of a sequencing vector; however, doing so will result in a less efficiently digested DNA fragment, which may result in decreased cloning efficiency in the pET 32a vector.
9. If different restriction sites were used in the design of the PCR primers, use their corresponding restriction enzymes accordingly.
10. If using another cloning system (i.e. In-Fusion™ (Clontech)), follow the manufacturer's instructions.
11. IPTG induction can also be performed overnight at 20 °C.
12. Upon purification, some thioredoxin-RVFV nucleoprotein fusion protein will be degraded by endogenous proteases in the bacteria. While the addition of protease inhibitors is recommended in similar protocols, the author did not notice any significant change in protein degradation when the protease inhibitors were used.
13. If extremely pure trx-N is required, an additional wash using 100 mM imidazole in PBS can be performed. While the purified protein will have considerably less contaminant, less protein will also be recovered (due to some trx-N being eluted during this final step).

References

1. Boshra H (2022) An overview of the infectious cycles of bunyaviruses. Viruses 14(10):1–14. https://doi.org/10.3390/v14102139
2. Sasaya T, Palacios G, Briese T et al (2023) ICTV Virus taxonomy profile: phenuiviridae. J Gen Virol 104(9):1–2. https://doi.org/10.1099/jgv.0.001893
3. Chevalier V, Pepin M, Plee L et al (2010) Rift Valley fever – a threat for Europe? Euro Surveill 15(10):19506. https://doi.org/10.2807/ese.15.10.19506-en
4. Hendrickx G, Lancelot R (2010) A perspective on emerging mosquito and phlebotomine-borne diseases in Europe. Euro Surveill 15(10):19503. https://doi.org/10.2807/ese.15.10.19503-en
5. Balkhy HH, Memish ZA (2003) Rift Valley fever: an uninvited zoonosis in the Arabian

peninsula. Int J Antimicrob Agents 21(2): 153–157. https://doi.org/10.1016/s0924-8579(02)00295-9

6. Boshra H, Lorenzo G, Busquets N et al (2011) Rift Valley fever: recent insights into pathogenesis and prevention. J Virol 85(13): 6098–6105. https://doi.org/10.1128/JVI.02641-10
7. Alkan C, Jurado-Cobena E, Ikegami T (2023) Advancements in Rift Valley fever vaccines: a historical overview and prospects for next generation candidates. NPJ Vaccines 8(171):1–16. https://doi.org/10.1038/s41541-023-00769-w
8. Sherman MB, Freiberg AN, Holbrook MR, Watowich SJ (2009) Single-particle cryo-electron microscopy of Rift Valley fever virus. Virology 387(1):11–15. https://doi.org/10.1016/j.virol.2009.02.038
9. Besselaar TG, Blackburn NK (1991) Topological mapping of antigenic sites on the Rift Valley fever virus envelope glycoproteins using monoclonal antibodies (1991). Arch Virol 121(1–4):111–124. https://doi.org/10.1007/BF01316748
10. Ly HJ, Ikegami T (2016) Rift Valley fever virus NSs protein functions and the similarity to other bunyavirus NSs proteins. Virol J 13(118):1–13. https://doi.org/10.1186/s12985-016-0573-8
11. Swanepoel R, Blackburn NK (1977) Demonstration of nuclear immunofluorescence in Rift Valley fever infected cells. J Gen Virol 34(3): 557–561. https://doi.org/10.1099/0022-1317-34-3-557
12. Yadani FZ, Kohl A, Préhaud C et al (1999) The carboxy-terminal acidic domain of Rift Valley fever virus NSs protein is essential for the formation of filamentous structures but not for the nuclear localization of the protein. J Virol 73(6):5018–5025. https://doi.org/10.1128/JVI.73.6.5018-5025.1999
13. Carnec X, Ermonval M, Kreher F et al (2014) Role of the cytosolic tails of Rift Valley fever virus envelope glycoproteins in viral morphogenesis. Virology 448:1–14. https://doi.org/10.1016/j.virol.2013.09.023
14. Lagerqvist N, Näslund J, Lundkvist A et al (2009) Characterisation of immune responses and protective efficacy in mice after immunisation with Rift Valley fever virus cDNA constructs. Virol J 6(6):1–10. https://doi.org/10.1186/1743-422X-6-6
15. Lorenzo G, Martín-Folgar R, Hevia E et al (2010) Protection against lethal Rift Valley fever virus (RVFV) infection in transgenic IFNAR(−/−) mice induced by different DNA vaccination regimens. Vaccine 17:2937–2944. https://doi.org/10.1016/j.vaccine.2010.02.018
16. Lorenzo G, Martín-Folgar R, Rodríguez F et al (2008) Priming with DNA plasmids encoding the nucleocapsid protein and glycoprotein precursors from Rift Valley fever virus accelerates the immune responses induced by an attenuated vaccine in sheep. Vaccine 26(41): 5255–5262. https://doi.org/10.1016/j.vaccine.2008.07.042
17. Boshra H, Lorenzo G, Rodriguez F et al (2011) DNA vaccine encoding ubiquitinated Rift Valley fever virus nucleoprotein provides consistent immunity and protects IFNAR (−/−) mice upon lethal virus challenge. Vaccine 29(27):4469–4475. https://doi.org/10.1016/j.vaccine.2011.04.043
18. Martín-Folgar R, Lorenzo G, Boshra H et al (2010) Development and characterization of monoclonal antibodies against Rift Valley fever virus nucleocapsid protein generated by DNA immunization. MAbs 2(3):275–284. https://doi.org/10.4161/mabs.2.3.11676

Chapter 14

Bunyaviruses: Expression of Recombinant Glycoproteins from Insect Cells

S. Matthijn De Boer

Abstract

High-density suspension cultures of insect cells offer a scalable and serum-free system for the expression of recombinant proteins. Rift Valley fever virus (RVFV), an arthropod-borne virus spread by mosquitoes, contains two envelop glycoproteins Gn and Gc. These glycoproteins are crucial for eliciting neutralizing antibodies that can offer protection against RVFV infection. Here, we describe the production of individual RVFV glycoproteins and empty RVFV particles using S2 *Drosophila* insect cells, followed by purification and analysis of the purified proteins.

Key words Insect cells, Expression, Glycoprotein, Purification, Strep-tag, Fusion protein, RVFV, VLP

1 Introduction

RVFV is a zoonotic mosquito-borne RNA virus belonging to the Phlebovirus genus of the Phenuiviridae family [12]. RVFV can cause massive death including abortion in cattle and sheep. Mortality rates in young lambs can exceed 95%. Human clinical cases are characterized by a sudden onset of high fever, severe headache, and myalgia. Over the years, the human case-fatality rates were estimated at 0.5–1.0%. Vertical transmission of RVFV in humans [13], and transmission via raw milk [14], have also been reported. RVFV is a serious zoonotic pathogen and listed as category A agent by the Centers for Disease Control. There are currently no licensed RVF vaccines available for humans, and commercial veterinary vaccines are only available in endemic areas. This highlights the need for additional safe and effective vaccines and antiviral therapies.

The RVFV genome (~12 kb) consists of a tri-segmented single-stranded RNA genome of negative and ambisense polarity. The three virion RNA (vRNA) molecules, designated small (S), medium (M), and large (L) encode the virus proteins. The RVFV glycoproteins, encoded by the M segment are exposed on the surface of the

Hani Boshra (ed.), *Bunyaviruses: Methods and Protocols*, Methods in Molecular Biology, vol. 2893,
https://doi.org/10.1007/978-1-0716-4338-9_14,

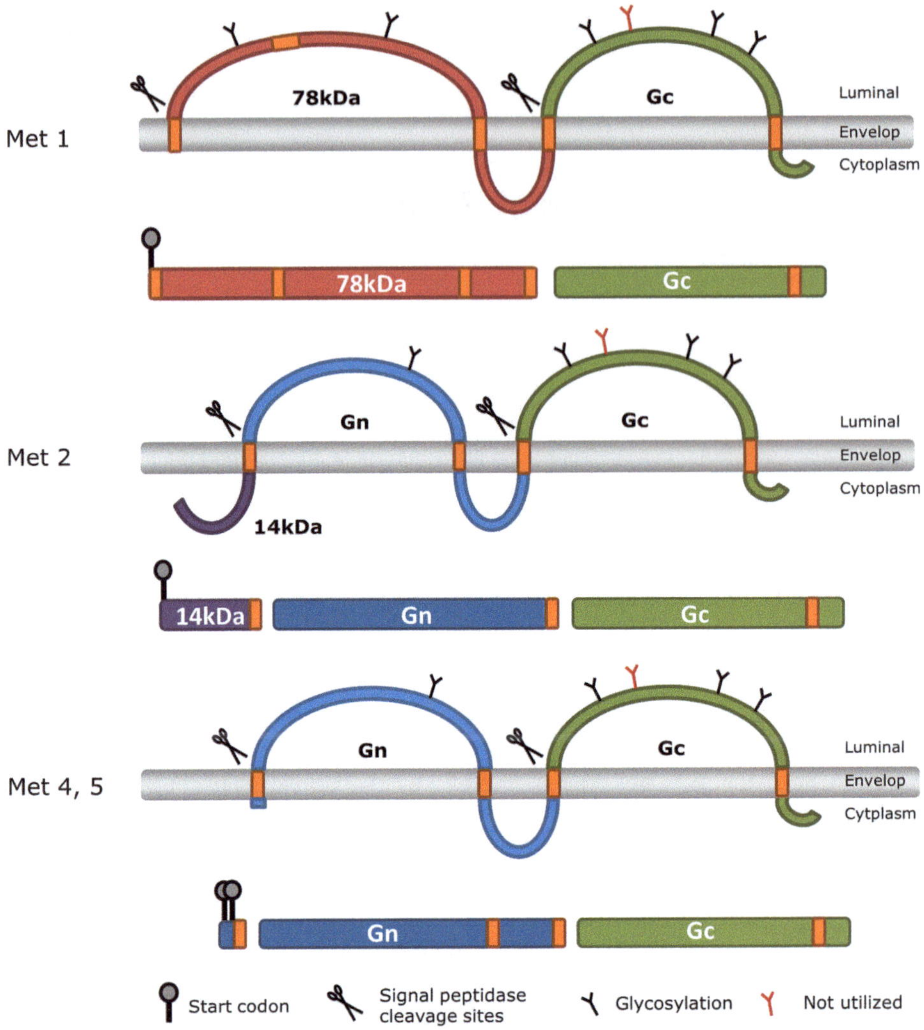

Fig. 1 Membrane topology and proteolytic processing of RVFV polyproteins. Membrane topology of the RVFV M-segment-encoded polyproteins. The translation products start from the in-frame AUG codons 1,2 and 4/5 of the open reading frame. The "Y" symbols indicate the predicted N-linked glycosylation sites. The red "Y" symbol is known not to be utilized. The 14-kDa, 78-kDa, Gn, and Gc proteins are colored in purple, red, blue, and green, respectively. The orange blocks represent the predicted transmembrane spanning regions

virus and are the first targets of the host immune defense system upon infection. RVFV has a single serotype and the targets for neutralizing antibodies are the Gn and Gc glycoproteins. Antibodies against the glycoproteins are the only established correlate of protection against RVFV infection.

The M segment contains five in-frame AUG codons (Fig. 1), of which four are used for synthesis of proteins [2]. A 78 kDa protein is transcribed from the first methionine, followed by a 14 kDa nonstructural protein from the second methionine. From the fourth and fifth start codons a polyprotein precursor is co-translationally expressed and cleaved into Gn and Gc [3]. Both the 78 kDa and 14 kDa proteins are nonessential for the replication of RVFV in cell culture, suggesting that these proteins have a function in vivo either in the mammalian or insect host [3, 8]. The Gn (~54 kDa) and Gc (~59 kDa) glycoproteins [3] form a heterodimer after processing of the polyprotein by host proteases in the endoplasmic reticulum (ER), and 720 Gn-Gc heterodimers are incorporated into the envelope forming the T = 12 icosahedral RVFV particle [4]. Co-expression of the RVFV Gn and Gc glycoproteins leads to the assembly of empty virus-like particles (VLPs). These VLPs can be produced and secreted from insect cells, with potential to serve as vaccine antigens [5, 6].

In order to study these glycoproteins, to setup a bioassay or generate antibodies against them, it is necessary to express these structural components of the virus particle. Such protein structures can be produced by different expression systems. In this chapter we describe the production of RVFV glycoproteins using the insect cell system. The advantage of using an insect cell system is the ease of scalability to several liters or more, the capacity for high density cell growth in suspension, the ability for protein glycosylation, and cultivation without the need for CO_2 or serum supplementation [1]. We selected the well described Drosophila expression system (InvitrogenTM) for expression and secretion of RVFV glycoproteins since it allows inducible expression and secretion under non-lytic conditions.

We have successfully expressed and purified the Gn ectodomain (Gn-e) individually and in the context of VLPs by co-expression of GnGc from the Drosophila expression system.

2 Materials

All solutions are prepared using ultrapure water (prepared by purifying deionized water) and analytical grade reagents. Prepare and store all reagents at room temperature (unless indicated otherwise). Diligently follow all waste disposal regulations when disposing waste materials.

2.1 Plasmids

1. pMT/BiP/V5-HisA (InvitrogenTM, Carlsbad, CA, USA).
2. Plasmids encoding the Gn-e or GnGc coding sequence.

2.2 Transfection and Expression

1. Drosophila Schneider (S2) cells [7].
2. 6 W cluster plates.
3. pCoBlast vector (Thermo Fisher Scientific).
4. Nanodrop2000 (Thermo Fisher Scientific).
5. Cellfectin reagent (Invitrogen).

6. Serum-free InsectXpress medium (Lonza, Westburg).
7. Eppendorf tubes (1.5 mL).
8. Polystyrene tubes (5 or 10 mL).
9. Blasticidin-S-HCl (InvitrogenTM).
10. 500 μM Copper sulphate.
11. cOmplete™, Mini Protease Inhibitor Cocktail (ROCHE).
12. Avidin.
13. 25 m^2 and 175 m^2 cell culture flasks (Greiner Bio-One B.V.).
14. Incubator 27 °C.

2.3 Purification Gn-e and VLPs

1. Microcentrifuge Micro CL 21R.
2. Ultra centrifuge (BECKMAN) and SW32Ti, SW41Ti, SW55Ti and JA-10 rotors (Beckman coulter).
3. 0.22 μm filter sterilized TN buffer; 10 mM Tris pH 7.4, 100 mM NaCl.
4. 0.22 um filter sterilized TEN Buffer; 10 mM Tris ph 7.4, 1 mM EDTA, 100 mM NaCl.
5. Weigh 20 g of sucrose and add TN buffer to a total of 100 g to prepare 20% w/w Sucrose suspensions in TN buffer. Dissolve the sugar under gentle heating on a magnet stirring device until dissolved, filter sterilize with 0.22 μm filter (store at 4 °C). For 30%, 40%, 50%, and 60% sucrose solution weigh 30, 40, 50, or 60 g of sucrose respectively and add TN buffer to a total of 100 g.
6. Sucrose Refractometer.
7. Ultra clear tubes, BECKMANN (344058); 1 × 3½ (25 × 89 mM).
8. ProBond™ nickel-chelating resin (Invitrogen™).
9. Imidazole.
10. 100 mL (desthio)Biotin elution buffer without imidazole; 100 mM Tris pH 8 (10 mL 1 M Tris pH 8.0), 150 mM NaCl (3 mL 5 M NaCl), 1 mM EDTA (200 μL 0.5 M EDTA pH 8.0), store at 4 °C storage.
11. Molecular mass cut-off of 30 kDa Amicon® Ultra-4 concentrator (MiliporeTM, Billerica, MA, USA).

2.4 SDS-PAGE & Western Blotting

1. 10% SDS-PAGE gels and blot buffer (Biorad system).
2. Polyvinylidene fluoride; 0,2 μM PVDF.
3. 100% Methanol.
4. Whattman paper 3 M.

5. Block buffer (Phosphate Buffered Saline with 0.05% TWEEN20 and 5% Protifar).
6. Wash buffer (PBS with 0.05% TWEEN 20).
7. 10× Phosphate Buffered Saline.
8. Rabbit polyclonal antibodies derived from Gn and Gc proteins.
9. Gn antiserum (α-Gn) synthetic peptide comprising residues 374-CFEHKGQYKGTMDSGQTKRE-393 [9].
10. Gc antiserum (α-Gc) a peptide comprising residues 975-VFE RGSLPQTRNDKTFAASK-994 [9]
11. Goat anti-rabbit horseradish peroxidase-conjugate.
12. Amersham ECL Western blotting detection reagents (GE Healthcare).
13. Tweezers and incubation trays (or 50 mL tubes.)

3 Methods

3.1 Plasmid Construction

Codon optimize the glycoprotein-coding sequence of the M segment of RVFV strain 35/74 (GenBank: JF784387.1) for expression in insect cells. The ectodomain of the Gn protein will be produced using the Drosophila expression system.

In the Drosophila expression system cassette, the expression vector contains the inducible metallothionein (MT) promoter and the Drosophila BiP secretion signal. For the expression of Gn the ectodomain (pMT-Gn-e) needs to be cloned, and for the VLPs the complete GnGc coding sequence (pMT-GnGc) needs to be cloned into the expression vector. In both cases, the sequence encoding the Gn signal peptide needs to be replaced by that coding for the BiP signal peptide, specifying the junction sequence "GLSLG-RSL-AEDPH" (BiP, linker, Gn ectodomain).

The sequence encoding the ectodomain of the Gn protein was fused to a sequence encoding a combined FLAG-tag/enterokinase (EK) cleavage site and three Strep-tags separated by glycine linkers to allow easy detection and purification of the monomeric protein. Specifying the junction sequence "*QCHTDPTG*-DYKDDDD-KAGPG-WSHPQFEK-GGGSGGGSGGGS-WSHPQFEK-GGGS GGGSGGGS-WSHPQFEK* " (*Gn ectodomain*-FLAG/EK-Strep-GC linker-Strep-GC linker, Strep" [17] (*see* **Note 8**).

In case you need expression in mammalian cells, clone a human codon-optimized GnGc gene into the pCAGGS expression vector [10], yielding pCAGGS-GnGc. All constructs were verified by DNA sequencing.

3.2 Expression and Purification of RVFV Gn Ectodomain

3.2.1 Transfection of Drosophila Schneider (S2) Cells

Co-transfect the pMT-Gn-e construct into Drosophila Schneider (S2) cells [7, 11] at a ratio of 19:1 with the Blasticidin resistance vector pCoBlast according to manufacturers' recommendations.

3.2.2 Transient Expression in Drosophila Schneider (S2) Cells

At day 0 seed S2 cells in 6 W cluster plates at (1×10^6 cells/mL, 3 mL serum-free InsectXpress medium per well at 27 °C). In order to transfect one 6 W cluster prepare the following. On day 1 pipette 5 μg plasmid DNA in 20 μL sterile distilled water in an Eppendorf tube (DNA mix) and add 10 μL Cellfectin to 10 μL sterile distilled water in a polyethyleen tube (transfection mix). Add the DNA mix to the transfection mix (also *see* **Notes 2** and **3**). Incubate this mixture for 15 min at room temperature. During this incubation mix every 3 min gently by tapping. After the 5-min incubation, add the mix to the cells (gently one drop at the time). At day 2 you add the induction medium containing Cupper sulfate (CuSO4) to a final concentration in the well of 500 μM. Harvested a sample of the medium containing the cells at multiple days to determine the optimum conditions for expression of your protein.

3.2.3 Multiple Induction Cycles for Expression in Drosophila Schneider (S2) Cells

At day 0 you seed S2 cells in a 6 W cluster plate at (1×10^6 cells/mL, 3 mL serum-free InsectXpress medium per well at 27 °C). At day 1, you prepare the DNA mix by pipetting 19 μg of the plasmid DNA with 1 μg or 0,5 μg or 0,25 μg of pCoBlast into a final volume of 20 μL distilled sterile water in an Eppendorf tube. Add the plasmid mixture to the transfection mix (10 μL Cellfectin with 10 μL sterile distilled water) and incubate at room temperature. Mix the transfection mixture gently by tapping every 3 min for 15 min. Add the transfection mix to the cells (gently 1 drop at the time). Replace the medium on day 4, 8, and 12 with 1 mL fresh S2 medium with 75 μM Blasticidin-S-HCL, these suspension cells do adhere to the bottom of your 6 W cluster plate, but pipette gently. Pellet the cells on day 14 by centrifugation for 5 min at 500 *g* and transfer the cells gently to a T25 flask after resuspension with 5 mL of fresh S2 medium containing 10 μM Blasticidin-S-HCL (*see* **Note 6**).

After successful selection, you still observe clear spherical cells, you harvested the cells by centrifugation (5 min at 500 *g*). The collected cells are counted and resuspend gently (1.3–1.5mln cells/mL) in 10 mL of InsectXpress medium with 10 uM Blasticidin-S-HCL and 500 μM CuSO4 (induction medium). After around 7 days of incubation at 27 °C the cell density will be sufficient to repeat the aforementioned steps. In most cases multiple induction cycles need to be performed with the S2 cell line in the presence of

Blasticidin-S-HCL. Allow the cells to expand up to multiple 175 m^2 cell culture flasks containing InsectXpress medium with 10 µM Blasticidin-S-HCL. In order to proceed to the purification, harvest the InsectXpress medium containing secreted protein from the T175 flasks. This will be around 7 days post induction. The collected medium is cleared of cell debris by a 5 min centrifugation at 10.000 *g*. For each liter of collected medium add 500 uL Avidin (10 mg/mL, *see* **Note 1**) and 1 tablet of complete mini, *see* **Note 4**. Add 8 mL of Streptactine Sepharose beads 50% to the cleared S2 medium and incubated overnight at 4 °C rotating. After incubation with the Streptactine Sepharose beads transfer the beads to a (glass) column for further purification (store at 4 °C or proceed to purification). Make sure that all air bubbles in the stacked beads of the column are removed.

3.2.4 Protein Purification

If possible, perform the complete purification process in a 4 °C cold room. The Gn ectodomain is affinity purified from the cell culture supernatant through its C-terminal 3xstrep-tag. The column containing the Streptactin Sepharose beads is first washed with 400 mL of sterile distilled water (4 °C). Then prime the Streptactin Sepharose beads with 50 mL elution buffer (without imidazole, 4 °C). For elution you incubated the beads with 10 ml elution buffer containing 250 mM imidazole for 10 min at 4 °C. Repeat the elution for two consecutive times and concentrate each elution (10 min, at 500 *g* to a maximum of 3 mg protein/mL) over (multiple) Amicon® Ultra-4 concentrators with a molecular mass cut-off of 30 kDa, *see* **Note 7**. Determine the protein concentration of the pooled concentrate in 2 µL sample using the Nanodrop2000 measuring the A280/A260 nM ratio following manufacture specifications. Store the suspension at 4 °C.

3.2.5 Purification of RVFV VLPs

Co-transfect the pMT-GnGc construct into Drosophila Schneider (S2) cells [7, 11] at a ratio of 19:1 with the Blasticidin resistance vector pCoBlast according to manufacturers' recommendations and induced as described before.

To purify RVFV VLPs, harvest culture supernatant of GnGc expressing cells 10 days after induction, and precleared the medium by low speed centrifugation (10,000 × g, 20 min, 4 °C). Pellet the sedimentation through a sucrose cushion. Typically, add 4 mL of 20% (w/w) sucrose solution in TN buffer to the centrifuge tube. Gently add 32 mL of precleared supernatant on top of 4 mL 20% (wt/wt) sucrose in TN buffer and centrifuge for 2 h at 100,000 x *g* at 4 °C in a Beckman SW32Ti rotor. Dissolve the pellet by gently adding 100 µL TN buffer and resting on ice including occasional gentle swirling. For further purification you can perform flotation experiments. For flotation experiments, gently mix concentrated VLPs with a TN-buffered 60% sucrose solution (total volume

1.4 mL) by swirling and used this as the bottom layer of a discontinuous sucrose gradient. Add subsequent layers of 50, 40, 30, and 20% (wt/wt) sucrose, of each 0.9 mL, on top of each other using a 1 mL tip, (*see* **Note 5**). Centrifugate the discontinuous sucrose gradient for 72 h at 160.000 × *g* at 4 °C in a Beckman SW55Ti rotor. After centrifugation collect 0.5 mL fractions using a 1 mL tip (tip cut off in order to gently pipette sucrose layers with high viscosity). Determine the sucrose concentration of the individual fractions with a sucrose refractometer. Analyze all fractions by sodium dodecyl sulfate polyacrylamide gel electrophoresis (SDS-PAGE) and Western blotting for the presence of RVFV glycoproteins.

3.3 SDS-PAGE and Western Blotting

Separated the proteins on 10% SDS-PAGE gels following manufacture specifications and subsequently transfer the separated proteins to polyvinylidene fluoride (PVDF, 0,2 μM) membranes. For transfer of the proteins from the SDS-PAGE gel to the PVDF membrane stack four Whattman papers 3 M, prewetted in blot buffer, followed by the SDS-PAGE gel. Prewet the PVDF membrane in 100% methanol (30 s) and gently place the PVDF paper on top of the SDS-PAGE gel using tweezers in one movement. Finish the stack with another four prewetted (blot buffer) Whattman papers 3 M. Blot for 90 min at max V, 80 mA. Remove the PVDF membrane after blotting was finished and place in a container with block buffer. Initiate the overnight blocking at 4 °C under continuous shaking. After overnight blocking, incubate the membrane (shaking) with rabbit polyclonal antisera raised against peptides derived from the Gn or the Gc protein for 1 h at room temperature in block buffer. Wash the membrane three times for 15 min with wash buffer. The Gn antiserum (α-Gn) was raised against a synthetic peptide comprising residues 374-CFEHKGQYKGTMDSGQTKRE-393. For obtaining the Gc antiserum (α-Gc) a peptide comprising residues 975-VFERGSLPQTRNDKTFAASK-994 was used [9]. Incubate the membrane (gently shaking) with the secondary antibody in block buffer for 1 h at room temperature followed by washing three times with wash buffer (15 min each) and one time with PBS for 15 min. As secondary antibody add goat antirabbit horseradish peroxidase-conjugate and for protein detection use the Amersham ECL Western blotting detection reagents following manufacture specifications (Fig. 2). For protein quantification of Gn and Gc you may use quantitative Western blotting using a twofold serial dilution of purified GST-GnGc fusion protein as a calibration marker (data not shown).

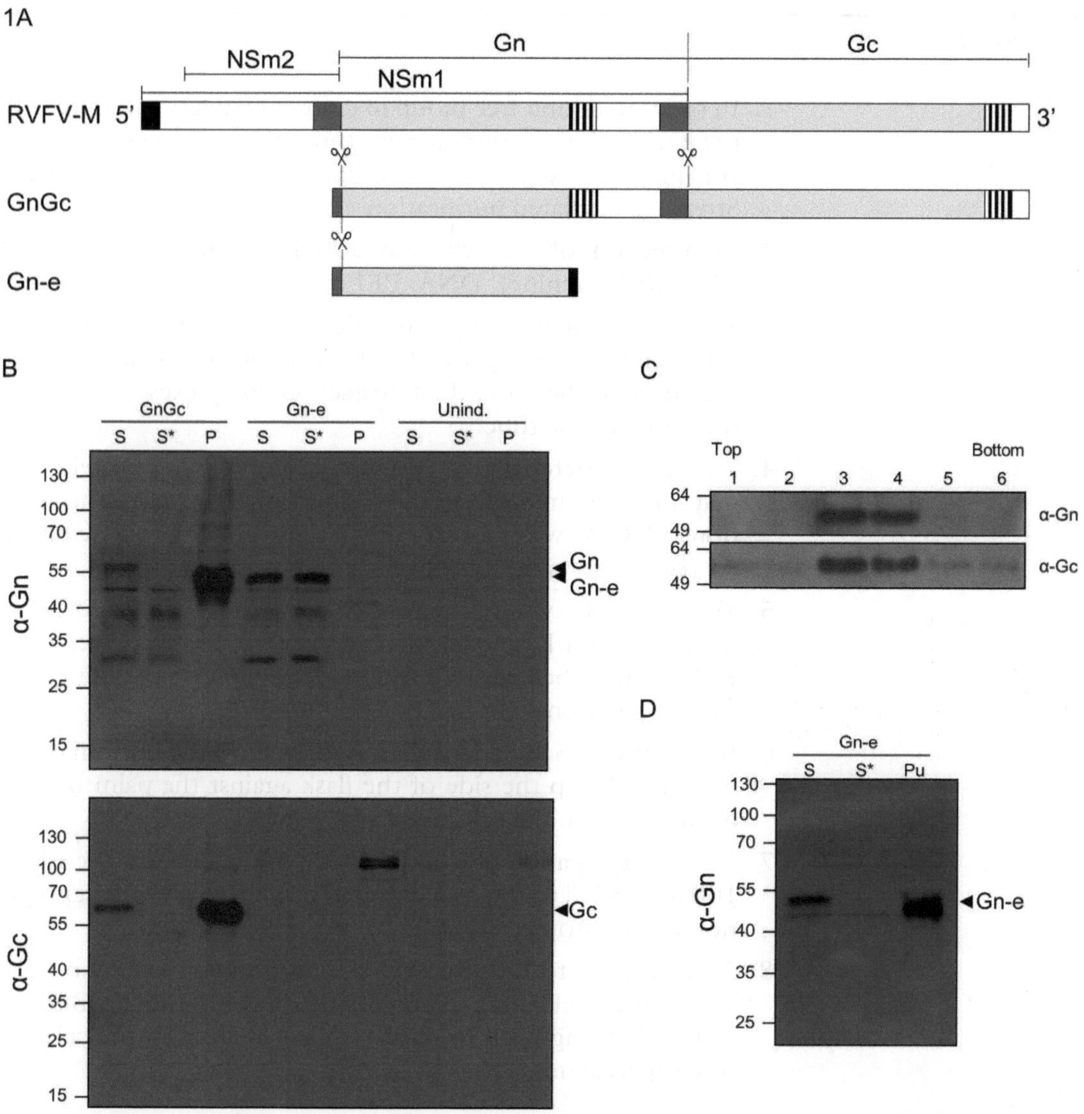

Fig. 2 Expression and analyses of the RVFV Gn and Gc glycoproteins. (**a**) Schematic representation of the RVFV M segment, encoding the Gn and Gc glycoproteins as well as two nonstructural proteins NSm1 and NSm2 [15]. The constructs, encoding parts of the M segment used for expression in Drosophila Schneider 2 (S2) cells, named GnGc and Gn-e are also depicted. The positions of a hydrophobic domain (black box), signal peptides (dark grey box), transmembrane domains (vertical-hatched box), and the signal peptide cleavage sites (scissors) are indicated. The expression constructs encode the GnGc or Gn ectodomain (Gn-e) with an N-terminal Drosophila BiP signal sequence (dark grey bar) and—in case of Gn—a C-terminal-fused tag (black bar). (**b**) Western blot analysis of culture supernatant of S2 cells expressing the GnGc polyprotein or Gn-e before and after ultracentrifugation. Cell culture supernatant before (S) and after (S*) pelletation and the pellet (P) were analyzed for the presence of Gn or Gc by Western blotting using an α-Gn (upper panel) or α-Gc (lower panel) peptide serum. Culture supernatant of uninduced S2 cells (unind.) was taken as a negative control. Subsequently co-floated of the virus-like particles in a discontinuous sucrose gradient are depicted in **c** (bottom of tube containing pellet to top of the tube). Purification of Gn-e depicted in panel **d**. Position of Gn, Gn-e, or Gc are indicated on the right (arrowheads); sizes of molecular weight markers are indicated on the left in kilodaltons (**b**–**d**) [16]

4 Notes

1. In order to inhibit free biotin in culture media to interfere with the purification of your proteins through Strep-tag purification you may add Avidin. The Avidin will not interfere with the Strep-tag mediated purification.
2. Transfection of S2 cells can also be performed with PEI (Polyethyleenimine), DNA: PEI ratio = 1:5.
3. To prevent inefficient transfection, always use polyethylene tubes and always pipette the transfection mixes directly into the liquid. Otherwise, the transfection complexes may stick to the inside of the tube.
4. To prevent bacterial growth and digestion of your protein, you may add sodium azide to the clarified medium at a concentration of 0.05% (w/v) for storage. We also added protease blockers (complete mini) and always stored at 4 °C.
5. To pipette the viscose sucrose solutions, we generated boxes with sterile 1 mL tips of which the first 5 millimeter is removed with a scissor before sterilization. This allows pipetting of the sucrose solutions.
6. In case the S2 Schneider cells are adjacent to the bottom of the flask, gently tap the side of the flask against the palm of your hand to release the cells.
7. In case aggregation of your purified proteins occurs. We added 100 mM NaCl to the concentrated protein to prevent aggregation during storage at 4 °C.
8. As alternative to the strep-tag purification you may also use a C-terminal 6xHis-tag in combination with the ProBond™ nickel-chelating resin to purify your protein from the cell culture supernatant.

References

1. Bernard AR, Kost TA, Overton L, Cavegn C, Young J, Bertrand M et al (1994) Recombinant protein expression in a Drosophila cell line: comparison with the baculovirus system. Cytotechnology 15(1–3):139–144
2. Suzich JA, Collett MS (1988) Rift Valley fever virus M segment: cell-free transcription and translation of virus-complementary RNA. Virology 164:478–486
3. Gerrard SR, Nichol ST (2007) Synthesis, proteolytic processing and complex formation of N-terminally nested precursor proteins of the Rift Valley fever virus glycoproteins. Virology 357:124–133
4. Huiskonen JT, Overby AK, Weber F, Grunewald K (2009) Electron cryo-microscopy and single-particle averaging of Rift Valley fever virus: evidence for GN-GC glycoprotein heterodimers. J Virol 83:3762–3769
5. Mandell RB, Koukuntla R, Mogler LJ, Carzoli AK, Freiberg AN, Holbrook MR, Martin BK, Staplin WR, Vahanian NN, Link CJ, Flick R (2009) A replication-incompetent Rift Valley fever vaccine: chimeric virus-like particles protect mice and rats against lethal challenge. Virology 397:187–189
6. Overby AK, Popov V, Neve EP, Pettersson RF (2006) Generation and analysis of infectious

virus-like particles of uukuniemi virus (bunyaviridae): a useful system for studying bunyaviral packaging and budding. J Virol 80:10428–10435

7. Schneider I (1972) Cell lines derived from late embryonic stages of Drosophila melanogaster. J Embryol Exp Morphol 27:353–365
8. Won S, Ikegami T, Peters CJ, Makino S (2006) NSm and 78-kilodalton proteins of Rift Valley fever virus are nonessential for viral replication in cell culture. J Virol 80:8274–8278
9. Filone CM, Heise M, Doms RW, Bertolotti-Ciarlet A (2006) Development and characterization of a Rift Valley fever virus cell-cell fusion assay using alphavirus replicon vectors. Virology 356:155–164
10. Niwa H, Yamamura K, Miyazaki J (1991) Efficient selection for high-expression transfectants with a novel eukaryotic vector. Gene 108:193–199
11. Johansen H, van der Straten A, Sweet R, Otto E, Maroni G, Rosenberg M (1989) Regulated expression at high copy number allows production of a growth-inhibitory oncogene product in Drosophila Schneider cells. Genes Dev 3:882–889
12. Adams MJ, Lefkowitz EJ, King AMQ, Harrach B, Harrison RL, Knowles NJ, Kropinski AM, Krupovic M, Kuhn JH, Mushegian AR, Nibert M, Sabanadzovic S, Sanfaçon H, Siddell SG, Simmonds P, Varsani A, Zerbini FM, Gorbalenya AE, Davison AJ (2017) Changes to taxonomy and the international code of virus classification and nomenclature ratified by the international committee on taxonomy of viruses. Arch Virol 162:2505–2538
13. Adam I, Karsany MS (2008) Case report: Rift Valley fever with vertical transmission in a pregnant Sudanese woman. J Med Virol 80:929
14. Arishi HM, Aqeel AY, Al Hazmi MM (2006) Vertical transmission of fatal Rift Valley fever in a newborn. Ann Trop Paediatr 26:251–253
15. Bouloy M, Flick R (2009) Reverse genetics technology for Rift Valley fever virus: current and future applications for the development of therapeutics and vaccines. Antivir Res 84:101–118
16. de Boer SM, Kortekaas J, Antonis AF, Kant J, van Oploo JL, Rottier PJ, Moormann RJ, Bosch BJ (2010) Rift Valley fever virus subunit vaccines confer complete protection against a lethal virus challenge. Vaccine 28:2330–2339
17. Kortekaas J, Antonis AF, Kant J, Vloet RP, Vogel A, Oreshkova N, de Boer SM, Bosch BJ, Moormann RJ (2012) Efficacy of three candidate Rift Valley fever vaccines in sheep. Vaccine 30(23):3423–3429

Chapter 15

Murine Models to Study Crimean-Congo Hemorrhagic Fever Hepatic Injury

Joseph W. Golden, Aura R. Garrison, and Xiankun Zeng

Abstract

Crimean-Congo hemorrhagic fever virus (CCHFV) is a member of the family *Nairoviridae* in the *Hareavirales* order and is an important human pathogen. Hepatic injury is a salient feature of CCHF human disease and can be recapitulated in murine models. Here, we described techniques to study liver damage caused by CCHFV in the mouse system. These techniques include a description of the mouse model used to study disease, and classical histopathological methods. Additionally, the use of in situ hybridization to evaluate liver injury is described in the context of viral infection. Immunofluorescence staining techniques to identify infected and/or damaged Kupffer cells is also explained. This work provides a primer for the comprehensive pathophysiological study of liver infection by a virus, but also has utility for noninfectious liver injury as well.

Key words Nairoviruses, Crimean-Congo hemorrhagic fever virus, Liver damage, Hepatic injury, Viral infection, Kupffer cells, Hepatocyte, In situ hybridization

1 Introduction

CCHFV is an important human pathogen that is both endemic and emerging throughout Asia, Africa, and Europe [1, 2]. Hepatic injury is a prominent component of CCHF and a significant factor contributing to morbidity and mortality [3–7]. Severe disease models for CCHFV have been developed in mice [8–10]. This work revealed that CCHFV produces acute disease only in mice lacking functional type I IFN signaling (IFN-I) either through STAT-1 deficiency [8] or deletion of the type I interferon receptor [9, 10]. These murine infection models recapitulate CCHFV-mediated hepatic injury observed in humans with marked liver pathology, including virus-mediated destruction of Kupffer cells [8, 11–13]. Concomitant with liver damage is the activation of inflammatory cytokines including TNF-α, IL-6, and IL-1 [11–13]. A transient IFN-I antibody blockade model was adapted to

Hani Boshra (ed.), *Bunyaviruses: Methods and Protocols*, Methods in Molecular Biology, vol. 2893,
https://doi.org/10.1007/978-1-0716-4338-9_15,

study CCHFV pathogenesis in transgenic mice [11, 12, 14–16]. For these studies, IFN-I signaling was blocked using a commercially available murine non-cell depleting monoclonal antibody (mAb) targeting the IFNAR-1 subunit of the mouse IFN-α/β receptor (MAR1-5A3) [17]. Similar to genetic knockout mice, IFN-I antibody blockaded mice develop a severe and fatal disease analogous to humans. As in humans, the primary target organ in mice is the liver. Here, CCHFV infection is described in these murine systems. Furthermore, downstream techniques are described to study the pathophysiological aspects of virus-mediated liver injury, including observation of infiltrating immune cells and Kupffer cell loss. Additionally, liver processing is described for the assessment of viral load. These techniques can be implemented to study CCHFV or other nairoviruses, such as Hazara virus. More broadly, several of these methodologies are useful for studies examining liver injury from both infectious and noninfectious causes.

2 Materials

2.1 Mouse Viral Challenge

1. 1 mL syringes.
2. Needles (22-27 g; 0.5 inch).
3. Sterile 1× phosphate buffered saline (PBS); pH 7.4.
4. Stock of Crimean-Congo hemorrhagic fever virus or other nairoviruses (*see* **Note 1**).
5. Septum cap surface sterilized with alcohol or autoclaved (*see* **Note 2**).
6. Mice (*see* **Note 3**).
7. MAR1-5A3 monoclonal antibody (Leninco Technologies or BioXCell).
8. 5% Micro-Chem Plus or equivalent (*see* **Note 4**).
9. Single channel pipettes and tips.

2.2 Liver Harvest

All components were sterilized either by the manufacturer or by autoclaving at a temperature of 121 °C for >30 m by using saturated steam under >15 psi of pressure.

1. Mouse Ketamine-Xylazine solution: 100 mg/mL Ketamine, 20 mg/mL Xylazine and 0.9% sterile saline (KAK).
2. Serum tubes.
3. Scissors.
4. Forceps.
5. 10% neutral buffered formalin.

6. 50 mL conical tubes.
7. Autoclave bags.

2.3 Histo/Paraffin Embedding and Tissue to Slide

1. Vacuum tissue processor.
2. Microtome.
3. Water bath.
4. Ice tray.
5. Charged glass slides.
6. Forceps.
7. Ethanol (100%, 95%, and 70%).
8. Xylene substitute.
9. Paraffin.

2.4 RNA In Situ Hybridization

1. Oven.
2. Rocking shaker.
3. Staining dish.
4. ImmEdge® Hydrophobic Barrier PAP Pen (Cat No. H-4000; Vector Laboratories).
5. ACD HybEZ™ II Hybridization System with ACD EZ-Batch Slide System (Cat No. 321710, 321717, and 321716; Advanced Cell Diagnostics).
6. ACD RNAscope™ CCHFV probe (Cat No. 494801; Advanced Cell Diagnostics).
7. ACD RNAscope™ 2.5 High Definition (HD)-RED Assay (Cat No. 322350; Advanced Cell Diagnostics).
8. Xylene substitute.
9. Ethanol (100%, 95%, and 70%).
10. Hematoxylin.
11. EcoMount.
12. Slide coverslips.

2.5 Immuno-fluorescence (CCHFV, Kupffer Cells, Inflammatory Cells, etc.)

1. Oven.
2. Rocking shaker.
3. Slide staining tray.
4. Tissue-Tek® Staining Dish.
5. Phosphate buffered saline, pH 7.4.
6. Tris-EDTA buffer (10 mM Tris Base, 1 mM EDTA Solution, 0.05% Tween 20, pH 9.0).
7. PBST (PBS +0.1% Tween-20).
8. CAS-Block with 5%.

9. Primary antibodies.
10. Fluorescence-dye conjugated secondary antibodies.
11. DAPI (4′,6-diamidino-2-phenylindole).
12. Fluorescence mounting medium.
13. Slide coverslips.
14. Confocal microscope.

2.6 Liver Homogenization

1. TissueLyser II.
2. TissueLyser II adapter 2 × 24 (Qiagen).
3. Stainless steel beads 5 mM (Qiagen).
4. Dulbecco's Modified Eagle Medium (DMEM) supplemented with 5% (10%) heat-inactivated fetal bovine serum (FBS), 1% nonessential amino acids, 10 U/mL Penicillin- 10 μg/mL Streptomycin, and 1% glutamine.
5. 2 mL Polypropylene/Eppendorf test tubes (or equivalent).
6. Single channel pipettes or multi-channel pipette, and tips.
7. Analytical scale.
8. Refrigerated microcentrifuge.

2.7 Quantification of Viral Load (Plaque Assay)

1. SW-13 cells (ATCC; CCL-105).
2. 6-well plates.
3. 96-well plates (*see* **Note 5**).
4. For dilutions: 1× Eagle's mimum essential medium (EMEM) supplemented with 5% heat-inactivated FBS, and 1% L-glutamine.
5. For overlay: 2× EMEM supplemented with 10% heat inactivated FBS, and 2% L-glutamine.
6. 1% low melt agarose in sterile H_2O.
7. 5% neutral red solution (0.33% w/v neutral red in PBS).
8. 1× PBS, pH 7.4.
9. 37 °C incubator with 5% CO_2.

3 Methods

3.1 Murine Infection

1. All work with infectious virus is conducted in a biosafety level 4 (BSL-4) or equivalent environment, which may vary depending on jurisdictional rules and regulations. Research with other nairoviruses may not require higher containment levels (always consult jurisdictional rules for biocontainment work before initiating studies).

2. In a biosafety cabinet (BSC) class 2 type A2, dilute the virus to 100 Plaque forming units (PFU)/0.2 mL in sterile PBS in a 50 mL tube (*see* **Note 6**).
3. Load the virus containing solution in sterile syringes with needles using a septum cap or other needle-safe method. Alternatively, syringes without needles can be filled and then equipped with sterile capped needles, ensure to purge the air from the needle before using. If transporting syringes, needles require secure caps. If recapping, utilize a single hand technique.
4. Challenge mice with the virus solution by the interperitoneally or subcutaneous route with 0.2 mL of virus material (*see* **Note 7**).
5. Inactivate virus solution by pouring contents into a pan of 5% Micro-Chem Plus, or institute required inactivation reagent.
6. If needed to block IFN-I, 2.5 mg of MAR1-5A3 is injected 24 h post-challenge by the IP route (*see* **Note 8**).
7. Monitor mice daily for signs of disease including hunched posture, lethargy, irregular breathing, and ruffled fur.

3.2 Liver and Spleen Harvest

1. At the designated time point(s) (*see* **Note 9**), deeply anesthetized mice by KAK and harvest blood by cardiac stick.
2. Confirm euthanasia by cervical dislocation.
3. Remove the abdominal and thoracic viscera en bloc, and gently dissect the liver and spleen away separately (*see* **Note 10**).
4. Remove a portion of the liver (~100–300 mg) and place in a tube to assess viral load (see Subheading 3.6).
5. Place the remaining intact liver (and spleen) section into a 10% neutral buffered formalin.
6. Leave the sections in formalin for 21 days (*see* **Note 11**).
7. Place carcasses into an autoclave-safe bag and autoclave (*see* **Note 12**).

3.3 Histology Processing

1. Select the larger median and lateral lobes of the liver, with particular attention to including any gross abnormalities. If possible, include a section of gall bladder within the median lobe.
2. Embed three to four sections of liver tissue and if taken, the spleen, in paraffin using an embedding mold.
3. After cooling, label the paraffin with study information to enable tracking.
4. After cooling the block, use a rotary microtome to cut tissue sections to a thickness of five microns.

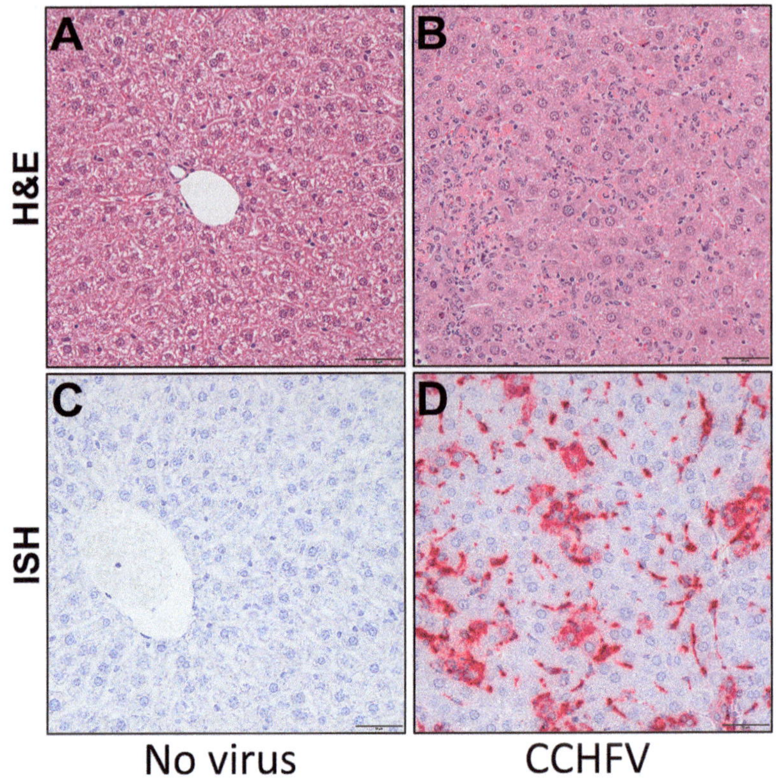

Fig. 1 Histological evaluation and CCHFV RNA detection of liver tissues **a–b**. Liver sections from an uninfected control mouse (**a**) and a CCHFV strain IbAr 10200-infected mouse (**b**) at day 4 post-exposure were stained by hematoxylin and eosin (H&E). **c–d**. Liver sections from an uninfected control mouse (**a**) and a CCHFV strain IbAr 10200-infected mouse at day 4 post-exposure (**b**) were stained for CCHFV RNA using in situ hybridization (ISH). Nuclei were counterstained with blue hematoxylin

5. Cut the ribbon of tissue from the block and float it on a tissue floatation water bath.
6. Stretch the ribbon of tissue using a forceps or brush to remove wrinkles, folds, and air bubbles.
7. Float tissue onto a clean glass slide and allowed to dry (*see* **Note 13**).
8. Deparaffinized the tissue sections and stain with hematoxylin and eosin (H&E) using standard procedures (Fig. 1).

3.4 RNA In Situ Hybridization

1. RNA in situ hybridization (ISH) is performed using an RNAscope™ 2.5 HD-RED Assay kit according to the manufacturer's protocol (Advanced Cell Diagnostics) (*see* **Note 14**). The RNAscope assay consists of Z-shaped probe pairs and a signal amplification system composed of preamplifiers,

amplifiers, and labelled probes (from AMP1 to AMP6). **Steps 2** through **12** delineate the procedures for this assay.

2. Bake the tissue slides for 1 h in 60 °C oven.
3. Deparaffinize the slides using xylene substitute and dehydrate in 100% ethanol.
4. Cover the entire section using RNAscope Hydrogen Peroxide to quench the endogenous peroxidase.
5. Conduct the heat-induced epitope retravel using the kit provided RNAscope Target Retrieval Reagent.
6. Conduct protease digestion using kit provided RNAscope® Protease Plus (*see* **Note 15**).
7. Expose the pretreated slides to the CCHFV ISH probes and incubate at 40 °C in a hybridization oven for 2 h.
8. Rinse the slide in washing buffer.
9. Amplify the ISH signal with kit-provided Pre-amplifiers, Amplifiers, probes conjugated to alkaline phosphatase and incubated with a Fast Red substrate solution for 10 min at room temperature.
10. Counterstain slides with hematoxylin.
11. Air-dry slide overnight and apply coverslip.
12. Observe the CCHFV ISH signal in infected liver cells, which primarily consist of Kupffer cells, hepatocytes, and endothelial cells of mouse liver tissues (Fig. 1) (*see* **Note 16**).

3.5 Immuno-fluorescence

1. Bake tissue slides for 1 h in 60 °C oven.
2. Deparaffinize the slide using xylene substitute and a series of ethanol washes.
3. Heat the tissue sections in Tris-EDTA buffer (pH 9) for 20 min for epitope retrieval.
4. Rinse slides in PBS.
5. Block the slides by overnight incubation in CAS-Block containing 5% normal goat or donkey serum at 4 °C.
6. Incubate slides with primary antibodies diluted in CAS-Block for 2 h at room temperature or overnight at 4 °C as follows: rabbit polyclonal anti-CCHFV NP (04-0011. IBT Bioservices) at a dilution of 1:200, goat polyclonal anti-CLEC4F antibody (Kupffer cell marker, PA5-47396, Thermo Fisher Scientific) at a dilution of 1:200, rabbit polyclonal anti-CD68 (macrophage marker, ab125212, Abcam) at a dilution of 1:200, rabbit polyclonal anti-myeloperoxidase (MPO, a neutrophil marker, A039829-2, Dako Agilent Pathology Solutions) at a dilution of 1:200, rat monoclonal anti-CD45 antibody (leukocytes marker, 562166, BD Biosciences) at a dilution of 1:100,

mouse monoclonal anti-Ki67 (cell proliferation marker, clone B56, BD Biosciences) at dilution of 1:200, or mouse monoclonal anti-CCHFV NP 9D5 (USAMRIID) at a dilution of 1: 500.

7. Rinse slides three times in PBST.
8. Stain slides with the secondary goat or donkey Alexa Fluor® 488-conjugated anti-rabbit antibody (Thermo Fisher Scientific) at a dilution of 1:500 and with goat or donkey Alexa Fluor® 561-conjugated anti-mouse or rat antibody (Thermo Fisher Scientific) at a dilution of 1:500 for 50 min at room temperature (*see* **Note 17**).
9. Counterstain the slides with DAPI and apply a coverslip using fluorescence mounting medium.
10. Capture images on a confocal microscope and process using an open-source ImageJ software (National Institutes of Health, Fig. 2) (*see* **Note 18**).

3.6 Liver Homogenization

1. All work with infectious virus is conducted in a biosafety level 4 (BSL-4) or equivalent environment, which may vary depending on jurisdictional rules and regulations. Research with other nairoviruses may not require higher containment levels (always consult jurisdictional rules for biocontainment work before initiating studies).
2. From the mouse, cut the liver tissue, removing excess adipose tissue, into 100–300 mg size pieces using forceps and scissors (or scalpel).
3. Place tissue into pre-labeled 2 mL microcentrifuge tube or cryovial.
4. Weigh the tissue in the tube using blank tubes as the tare weight.
5. Place a single stainless 5 mM steel bead into each tube containing the liver section.
6. Add 0.75–1 mL of homogenization media to each tube (*see* **Note 19**).
7. Set the TissueLyser II to 30 Hz and time to 1 min (*see* **Note 20**).
8. Place samples into the TissueLyser II adapter 2 × 24 (comprised of a Top plate, Bottom plate, and Tube holder). Ensure the samples are appropriately balanced across adapters (analogous to a centrifuge).
9. Place the adapters into TissueLyser II by pulling the locking pin upward and out of its slot turning it 90°. Turn the handwheel counterclockwise until the maximum clamping range is reached (fully open). Rotate back the locking pin by 90° until it

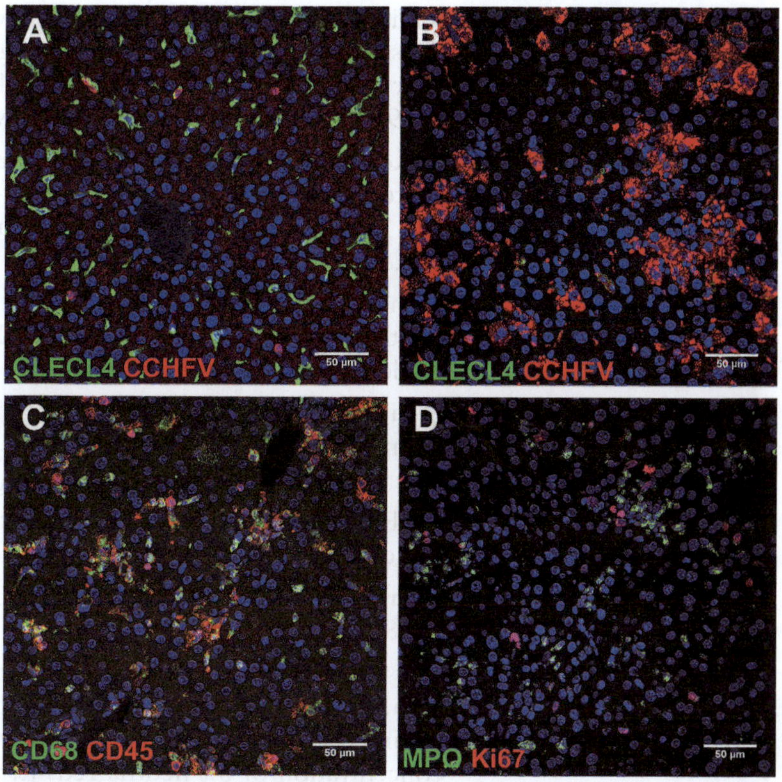

Fig. 2 Immunofluorescence staining on CCHFV infected liver tissues. **a–b**. Liver sections from CCHFV strain IbAr 10200-infected mice treated with isotype control (**a, c**, and **d**) or anti-CCHFV NP (**b**) at day 4 post-exposure were stained with anti-CLEC4F (green in **a** and **b**) and anti-CCHFV NP antibodies (red in **a** and **b**), anti-CD68 (green in **c**, macrophages) and anti-CD45 antibodies (red in **c**, leukocytes), or anti-MPO (green in **d**, neutrophil granulocytes) and anti-Ki67 (red in **d**, proliferating cells) antibodies. Nuclei were counterstained blue with 4′,6-diamidino-2-phenylindole (DAPI)

locks in place. Insert the adapter set and press it lightly into the indentation in the clamp and tighten.

10. Initiate homogenization and after the cycle is completed, rotate the adapters 180° and repeat the homogenization.
11. Remove samples from TissueLyser adaptors inside a BSC and place samples in a refrigerated microcentrifuge set to 4 °C.
12. Clarify the samples by centrifugation at 5000 × g for 10 min.
13. In a BSC class 2 type A2, remove the supernatant from the tube and place in a new pre-labeled tube of appropriate size.
14. Samples can be frozen for downstream analysis (*see* **Note 21**).
15. Disinfect the equipment with 5% Microchem.

3.7 Viral Load (Plaque Assay)

1. Seed SW-13 cells into 6-well plates, with enough plates to run the samples in duplicate, a plaque negative plate (no virus) and a virus plaque positive control (stock virus). Cells should be 80–90% confluent at the time of the viral titration.
2. Make a dilution series of the clarified liver homogenate (*see* **Note 22**) from 10^{-1} to 10^{-12} in 96-well plates (final volume after dilution should be 100 μL of medium/well) (*see* **Note 23**).
3. Remove the medium from SW-13 cell monolayers and add the diluted samples in duplicate in six well plates (100 μL of sample is added per well).
4. Incubate the plates at 37 °C with 5% CO_2 for 1 h and rock them ~15 m to prevent drying of the monolayer.
5. Add 2 mL of primary overlay consisting of a 1:1 volume of 2× EMEM and 1% low melt agarose to each well, final concentration of 1× EMEM and 0.5% low melt agarose (*see* **Note 24**).
6. Incubate the plates for 3 days at 37 °C with 5% CO_2 (humidity 70–80%).
7. Use one of two secondary overlay options (*see* **Note 25**): add 2 mL per well of solid secondary overlay after 2 days (1:1 volume of 2× EMEM and 1% low melt agarose, and final concentration of 5% neutral red) (*see* **Note 26**), or 2 mL per well of a liquid overlay after 3 days (5% neutral red in PBS to each well).
8. Solid overlay option, visualize plaques after overnight incubation. Liquid overlay option, remove liquid overlay and visualize plaques after a 1–2 h incubation (*see* **Note 27**).
9. Calculated the PFU/mL and then divided by the weight (grams) of the respective liver section to determine the PFU/gram. Record the plaque counts.

4 Notes

1. CCHFV strains (Kosova Hoti, Afg09-2990, IbAr 10200; Turkey 2004, UG3010) and Hazara virus strain JC280 have been tested in the IFN-I blockade model. Strain Kosova Hoti is not lethal in the IFN-I blockade model, but mice lose weight [11].
2. Any non-glass tubes with a rubber septum for syringes use can be used; Syringa Lab Supplies has both 50 mL and 15 mL conical caps (cat#1F920 for 50 mL) that work well but ensure the caps match the conical tube manufacturer used. The 15 ml caps are more difficult to secure on the 15 mL conical tubes.
3. If using type I interferon receptor deficient mice or STAT1 deficient mice, it is not necessary to block IFN-I because it is

already genetical absent. For other mice, MAR1-5A3 treatment will be needed to disrupt function of IFN-I signaling. This product is commercially available from multiple sources.

4. Our institute uses a solution of 5% microchem plus (diluted in water) (National Chemcial Laboratories, Inc.; EPA registration number 1839-95-2296) to inactivate enveloped viruses. Other inactivation chemicals may be required for different laboratories.
5. Any dilution arrangement is acceptable, 96-well plates are readily available. Dilution titer boxes are also commercially available.
6. The virus is typically diluted 1:10 before making the challenge stock to ensure pipetting >10 μL in order to maintain consistency between studies.
7. All animal work requires approval from an institutional animal care and use committee with predefined euthanasia criteria. For intraperitoneal injections, mice are injected in two spots with half the inoculum per spot.
8. MAR1-5A3 can alternatively be added at the time of challenge or in two doses (2 mg day −1 and 0.5 mg day +1).
9. Often the day 4 timepoint is used in mice infected with CCHFV to compare the efficacy of drugs against control animals. This is a time point where acute liver injury can be detected in infected, non-drug treated mice and used to assess drug-mediated protection.
10. The spleen is taken in addition to the liver and processed on the same slide. This gives added information regarding the histopathological impact of the viral infection on the host.
11. Inactivation time of biological select agents in formalin may vary depending on jurisdiction and laboratory.
12. Biological select agent handling may require logging of infected animals for tracking purposes. This will vary for various jurisdictions. Other nairoviruses, such as Hazara virus, are not select agents.
13. Overnight at room temperature, or drying oven <60 °C for 30–60 min.
14. Due to high divergence of different strains of CCHFV, a probe targeting a specific strain of CCHFV is needed for CCHFV ISH.
15. To preserve tissue integrity and cell morphology, proteinase digestion of mouse liver and spleen tissues using RNAscope® Protease Plus should be no more than 16 min at room temperature.

16. RNAscope is a highly sensitive and specific ISH platform, but nonspecific staining still occurs in most samples. It is important to differentiate specific ISH signal from nonspecific signal for the samples with few CCHFV infected cells, thus a noninfected control sample is encouraged.
17. The incubation time of secondary antibodies should not exceed 1 h at room temperature to avoid cross binding of secondary antibody to the primary antibody of the other species.
18. Some formalin fixed paraffin embedded tissues have background autofluorescence that is usually not as bright as real fluorescence signal. Commercial autofluorescence quencher or Sudan Black B can be used to reduce autofluorescence.
19. Homogenization medium can be cell culture medium or PBS. If there is no need for viable virus, an RNA extraction reagent. Inactivation of biological select agents in RNA extraction reagents may need to be verified depending on jurisdiction and laboratory.
20. Other homogenization systems will also work, such as the gentleMACS™ Dissociator and M-tube.
21. Clarified liver homogenate can be added to added to an appropriate RNA extraction reagent and used for qRT-PCR assays. Inactivation of biological select agents in RNA extraction reagents may need to be verified depending on jurisdiction and laboratory.
22. Liver enzymes may damage the virus thus impacting titer. Livers are therefore processed all at once and the plaque assay conducted promptly thereafter. If collecting multiple time points, freeze down the unprocessed livers at −70 to −80 °C and thaw for processing once all samples are collected.
23. Include a positive control sample consisting of virus stock, and a mock sample consisting of just media. The positive control is critical to ensure that the plaque assay functioned correctly, and plaques could be detected, the mock sample ensures the monolayers remain fully intact throughout the assay.
24. A 2× concentration of agarose (1% in H2O) is heated in a microwave until melted and cooled in a 45 °C water bath prior to mixing with medium. The medium is kept at 37 °C. A disposable plastic Erlenmeyer flask or similar type of sterile plastic bottle is used to make the overlay mixture. Continuous swirling helps prevent solidification prior to adding to the plates. If a large number of plates are being used, aliquots of the agarose and media should remain in the water bath and mixed just prior to use to decrease the chance of solidification. Do not microwave the medium and agarose mixture if it has

solidified. Let overlay solution completely solidify prior to moving the plates and placing them back into the incubator. This prevents cells from being lifted off the plates.

25. Oxidation from repeated use may impact neutral red effectiveness, therefore the smallest size bottles are used.
26. For the solid overlay option a 2× concentration of agarose (1% in H2O) is heated in a microwave until melted and cooled in a 45 °C water bath prior to mixing with 2×EMEM with 10% neutral red (for a 5% final concentration). The medium is kept at 37 °C. A disposable plastic Erlenmeyer flask or similar type of sterile plastic bottle is used to make the overlay mixture. Continuous swirling helps prevent solidification prior to adding to the plates. If the mixture starts to solidify do not add it to the plates, it will distort the visibility of the plaques.
27. The solid overlay option will have better contrast compared to the liquid overlay. Starting day 2 post infection the appearance of faint opaque spots may be seen from the underside of the wells with some virus strains. This can be a good indication of overlay timing. Extended incubation with neutral red beyond 24 h can deteriorate the monolayer.

Acknowledgement

This work was supported in part by a Military Infectious Disease Research Program (MIDRP) awarded to J.W.G. The opinions, interpretations, conclusions, and recommendations presented are those of the authors and are not necessarily endorsed by the U.S. Army.

References

1. Bente DA, Forrester NL, Watts DM, McAuley AJ, Whitehouse CA, Bray M (2013) Crimean-Congo hemorrhagic fever: history, epidemiology, pathogenesis, clinical syndrome and genetic diversity. Antivir Res 100(1):159–189. https://doi.org/10.1016/j.antiviral.2013.07.006
2. Hawman DW, Feldmann H (2023) Crimean-Congo haemorrhagic fever virus. Nat Rev Microbiol 21(7):463–477. https://doi.org/10.1038/s41579-023-00871-9
3. Ergonul O, Celikbas A, Baykam N, Eren S, Dokuzoguz B (2006) Analysis of risk-factors among patients with Crimean-Congo haemorrhagic fever virus infection: severity criteria revisited. Clin Microbiol Infect 12(6): 551–554. https://doi.org/10.1111/j.1469-0691.2006.01445.x
4. Negredo A, de la Calle-Prieto F, Palencia-Herrejon E, Mora-Rillo M, Astray-Mochales J, Sanchez-Seco MP, Bermejo Lopez E, Menarguez J, Fernandez-Cruz A, Sanchez-Artola B, Keough-Delgado E, Ramirez de Arellano E, Lasala F, Milla J, Fraile JL, Ordobas Gavin M, Martinez de la Gandara A, Lopez Perez L, Diaz-Diaz D, Lopez-Garcia MA, Delgado-Jimenez P, Martin-Quiros A, Trigo E, Figueira JC, Manzanares J, Rodriguez-Baena E, Garcia-Comas L, Rodriguez-Fraga O, Garcia-Arenzana N, Fernandez-Diaz MV, Cornejo VM, Emmerich P, Schmidt-Chanasit J, Arribas JR, Crimean Congo Hemorrhagic Fever@Madrid Working G (2017) Autochthonous Crimean-Congo Hemorrhagic fever in Spain. N Engl J

Med 377(2):154–161. https://doi.org/10.1056/NEJMoa1615162

5. Conger NG, Paolino KM, Osborn EC, Rusnak JM, Gunther S, Pool J, Rollin PE, Allan PF, Schmidt-Chanasit J, Rieger T, Kortepeter MG (2015) Health care response to CCHF in US soldier and nosocomial transmission to health care providers, Germany 2009. Emerg Infect Dis 21(1):23–31. https://doi.org/10.3201/eid2101.141413
6. Swanepoel R, Gill DE, Shepherd AJ, Leman PA, Mynhardt JH, Harvey S (1989) The clinical pathology of Crimean-Congo hemorrhagic fever. Rev Infect Dis 11(Suppl 4):S794–S800
7. Burt FJ, Swanepoel R, Shieh WJ, Smith JF, Leman PA, Greer PW, Coffield LM, Rollin PE, Ksiazek TG, Peters CJ, Zaki SR (1997) Immunohistochemical and in situ localization of Crimean-Congo hemorrhagic fever (CCHF) virus in human tissues and implications for CCHF pathogenesis. Arch Pathol Lab Med 121(8):839–846
8. Bente DA, Alimonti JB, Shieh WJ, Camus G, Stroher U, Zaki S, Jones SM (2010) Pathogenesis and immune response of Crimean-Congo hemorrhagic fever virus in a STAT-1 knockout mouse model. J Virol 84(21):11089–11100. https://doi.org/10.1128/JVI.01383-10
9. Zivcec M, Metcalfe MG, Albarino CG, Guerrero LW, Pegan SD, Spiropoulou CF, Bergeron E (2015) Assessment of inhibitors of pathogenic Crimean-Congo Hemorrhagic fever virus strains using virus-like particles. PLoS Negl Trop Dis 9(12):e0004259. https://doi.org/10.1371/journal.pntd.0004259
10. Bereczky S, Lindegren G, Karlberg H, Akerstrom S, Klingstrom J, Mirazimi A (2010) Crimean-Congo hemorrhagic fever virus infection is lethal for adult type I interferon receptor-knockout mice. J Gen Virol 91(Pt 6):1473–1477. https://doi.org/10.1099/vir.0.019034-0
11. Golden JW, Zeng X, Cline CR, Smith JM, Daye SP, Carey BD, Blancett CD, Shoemaker CJ, Liu J, Fitzpatrick CJ, Stefan CP, Garrison AR (2022) The host inflammatory response contributes to disease severity in Crimean-Congo hemorrhagic fever virus infected mice. PLoS Pathog 18(5):e1010485. https://doi.org/10.1371/journal.ppat.1010485
12. Lindquist ME, Zeng X, Altamura LA, Daye SP, Delp KL, Blancett C, Coffin KM, Koehler JW, Coyne S, Shoemaker CJ, Garrison AR, Golden JW (2018) Exploring Crimean-Congo hemorrhagic fever virus-induced hepatic injury using antibody-mediated type I interferon blockade in mice. J Virol. https://doi.org/10.1128/JVI.01083-18
13. Zivcec M, Safronetz D, Scott D, Robertson S, Ebihara H, Feldmann H (2013) Lethal Crimean-Congo hemorrhagic fever virus infection in interferon alpha/beta receptor knockout mice is associated with high viral loads, proinflammatory responses, and coagulopathy. J Infect Dis 207(12):1909–1921. https://doi.org/10.1093/infdis/jit061
14. Garrison AR, Shoemaker CJ, Golden JW, Fitzpatrick CJ, Suschak JJ, Richards MJ, Badger CV, Six CM, Martin JD, Hannaman D, Zivcec M, Bergeron E, Koehler JW, Schmaljohn CS (2017) A DNA vaccine for Crimean-Congo hemorrhagic fever protects against disease and death in two lethal mouse models. PLoS Negl Trop Dis 11(9):e0005908. https://doi.org/10.1371/journal.pntd.0005908
15. Golden JW, Fitzpatrick CJ, Suschak JJ, Clements TL, Ricks KM, Sanchez-Lockhart M, Garrison AR (2023) Induced protection from a CCHFV-M DNA vaccine requires CD8(+) T cells. Virus Res 334:199173. https://doi.org/10.1016/j.virusres.2023.199173
16. Golden JW, Shoemaker CJ, Lindquist ME, Zeng X, Daye SP, Williams JA, Liu J, Coffin KM, Olschner S, Flusin O, Altamura LA, Kuehl KA, Fitzpatrick CJ, Schmaljohn CS, Garrison AR (2019) GP38-targeting monoclonal antibodies protect adult mice against lethal Crimean-Congo hemorrhagic fever virus infection. Sci Adv 5(7):eaaw9535. https://doi.org/10.1126/sciadv.aaw9535
17. Sheehan KC, Lai KS, Dunn GP, Bruce AT, Diamond MS, Heutel JD, Dungo-Arthur C, Carrero JA, White JM, Hertzog PJ, Schreiber RD (2006) Blocking monoclonal antibodies specific for mouse IFN-alpha/beta receptor subunit 1 (IFNAR-1) from mice immunized by in vivo hydrodynamic transfection. J Interf Cytokine Res 26(11):804–819. https://doi.org/10.1089/jir.2006.26.804

Chapter 16

A Review on Reliable and Standardized Animal Models to Study the Pathogenesis of Schmallenberg Virus in Ruminant Natural Host Species

Ludovic Martinelle and Claude Saegerman

Abstract

In the late summer of 2011, the Netherlands reported a cluster of reduced milk yield, fever, and diarrhea in dairy cattle. In March 2012, congenital malformations appeared, and Schmallenberg virus (SBV) was identified, becoming one of the few orthobunyaviruses distributed in Europe. Initially, little was known about the pathogenesis and epidemiology of these viruses in the European context, so assumptions were largely extrapolated from related viruses and other regions worldwide. To study SBV's pathogenesis and its ability to cross the placental barrier, standardized and repeatable models that mimic clinical signs observed in the field are essential. This review discusses some of the latest experimental designs for infectious disease challenges involving SBV, covering infectious doses, routes of infection, inoculum preparation, and origin. Special attention is given to the placental crossing associated with SBV.

Key words Schmallenberg, Culicoides, Vector-borne disease, Experimental challenge, Infection, Arboviruses

1 Introduction

Amongst pathogens, RNA viruses were a major source of emerging diseases during the last 30 years [1]. High mutation rate and in case of segmented genome, reassortment are responsible for genetic adaptability and variability of these viruses.

Schmallenberg virus (SBV) was responsible for a major outbreak in Mainland Europe in the past 15 years. This outbreak was singular in several ways: the disease was previously unknown; the emergence still has unexplained aspects; the virus displayed the ability to cross the placental barrier. Moreover, these events confirmed that Palearctic endemic *Culicoides* species contribute to the spread of SBV and to the epizootic aspect of the diseases.

Until recent nomenclature changes implemented by the International Committee on Taxonomy of Viruses [2–4] Schmallenberg

Hani Boshra (ed.), *Bunyaviruses: Methods and Protocols*, Methods in Molecular Biology, vol. 2893,
https://doi.org/10.1007/978-1-0716-4338-9_16,

virus was part of the Bunyaviridae family, genus Orthobunyavirus, grouped within the serogroup Simbu along with at least 27 other virus species. The members of the Simbu serogroup show cross-reactions to the complement fixation test but are distinguished by seroneutralization [5] and by genetic sequence analysis. Yet still part of the Orthobunyavirus genus, SBV is now considered as a species of its own, *Orthobunyavirus schmallenbergense*, along with Akabane virus (*Orthobunyavirus akabaneense*) and 136 other species [3, 6]. These belong to the new order Elliovirales (formerly Bunyavirales), family Peribunyaviridae (formerly Bunyaviridae), which comprises seven other genera.

By the time of SBV emergence, the virus was lacking pathogenesis and epidemiology data in the European context (naïve ruminants, Palearctic vectors, European climate) and most assumptions were extrapolated based on scientific knowledge on other related viruses and/or other regions of the World.

To study and determine the pathogenesis, the dynamic of SBV, to clarify the ability to cross the placental barrier, standardized, repeatable models ideally displaying field-like outcome are required.

2 Studying the Pathogenesis and Immune Response of SBV in Natural Ruminant Host Species

Experimental infections of mammalian hosts proved to be a highly valuable tool to study pathogenicity, virulence, pathogenesis, and transplacental infections since the dawn of the study of infectious diseases [7]. Design of in vivo models evolved and were usefully complemented with in vitro and in silico approaches to better comprehend the host–pathogen interactions.

To study the pathogenesis of SBV in ruminants, including teratogenic potential, experimental models reproducing the disease had to be found.

To date there are no lab-adapted colonies of Palearctic *Culicoides*. Given the feeding behavior of *Culicoides*, investigating the most adapted route of inoculation is of prime importance to ensure standardization and repeatability of challenge experiments. Amongst other important pathogenesis factors to consider when designing experimental infections, the origin of the inoculum and its passage history has to be carefully evaluated. Indeed, the number of passages, the cell culture system used to grow the inoculum or by contrast, infectious blood or serum source are central to achieve experimental infection matching virological, clinical, and serological parameters of field infection with wild-type viruses.

2.1 A Proper Experimental Infection Needs a Proper Inoculum

An adequate inoculum to use in infectious challenges to study viral pathogenesis should be:

1. Safe, meaning it should have been screened for contaminations, adventitial agents, or other pathogens [8].
2. Easily available, practical and standardized.
3. Contain a virus the closest from wild-type virus found in the field displaying similar replication and virulence properties.

2.2 Infectious Blood Versus Cell-Passaged Inoculum

An infectious inoculum can be used for several purposes, from the investigation of in vivo characteristics of recently discovered viruses [9] to vaccine efficacy requirements or to investigate certain specific aspects of the pathogenesis [10]. In the context of an emerging pathogen with an epizootic potential, it is obvious that time is of the essence, and since standardization is not critical in the very first steps of in vivo characterization, using an infectious animal product such as blood or serum could be both faster and more secure to reproduce expected clinical signs and viremia. Nevertheless, it appears that in most of the most recent experimental infections involving SBV, culture grown inocula were used in most challenges (Table 1). The main reasons to use cell-passaged virus can be summarized as follows:

1. The original isolate or any strain of particular interest can be distributed throughout the world, leading to great improvement of standardization.
2. Viral amplification by cell-passages allows a high increase in viral titer, subsequently allowing to inoculate lower volumes.
3. Screening for contamination or other pathogens is easier in cell culture and eliminate some veterinary public health concerns about using ruminant blood to infect other ruminants.
4. Virulence in cell culture can be easily standardized.

Clinical signs of acute SBV infection in adult cattle are reported to be generally mild and short-lived. These include loss of appetite, hyperthermia, diarrhea, and reduction in milk production [16]. As a matter of fact, regarding Bluetongue Disease Virus (BTV)—another culicoide-borne virus—it was reported that the inoculation of infectious material from field isolates rarely produce a clinical picture as severe as in natural infection [17]. A hypothesis would be that the *Culicoides* saliva might act as a catalyzer to enhance the ability of BTV to produce severe clinical signs [18]. Moreover, in the field and no matter the care and experience of the farmers there are always some animals that would be more sensitive to viral diseases within the herd due to individual poor immune defenses, genetic variability, nutritional aspects or metabolic status. In an experimental context, however, the health and sanitary status of every single animal is very strictly controlled and monitored,

Table 1
Inocula characteristics used in five recent experimental infection studies on SBV (as searched on PubMed with keywords "experimental infection Schmallenberg")

Virus	Type of inoculum	Host Species	Cell type	No. of passages	Inoculation route	Volume (ml)	Titer (TCID50/ml)	References
SBV	Cell-passaged	ECE	KC + BHK-21 + HmLu-1	1 + 5 + 2	Yolk sac	0.2	5 × 10^2–5 × 10^6.4	[11]
SBV	Serum (cattle)	Goats	/	/	SC	1	/	[12]
SBV	Cell-passaged	IFNAR mice	KC + BHK-21	1 + 1	SC	0.1	10^3	[13]
SBV	Cell-passaged	IFNAR mice	KC + BHK-22	1 + 2	SC	0.1	10^8	[14]
SBV	Cell-passaged	Sheep	KC + BHK-21	1 + 1	SC	1	2x10^3	[15]

inducing a bias toward the selection of very healthy and vigorous animals most likely able to face and recover from viral infection.

The milder nature of natural SBV infection in cattle hinders the proper clinical comparison between field and culture-grown inocula. Indeed, in cattle, Wernike et al. reported a reduced viral replication of culture-grown SBV when compared to natural host-passaged inoculum [19]. By contrast, 1 year later the same team concluded to the suitability of both infectious serum and low passage cell culture material for SBV experimental challenges in sheep [20]. Besides the passage history, the origin of the isolated virus seems to be of importance as central nervous system originating virus failed to reproduce RNAemia in inoculated animals [20]. Successive serial passages in cell-culture indeed are well known to usually result in decreased virulence. However, regarding SBV Varela et al. reported an increased pathogenicity in a SBV strain passaged 32 times in INF-incompetent sheep CPT-Tert cells, associated with a faster spread of the virus in the brain of suckling mice [21]. SBV was demonstrated to grow efficiently in several cell lines including sheep CPT-Tert, bovine BFAE, human 293 T, dog MDCK, hamster BHK-21, BSR, KC, and VERO cells [21, 22]. Whereas serial passages in CPT-Tert led to the accumulation of a variety of mutations mostly in the M and S segments, the porcine cell line SK-6 proved to be highly susceptible and to allow the genetic stability of SBV throughout successive passages [23]. Therefore, depending on the cell line used to grow SBV, serial passages can lead to attenuation, increased virulence, or efficient propagation with a low frequency of nucleotide exchanges.

2.3 A Matter of Doses and Routes

When it comes to arboviruses, the choice of the route of inoculation can be driven by two main considerations:

1. The need for a route that will mimicry the most the behavior of the vector in the wild. Usually, hematophagous arthropods are either telmophagous or solenophagous; depending on the vector species the route might be intradermal (ID), subcutaneous (SC), or intravenous (IV). In experimental infections given the size of the arthropods and the size of their mouthparts the inoculated viral load and volume are usually way higher than the ones inoculated through naturally occurring feeding [24]. Another drawback is the lack of vector saliva components, which can modify the structure and infectivity of *Bunyaviridae* viral particles [25].
2. The need for a route that will ensure the virus to reach blood stream. Quite obviously this is the intravenous route. Since vector saliva components can enhance the infectivity of arboviruses there is a risk that the inoculation of the virus alone or at a distal site from the vector feeding site would result in a failed

infection [26]. Therefore, the option to by-pass the skin to readily reach the bloodstream may be relevant.

Several authors used mixed routes to overcome the respective disadvantages of each approach (Table 1). In a previous study, we compared intranasal, intradermal, and subcutaneous routes for experimental infections with SBV [27]. Intradermal is an interesting yet underused route: indeed most hematophagous arthropods do not make it through the skin and their mouthparts only allow them to feed intradermally. Most of the cellular and fluid exchanges between the skin and the blood do occur in the dermis [28]. In addition, there are some evidences suggesting that intradermal inoculation can be more appropriate to reproduce many aspects of natural infection, including clinical disease, viral and immune responses [29]. However, to perform an actual intradermal inoculation the volume to be injected has to be limited, the dermis being mostly composed of a network of collagen fibres. Therefore, it is required to multiply the inoculation sites to reach the desired total inoculum volume and infectious titer. To realize the inoculation itself the most practical tools are the Dermojet® (Akra Dermojet) or a special syringe for intradermal injections (used to perform bovine tuberculosis skin tests as an example). These devices allow usually volumes between 0.1 and 0.4 mL, thus the need of multiple injections to reach the common 1–4 mL inoculation volume used in ruminant infectious challenges experiments (Table 1). Moreover, with both systems the inoculum must be transferred from its original vial to a small tank part of the body of the dermojet or to a special cartridge to be used with the intradermal syringe. This extra step increases the number of handlings, which should be limited especially in the case of BSL3 pathogens. The intranasal route was mostly investigated to test whether a potential direct contamination between sheep could be achieved.

Broadly speaking, several authors reported a direct link between the inoculated viral doses and the onset of clinical signs and viremia, i.e. the higher the dose the sooner the clinical signs and viral RNA detection [30–32]. In another study we evaluated four tenfold dilutions of a SBV infectious serum inoculum on ewes [33]. The undiluted original inoculum had a titer of 2×10^3 $TCID_{50}$/mL. It appears there is a critical dose to be inoculated to successfully reproduce field-like virological and immunological parameters, and once this threshold is over there no dose-dependent effect anymore. Indeed, in the productively infected animals no statistical differences between the different inoculation doses were found in the duration or quantity of viral RNA circulating in blood, nor in the amount of viral RNA present in virus positive lymphoid organs.

Inoculation by the bite of Culicoides was reported to be more efficient than intradermal inoculation, especially by delaying the early immune response of the host despite a generally lower inoculated viral dose when compared to needle inoculation [34]. Pharmacological agents contained in Culicoides saliva might affect the host's immune response by anti-proliferative effects on leucocytes [35] or a reduced INF alpha/beta expression, as demonstrated with vesicular stomatitis virus and mosquito saliva [36].

Nonetheless, the use of *Culicoides* to perform experimental challenges remains highly limited by practical constraints: to date besides *C. nubeculosus, C. riethi,* and *C. sonorensis* no other *Culicoides* species was successfully established as lab-adapted colonies [37, 38], the alternative being insects caught in the wild. In addition, prior to the infectious challenge on the ruminant host the infection of culicoides is particularly tricky given the size of the insect and the exact amount of virus delivered to each ruminant cannot be known.

Altogether the subcutaneous route seems to represent the best compromise for SBV experimental challenges. The dose itself has to be sufficient but there is no gain in using massive viral load.

2.4 Beware of Contaminations!

History of experimental challenges with RNA viruses is scarred with incidents of contamination of biological samples. For instance BTV modified live vaccines or inocula were reported to be contaminated in several occurrences [39–45].

So far literature does not report experimental infections with an SBV inoculum that was contaminated by another virus belonging to the same or a different family. Broadly speaking, contamination routes are most likely related to (i) laboratory contamination during sample preparation or (ii) natural multiple infection of the original donor animal [46]. Given the potential dramatic consequences of such contamination incidents, inocula should indeed be tested for major pathogens affecting the host species used in challenge experiments. Extensive screening could however be considered on a case-by-case basis.

3 Placental Crossing and Teratogenesis

Peribunyaviridae is a family of viruses able to cross the placental barrier, infect the fetus, and potentially cause teratogenic effects in the central nervous system or muskuloskeletal defects (arthrogryposis) [47]. It is generally considered that the age of the fetus (in gestational age) is a key criterion to determine the extent of the congenital defects as differentiating nervous tissues are important targets for SBV: usually the younger the fetus, the more severe the lesions [47].

However, to colonize the fetus viruses need a way in; therefore it is considered that SBV in utero infection can only occur once the first placentomes are established, around day 30 of pregnancy in cattle and slightly earlier in sheep [48–50].

Small ruminants and cow have slightly different definitive placental structure. In 1909 Grosser classified the mammalian placentas according to the number of tissue layers between fetal and maternal blood after implantation. The ruminants were considered to have a syndesmochorial placenta, i.e. where the uterine epithelium is removed and the chorion is in contact with the maternal connective tissue [51]. Other authors reclassified it as epitheliochorial since the uterine epithelium subsisted. From these historical and oversimplified categories, the ruminant placenta was later considered to be characterized by the migration of the fetal chorionic binucleate cells and their fusion with the uterine epithelial cells. The ruminant placenta is a mix between the syndesmochorial and epitheliochorial placentation and the uterine epithelium subsists as a fetomaternal syncytium [52]. In small ruminants more than 95% of the placenta forms a fetomaternal syncytium. In cow multinucleate cells appear only transiently once the uterine epithelium has regrown after implantation. This type of placenta is called synepitheliochorial [53].

At implantation several changes occur: the papillae in the uterine glands immobilize the conceptus and it starts to elongate (cow: 15 days post coitum (dpc); sheep: 13–16 dpc). Subsequently the cells of the trophectoderm and the uterine epithelium get interdigitated, binucleate cells start to be seen. Then binucleate cells start to differentiate and to migrate (cow: 20–22 dpc; sheep: 16–18 dpc). Fetal villi develop in the caruncular areas starting at 24–26 dpc in small ruminants and 28–30 dpc in cow, thus defining the end of the implantation and the start of the placental development [54]. Table 2 summarizes some of the essential events during the prenatal development in cattle and sheep.

In a previous study we challenged pregnant ewes with SBV at 45 and 60 days of pregnancy [55]. We reported the birth of 22 alive lambs, amongst which none had any anti-SBV neutralizing antibodies prior to colostrum intake.

Thus, the experimental infection took place within the critical timeframe, between 30 and 70 days of pregnancy for sheep (Fig. 1, [50]). The prenatal period can be divided into four main periods: (i) fertilization, (ii) blastogenesis, (iii) embryogenesis, and (iv) fetogenesis [56]. The embryo sprouts and develops tissues and organ structures from the three original germ layers (ecto-, meso-, and endoderm). By the end of the embryogenesis the conceptus became a "miniature" version of the adult animal, displaying all its specific features. Once the organs are differentiated the embryo becomes a fetus [57]. The fetal phase is characterized by a fast growth of the conceptus. In cattle and sheep the fetogenesis starts around 45 and 38 dpc, respectively [58]. Therefore, the critical timeframe for SBV infection overlaps the end of the embryo

Table 2
Key events in sheep and cow embryos/fetuses with particular emphasis on nervous and immune systems. Compiled from [58, 76–81]

Event	Timing in cow (dpc)	Timing in sheep (dpc)
Blastocyst hatching from zona pellucida	9	9
Elongation of the blastocyst, establishment of the primitive streak, emergence of the notochord	17–18	13–14
Appearance of neural folds, closure of the neural groove	17–19	15–16
Implantation begins	16–19	15–18
Neurula	20–21	17
Neural tube complete; optic and otic vesicles present	21–23	19–20
Placentation begins	22–23	17–22
Three brain vesicles visible	24–25	17
Placentoma are detectable	32–36	21
Lymphoid development of the thymus	42	36
Spleen development	55	43–44
Peripheral lymph nodes	60	45
IgM containing cells	59	65
Myelin sheath acquisition (starting)	60	54–63
IgG containing cells	145	87

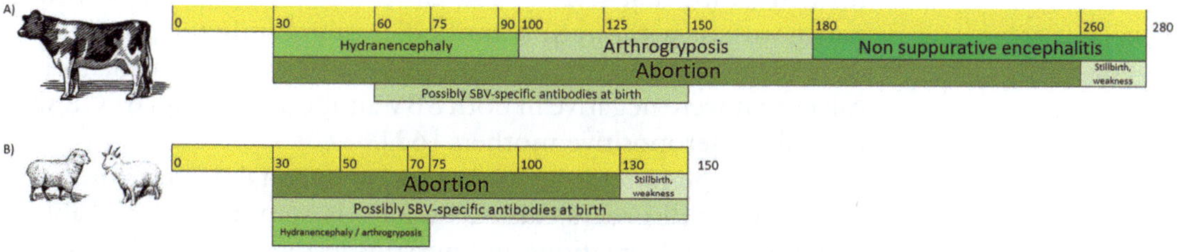

Fig. 1 In utero potential SBV-induced defects following infection of the pregnant dams along the whole gestation time for cattle (**a**) and small ruminants (**b**) [50, 55, 65, 74, 75]

stage and the beginning of the fetal stage. Moreover, although in ruminants γ-globulins are unable to go through the placental barrier from the mother to the fetus it is admitted that sheep fetuses become sequentially and increasingly immunocompetent to a larger variety of antigens throughout the pregnancy [59, 60]. The critical timeframe for SBV infection spans over the course of several important events in immune system development (Table 2). Although the sequence of antigens to be successively and progressively

Table 3
Summary of some of the most common central nervous and musculoskeletal lesions following in utero infection with bovine virus diarrhea virus (BVDV), SBV, BTV, Akabane virus (AKAV), or Aino virus (AV)

Lesion	Definition	BVDV	SBV	BTV	AKAV/AV
Hydranencephaly	Extensive loss of cerebral tissue with replacement by clear fluid	+	+	+	+
Porencephaly	Cystic fluid filled cavities in the brain tissue	+	+	+	+
Hydrocephalus	Dilation of the lateral ventricles by cerebrospinal fluid	+	+	+	−
Microencephaly	Reduced size of the cerebrum	+	+	+	+
Cerebellar hypoplasia	Reduced size of the cerebellum	+	+	+	
Kyphosis	Dorsal vertebral column curvature	−	+	−	−
Lordosis	Ventral vertebral column curvature	−	+	−	−
Scoliosis	Lateral vertebral column curvature	−	+	−	−
Torticollis	Twisted cervical vertebral column curvature	−	+	−	−
Arthrogryposis	Joint contraction of the limbs	−	+	±	+

Adapted from [82]

recognized by the fetal ruminant through pregnancy seems to be quite conserved between individuals, these antigens can be recognized starting with a difference of a few days between individuals [61]. This individual variability could explain the findings of malformed lambs that were found either SBV viropositive or vironegative with or without SBV antibodies, suggesting the possibility of an in utero clearance of the virus. Moreover, most of the malformed calves that were negative in both SBV antibodies and RTqPCR were born from seropositive mothers [62].

The range of teratogenic lesions, congenital defects, and other reproductive disorders caused by SBV is summarized in Table 3, along with malformations and nervous lesions induced by some of the most common viruses inducing such lesions in ruminants.

In our experiments timing of inoculation was optimal to achieve transplacental infection of the fetus regarding data available from the literature, yet no malformations could be seen. No antibodies against the virus used to infect the mothers could be detected as well. These striking results might even question the success of the infection, notwithstanding the positive RNA detection in the mothers. Our report of lesions and serology results, similar to another experiment on goats [63], in addition to the detection of SBV nucleic acids in organs of several lambs and many extraembryonic structures [55] provide support to an actual transplacental infection. In addition, in another study [64] we managed to isolate SBV from fetal envelopes in the animals at birth, thus 90 and 105 days post infection. The very low ratio of precolostral seroconversion in immunocompetent fetuses was also confirmed following the infection of pregnant cattle with SBV [65].

SBV vertical transmission rate seems to be lower when compared for instance to BTV, especially in cattle [66]. The rate of malformations caused by SBV was reported to be about 0.5% in cattle [67] although the rate of intrauterine infection—based on serological results of the calves prior to colostrum intake—was reported to be about 28% [48]. Other authors documented field data about congenital malformations affecting 3% of the calves but 8–10% of the lambs in farms at the beginning of the SBV epizootic [68, 69]. In Belgium based on a survey targeting farmers we also found an estimated 10% of malformed sheep in SBV positive flocks [70].

4 Final Thoughts and Prospects

The number of ruminants used in experimental infections is chosen based on welfare and statistical concerns but also quite unfortunately on economic and practical grounds [7]. We performed our experimental infections with SBV in the BSL2+/BSL3 facilities of Sciensano (Ukkel, Belgium) depending on the phase of the experiment. To date, the Belgian Service of Biosafety and Biotechnology as well as the Belgian law do not assign a particular pathogen class for SBV. Our biosafety measures for SBV were based on the analogy with AKAV, also classified as a class 3 pathogen [71]. Domestic ruminants being herd animals, they need to be housed in groups or at least not individually. Euthanasia methods have to be the most humane as possible and clear end points have to be defined. Given the scarcity of clinical signs caused by SBV in the field and the individual variations in the response to the infection the number of animals to be included has to be chosen very carefully to comply with the Reduction objective (Three Rs concept). Nonetheless the number of animals has to sufficient to limit the risk of not being able to provide useful data in the context of the ongoing scientific investigation. This is particularly difficult for experimental infection of pregnant ruminants with low malformation rates following transplacental transmission.

The most objective parameter to assess a vaccine efficacy against a virus and especially a RNA vector-borne virus is the evaluation of the viral RNA detection by RTqPCR in the host target [72]. SBV virulence was demonstrated to vary depending on the ruminant host whether it is cattle, sheep, or goat. In addition, pregnancy length differs between cattle and small ruminants while the placentation and the development of the fetal envelopes present slight differences [73]. Consequently, to study any of the specific aspects related to a ruminant species there are no other animal models or any alternative able to mimic the natural situation in a proper way [7].

According to our experiments, the subcutaneous route with an inoculum passaged a limited number of times on cell culture seems to represent the best compromise between a high probability to reproduce an infection similar to what happens in the field and logistics concerning the preparation/storage/management of the inoculum. To prevent the loss of viral variability and limit the risks of attenuation, SBV could benefit from an isolation on the highly susceptible SK-6 cell line [23]. Screening for concomitant pathogens should be considered on a case by case basis, if required. The dose should be chosen based on literature data yet no advantage is provided by inoculating a massive viral load.

Since some data from other authors suggest a better reproduction of the diseases with intradermal inoculation, it could be further investigated, especially if more user-friendly devices would be available. A breakthrough would be the successful adaptation of a colony of Palearctic SBV vector *Culicoides* species (*C. obsoletus/scoticus, pulicaris*) to laboratory conditions and subsequent use in infectious challenges.

In conclusion, targeting ruminant host species in experimental infections especially with BSL3 *Culicoides* borne pathogens is very expensive, time-consuming, and subject to stringent animal welfare constraints and critical sample size analysis to meet optimal statistical requirements. However, ruminant model remains unavoidable to assess the disease impact and to study the pathogenesis of emerging vector-borne viruses.

References

1. Carrasco-Hernandez R, Jacome R, Lopez Vidal Y, Ponce de Leon S (2017) Are RNA viruses candidate agents for the next global pandemic? A review. ILAR J 58(3):343–358. https://doi.org/10.1093/ilar/ilx026
2. Adams MJ, Lefkowitz EJ, King AMQ, Harrach B, Harrison RL, Knowles NJ, Kropinski AM, Krupovic M, Kuhn JH, Mushegian AR, Nibert M, Sabanadzovic S, Sanfacon H, Siddell SG, Simmonds P, Varsani A, Zerbini FM, Gorbalenya AE, Davison AJ (2017) Changes to taxonomy and the international code of virus classification and nomenclature ratified by the international committee on taxonomy of viruses. Arch Virol 162(8):2505–2538. https://doi.org/10.1007/s00705-017-3358-5
3. ICTV (2024) History of the taxon: species: Orthobunyavirus schmallenbergense (2023 Release, MSL #39). https://ictv.global/taxonomy/taxondetails?taxnode_id=202300126&taxon_name=Orthobunyavirus%20schmallenbergense. Accessed 02/07/2024
4. Hughes HR, Adkins S, Alkhovskiy S, Beer M, Blair C, Calisher CH, Drebot M, Lambert AJ, de Souza WM, Marklewitz M, Nunes MRT, Shi 石晓宏 X, Ictv Report C (2020) ICTV virus taxonomy profile: peribunyaviridae. J Gen Virol 101(1):1–2. https://doi.org/10.1099/jgv.0.001365
5. Kinney RM, Calisher CH (1981) Antigenic relationships among Simbu serogroup (Bunyaviridae) viruses. Am J Trop Med Hyg 30(6):1307–1318
6. De Regge N (2017) Akabane, Aino and Schmallenberg virus-where do we stand and what do we know about the role of domestic ruminant hosts and Culicoides vectors in virus

transmission and overwintering? Curr Opin Virol 27:15–30. https://doi.org/10.1016/j.coviro.2017.10.004

7. Coetzee P, van Vuuren M, Venter EH, Stokstad M (2014) A review of experimental infections with bluetongue virus in the mammalian host. Virus Res 182:21–34. https://doi.org/10.1016/j.virusres.2013.12.044
8. Speder B (2014) Regulatory requirements for viral-challenge studies: influenza case study. London
9. Breard E, Schulz C, Sailleau C, Bernelin-Cottet C, Viarouge C, Vitour D, Guillaume B, Caignard G, Gorlier A, Attoui H, Gallois M, Hoffmann B, Zientara S, Beer M (2018) Bluetongue virus serotype 27: experimental infection of goats, sheep and cattle with three BTV-27 variants reveal atypical characteristics and likely direct contact transmission BTV-27 between goats. Transbound Emerg Dis 65(2):e251–e263. https://doi.org/10.1111/tbed.12780
10. MacLachlan NJ, Nunamaker RA, Katz JB, Sawyer MM, Akita GY, Osburn BI, Tabachnick WJ (1994) Detection of bluetongue virus in the blood of inoculated calves: comparison of virus isolation, PCR assay, and in vitro feeding of Culicoides variipennis. Arch Virol 136(1–2):1–8
11. Collins AB, Mee JF, Kirkland PD (2018) Pathogenicity and teratogenicity of Schmallenberg virus and Akabane virus in experimentally infected chicken embryos. Vet Microbiol 216:31–37. https://doi.org/10.1016/j.vetmic.2018.01.009
12. Laloy E, Breard E, Trapp S, Pozzi N, Riou M, Barc C, Breton S, Delaunay R, Cordonnier N, Chateau-Joubert S, Crochet D, Gouzil J, Hebert T, Raimbourg M, Viarouge C, Vitour D, Durand B, Ponsart C, Zientara S (2017) Fetopathic effects of experimental Schmallenberg virus infection in pregnant goats. Vet Microbiol 211:141–149. https://doi.org/10.1016/j.vetmic.2017.10.011
13. Tauscher K, Wernike K, Fischer M, Wegelt A, Hoffmann B, Teifke JP, Beer M (2017) Characterization of Simbu serogroup virus infections in type I interferon receptor knock-out mice. Arch Virol 162(10):3119–3129. https://doi.org/10.1007/s00705-017-3475-1
14. Boshra HY, Charro D, Lorenzo G, Sanchez I, Lazaro B, Brun A, Abrescia NG (2017) DNA vaccination regimes against Schmallenberg virus infection in IFNAR(-/-) mice suggest two targets for immunization. Antiviral Res 141:107–115. https://doi.org/10.1016/j.antiviral.2017.02.013
15. Poskin A, Verite S, Comtet L, Van der Stede Y, Cay B, De Regge N (2015) Persistence of the protective immunity and kinetics of the isotype specific antibody response against the viral nucleocapsid protein after experimental Schmallenberg virus infection of sheep. Vet Res 46:119. https://doi.org/10.1186/s13567-015-0260-6
16. Dagnaw M, Solomon A, Dagnew B (2024) Serological prevalence of the Schmallenberg virus in domestic and wild hosts worldwide: a systematic review and meta-analysis. Front Vet Sci 11:1371495. https://doi.org/10.3389/fvets.2024.1371495
17. Drolet BS, Reister LM, Lehiy CJ, Van Rijn PA, Bowen RA (2015) Effect of Culicoides sonorensis salivary proteins on clinical disease outcome in experimental bluetongue virus serotype 8 infection of Dorset sheep. Vet Ital 51(4):379–384. https://doi.org/10.12834/VetIt.496.2398.1
18. Darpel KE, Langner KF, Nimtz M, Anthony SJ, Brownlie J, Takamatsu HH, Mellor PS, Mertens PP (2011) Saliva proteins of vector Culicoides modify structure and infectivity of bluetongue virus particles. PLoS One 6(3):e17545. https://doi.org/10.1371/journal.pone.0017545
19. Wernike K, Eschbaumer M, Breithaupt A, Hoffmann B, Beer M (2012) Schmallenberg virus challenge models in cattle: infectious serum or culture-grown virus? Vet Res 43:84. https://doi.org/10.1186/1297-9716-43-84
20. Wernike K, Hoffmann B, Breard E, Botner A, Ponsart C, Zientara S, Lohse L, Pozzi N, Viarouge C, Sarradin P, Leroux-Barc C, Riou M, Laloy E, Breithaupt A, Beer M (2013) Schmallenberg virus experimental infection of sheep. Vet Microbiol 166(3–4):461–466. https://doi.org/10.1016/j.vetmic.2013.06.030
21. Varela M, Schnettler E, Caporale M, Murgia C, Barry G, McFarlane M, McGregor E, Piras IM, Shaw A, Lamm C, Janowicz A, Beer M, Glass M, Herder V, Hahn K, Baumgartner W, Kohl A, Palmarini M (2013) Schmallenberg virus pathogenesis, tropism and interaction with the innate immune system of the host. PLoS Pathog 9(1):e1003133. https://doi.org/10.1371/journal.ppat.1003133
22. Kraatz F, Wernike K, Hechinger S, Konig P, Granzow H, Reimann I, Beer M (2015) Deletion mutants of Schmallenberg virus are avirulent and protect from virus challenge. J Virol 89(3):1825–1837. https://doi.org/10.1128/JVI.02729-14
23. Hofmann MA, Mader M, Fluckiger F, Renzullo S (2015) Genetic stability of

Schmallenberg virus in vivo during an epidemic, and in vitro, when passaged in the highly susceptible porcine SK-6 cell line. Vet Microbiol 176(1–2):97–108. https://doi.org/10.1016/j.vetmic.2015.01.010

24. Venter GJ, Mellor PS, Wright I, Paweska JT (2007) Replication of live-attenuated vaccine strains of bluetongue virus in orally infected South African Culicoides species. Med Vet Entomol 21(3):239–247. https://doi.org/10.1111/j.1365-2915.2007.00687.x

25. Horne KM, Vanlandingham DL (2014) Bunyavirus-vector interactions. Viruses 6(11):4373–4397. https://doi.org/10.3390/v6114373

26. Le Coupanec A, Babin D, Fiette L, Jouvion G, Ave P, Misse D, Bouloy M, Choumet V (2013) Aedes mosquito saliva modulates Rift Valley fever virus pathogenicity. PLoS Negl Trop Dis 7(6):e2237. https://doi.org/10.1371/journal.pntd.0002237

27. Martinelle L, Poskin A, Dal Pozzo F, Mostin L, Van Campe W, Cay AB, De Regge N, Saegerman C (2017) Three different routes of inoculation for experimental infection with Schmallenberg virus in sheep. Transbound Emerg Dis 64(1):305–308. https://doi.org/10.1111/tbed.12356

28. Nicolas JF, Guy B (2008) Intradermal, epidermal and transcutaneous vaccination: from immunology to clinical practice. Expert Rev Vaccines 7(8):1201–1214. https://doi.org/10.1586/14760584.7.8.1201

29. Umeshappa CS, Singh KP, Channappanavar R, Sharma K, Nanjundappa RH, Saxena M, Singh R, Sharma AK (2011) A comparison of intradermal and intravenous inoculation of bluetongue virus serotype 23 in sheep for clinico-pathology, and viral and immune responses. Vet Immunol Immunopathol 141(3–4):230–238. https://doi.org/10.1016/j.vetimm.2011.03.005

30. Alexandersen S, Quan M, Murphy C, Knight J, Zhang Z (2003) Studies of quantitative parameters of virus excretion and transmission in pigs and cattle experimentally infected with foot-and-mouth disease virus. J Comp Pathol 129(4):268–282

31. Howey R, Quan M, Savill NJ, Matthews L, Alexandersen S, Woolhouse M (2009) Effect of the initial dose of foot-and-mouth disease virus on the early viral dynamics within pigs. J R Soc Interface 6(39):835–847. https://doi.org/10.1098/rsif.2008.0434

32. Quan M, Murphy CM, Zhang Z, Alexandersen S (2004) Determinants of early foot-and-mouth disease virus dynamics in pigs. J Comp Pathol 131(4):294–307. https://doi.org/10.1016/j.jcpa.2004.05.002

33. Poskin A, Martinelle L, Mostin L, Van Campe W, Dal Pozzo F, Saegerman C, Cay AB, De Regge N (2014) Dose-dependent effect of experimental Schmallenberg virus infection in sheep. Vet J 201(3):419–422. https://doi.org/10.1016/j.tvjl.2014.05.031

34. Pages N, Breard E, Urien C, Talavera S, Viarouge C, Lorca-Oro C, Jouneau L, Charley B, Zientara S, Bensaid A, Solanes D, Pujols J, Schwartz-Cornil I (2014) Culicoides midge bites modulate the host response and impact on bluetongue virus infection in sheep. PLoS One 9(1):e83683. https://doi.org/10.1371/journal.pone.0083683

35. Bishop JV, Mejia JS, Perez de Leon AA, Tabachnick WJ, Titus RG (2006) Salivary gland extracts of Culicoides sonorensis inhibit murine lymphocyte proliferation and no production by macrophages. Am J Trop Med Hyg 75(3):532–536

36. Limesand KH, Higgs S, Pearson LD, Beaty BJ (2003) Effect of mosquito salivary gland treatment on vesicular stomatitis New Jersey virus replication and interferon alpha/beta expression in vitro. J Med Entomol 40(2):199–205

37. Boorman J (1974) The maintenance of laboratory colonies of Culicoides variipennis (Coq.), C. nubeculosus (Mg.) and C. riethi Kieff. (Diptera, Ceratopogonidae). Bull Entomol Res 64(3):371–377. https://doi.org/10.1017/S0007485300031254

38. Veronesi E, Antony F, Gubbins S, Golding N, Blackwell A, Mertens PP, Brownlie J, Darpel KE, Mellor PS, Carpenter S (2013) Measurement of the infection and dissemination of bluetongue virus in culicoides biting midges using a semi-quantitative rt-PCR assay and isolation of infectious virus. PLoS One 8(8): e70800. https://doi.org/10.1371/journal.pone.0070800

39. Evermann JF, McKeirnan AJ, Wilbur LA, Levings RL, Trueblood ES, Baldwin TJ, Hughbanks FG (1994) Canine fatalities associated with the use of a modified live vaccine administered during late stages of pregnancy. J Vet Diagn Invest 6(3):353–357

40. Wilbur LA, Evermann JF, Levings RL, Stoll IR, Starling DE, Spillers CA, Gustafson GA, McKeirnan AJ (1994) Abortion and death in pregnant bitches associated with a canine vaccine contaminated with bluetongue virus. J Am Vet Med Assoc 204(11):1762–1765

41. Breard E, Belbis G, Hamers C, Moulin V, Lilin T, Moreau F, Millemann Y, Montange C, Sailleau C, Durand B, Desprat A, Viarouge C, Hoffmann B, de

Smit H, Goutebroze S, Hudelet P, Zientara S (2011) Evaluation of humoral response and protective efficacy of two inactivated vaccines against bluetongue virus after vaccination of goats. Vaccine 29(13):2495–2502. https://doi.org/10.1016/j.vaccine.2010.12.105

42. Martinelle L, Dal Pozzo F, Sarradin P, Van Campe W, De Leeuw I, De Clercq K, Thys C, Thiry E, Saegerman C (2016) Experimental bluetongue virus superinfection in calves previously immunized with bluetongue virus serotype 8. Vet Res 47(1):73. https://doi.org/10.1186/s13567-016-0357-6
43. Eschbaumer M, Wackerlin R, Savini G, Zientara S, Sailleau C, Breard E, Beer M, Hoffmann B (2011) Contamination in bluetongue virus challenge experiments. Vaccine 29(26):4299–4301. https://doi.org/10.1016/j.vaccine.2011.04.049
44. Dal Pozzo F, Martinelle L, Thys C, Sarradin P, De Leeuw I, Van Campe W, De Clercq K, Thiry E, Saegerman C (2013) Experimental co-infections of calves with bluetongue virus serotypes 1 and 8. Vet Microbiol 165(1–2):167–172. https://doi.org/10.1016/j.vetmic.2013.01.016
45. Rasmussen LD, Savini G, Lorusso A, Bellacicco A, Palmarini M, Caporale M, Rasmussen TB, Belsham GJ, Botner A (2013) Transplacental transmission of field and rescued strains of BTV-2 and BTV-8 in experimentally infected sheep. Vet Res 44:75. https://doi.org/10.1186/1297-9716-44-75
46. Vandenbussche F, Sailleau C, Rosseel T, Desprat A, Viarouge C, Richardson J, Eschbaumer M, Hoffmann B, De Clercq K, Breard E, Zientara S (2015) Full-genome sequencing of four bluetongue virus serotype 11 viruses. Transbound Emerg Dis 62(5):565–571. https://doi.org/10.1111/tbed.12178
47. Doceul V, Lara E, Sailleau C, Belbis G, Richardson J, Breard E, Viarouge C, Dominguez M, Hendrikx P, Calavas D, Desprat A, Languille J, Comtet L, Pourquier P, Eleouet JF, Delmas B, Marianneau P, Vitour D, Zientara S (2013) Epidemiology, molecular virology and diagnostics of Schmallenberg virus, an emerging orthobunyavirus in Europe. Vet Res 44:31. https://doi.org/10.1186/1297-9716-44-31
48. Garigliany MM, Bayrou C, Kleijnen D, Cassart D, Desmecht D (2012) Schmallenberg virus in domestic cattle, Belgium, 2012. Emerg Infect Dis 18(9):1512–1514. https://doi.org/10.3201/eid1809.120716
49. Parsonson IM, McPhee DA, Della-Porta AJ, McClure S, McCullagh P (1988) Transmission of Akabane virus from the ewe to the early fetus (32 to 53 days). J Comp Pathol 99(2):215–227
50. Charles JA (1994) Akabane virus. Vet Clin North Am Food Anim Pract 10(3):525–546
51. Amoroso EC (1961) Histology of the placenta. Br Med Bull 17:81–90
52. Wooding FB (1992) Current topic: the synepitheliochorial placenta of ruminants: binucleate cell fusions and hormone production. Placenta 13(2):101–113
53. Wooding P, Burton G (2008) Comparative placentation. Springer, Cambridge
54. King GJ, Atkinson BA, Robertson HA (1979) Development of the bovine placentome during the second month of gestation. J Reprod Fertil 55(1):173–180
55. Martinelle L, Poskin A, Dal Pozzo F, De Regge N, Cay B, Saegerman C (2015) Experimental infection of sheep at 45 and 60 days of gestation with Schmallenberg virus readily led to placental colonization without causing congenital malformations. PLoS One 10(9): e0139375. https://doi.org/10.1371/journal.pone.0139375
56. Szabo KT (1989) Congenital malformations in laboratory and farm animals. Academic Press
57. Coppock RW, Dziwenka MM (2017) Chapter 72 – teratogenesis in livestock. In: Press A (ed) Reproductive and developmental toxicology, 2nd edn. Elsevier, pp 1391–1408. https://doi.org/10.1016/B978-0-12-804239-7.00072-X
58. Evans HE, Sack WO (1973) Prenatal development of domestic and laboratory mammals: growth curves, external features and selected references. Zentralbl Veterinarmed C 2(1):11–45
59. Schultz RD, Dunne HW, Heist CE (1973) Ontogeny of the bovine immune response. Infect Immun 7(6):981–991
60. Silverstein AM, Uhr JW, Kraner KL, Lukes RJ (1963) Fetal response to antigenic stimulus. II. Antibody production by the fetal lamb. J Exp Med 117:799–812
61. Fahey KJ, Morris B (1978) Humoral immune responses in foetal sheep. Immunology 35(4):651–661
62. De Regge N, van den Berg T, Georges L, Cay B (2013) Diagnosis of Schmallenberg virus infection in malformed lambs and calves and first indications for virus clearance in the fetus. Vet Microbiol 162(2–4):595–600. https://doi.org/10.1016/j.vetmic.2012.11.029
63. Coetzee P, Stokstad M, Myrmel M, Mutowembwa P, Loken T, Venter EH, Van Vuuren M (2013) Transplacental infection in

goats experimentally infected with a European strain of bluetongue virus serotype 8. Vet J 197(2):335–341. https://doi.org/10.1016/j.tvjl.2013.01.005

64. Poskin A, Martinelle L, Van der Stede Y, Saegerman C, Cay B, De Regge N (2017) Genetically stable infectious Schmallenberg virus persists in foetal envelopes of pregnant ewes. J Gen Virol 98(7):1630–1635. https://doi.org/10.1099/jgv.0.000841
65. Consortium E (2014) Schmallenberg virus Technical and scientific studies. Final report. Wageningen
66. Wernike K, Elbers A, Beer M (2015) Schmallenberg virus infection. Rev Sci Tech 34(2):363–373
67. Veldhuis AM, Carp-van Dijken S, van Wuijckhuise L, Witteveen G, van Schaik G (2014) Schmallenberg virus in Dutch dairy herds: potential risk factors for high within-herd seroprevalence and malformations in calves, and its impact on productivity. Vet Microbiol 168(2–4):281–293. https://doi.org/10.1016/j.vetmic.2013.11.021
68. Dominguez M, Gache K, Touratier A, Perrin JB, Fediaevsky A, Collin E, Breard E, Sailleau C, Viarouge C, Zanella G, Zientara S, Hendrikx P, Calavas D (2014) Spread and impact of the Schmallenberg virus epidemic in France in 2012-2013. BMC Vet Res 10:248. https://doi.org/10.1186/s12917-014-0248-x
69. Dominguez M, Hendrikx P, Zientara S, Calavas D, Jay M, Touratier A, Languille J, Fediaevsky A (2012) Preliminary estimate of Schmallenberg virus infection impact in sheep flocks – France. Vet Rec 171(17):426. https://doi.org/10.1136/vr.100883
70. Saegerman C, Martinelle L, Dal Pozzo F, Kirschvink N (2014) Preliminary survey on the impact of Schmallenberg virus on sheep flocks in south of Belgium. Transbound Emerg Dis 61(5):469–472. https://doi.org/10.1111/tbed.12047
71. Belgian Biosafety Server (2018) Service Biosafety and Biotechnology (SBB) https://www.biosafety.be/content/tools-belgian-classification-micro-organisms-based-their-biological-risks. Accessed 15/08/2018
72. Eschbaumer M, Wackerlin R, Rudolf M, Keller M, Konig P, Zemke J, Hoffmann B, Beer M (2010) Infectious blood or culture-grown virus: a comparison of bluetongue virus challenge models. Vet Microbiol. S0378-1135(10)00227-0 [pii]. https://doi.org/10.1016/j.vetmic.2010.05.004
73. Spencer TE, Forde N, Lonergan P (2016) Insights into conceptus elongation and establishment of pregnancy in ruminants. Reprod Fertil Dev 29(1):84–100. https://doi.org/10.1071/RD16359
74. Martinelle L, Dal Pozzo F, Kirschvink N, De la Grandiere MA, Thiry E, Saegerman C (2012) Le virus Schmallenberg ou l'émergence du premier Orthobunyavirus du sérogroupe Simbu en Europe. Ann Med Vet 156:7–24
75. Afonso A, Abrahantes JC, Conraths F, Veldhuis A, Elbers A, Roberts H, Van der Stede Y, Meroc E, Gache K, Richardson J (2014) The Schmallenberg virus epidemic in Europe-2011-2013. Prev Vet Med 116(4):391–403. https://doi.org/10.1016/j.prevetmed.2014.02.012
76. Assis Neto AC, Pereira FT, Santos TC, Ambrosio CE, Leiser R, Miglino MA (2010) Morpho-physical recording of bovine conceptus (Bos indicus) and placenta from days 20 to 70 of pregnancy. Reprod Domest Anim 45(5):760–772. https://doi.org/10.1111/j.1439-0531.2009.01345.x
77. Spencer TE, Johnson GA, Bazer FW, Burghardt RC (2004) Implantation mechanisms: insights from the sheep. Reproduction 128(6):657–668. https://doi.org/10.1530/rep.1.00398
78. Bryden MM, Evans HE, Binns W (1972) Embryology of the sheep. I. Extraembryonic membranes and the development of body form. J Morphol 138(2):169–185. https://doi.org/10.1002/jmor.1051380204
79. Jordan RK (1976) Development of sheep thymus in relation to in utero thymectomy experiments. Eur J Immunol 6(10):693–698. https://doi.org/10.1002/eji.1830061007
80. Khaksary-Mahabady M, Khazaeel K, Pourmahdi Borujeni M, Yazdanjoo B (2018) Morphometric development of sheep (Ovis aries) lymph nodes in fetal period. Vet Res Forum 9(2):121–128. https://doi.org/10.30466/VRF.2018.30833
81. Maddox JF, Mackay CR, Brandon MR (1987) Ontogeny of ovine lymphocytes. II. An immunohistological study on the development of T lymphocytes in the sheep fetal spleen. Immunology 62(1):107–112
82. Agerholm JS, Hewicker-Trautwein M, Peperkamp K, Windsor PA (2015) Virus-induced congenital malformations in cattle. Acta Vet Scand 57:54. https://doi.org/10.1186/s13028-015-0145-8

Chapter 17

Schmallenberg Virus, from the Necropsy Room to the Laboratory

Calixte Bayrou, Daniel Desmecht, and Anne-Sophie Van Laere

Abstract

In a world facing significant climate changes, arboviruses are expanding into new regions. In recent decades, Western Europe has experienced multiple outbreaks of epizootic arboviruses in ruminants, including the Schmallenberg virus. This virus emerged in 2011 and spread rapidly across the continent, causing severe malformations in the central nervous systems of ruminant fetuses. As such, it serves as a strong example of the potential threat posed to other species and provides a valuable model for studying vertical transmission and viral damage to the central nervous system of fetuses. This chapter presents a detailed methodology for studying the virus from natural cases. It begins with the systematic determination of the virus's distribution within the fetus and the associated lesions and continues with the biomolecular characterization of the virus.

Key words Schmallenberg virus, Necropsy, Histopathology, Polymerase chain reaction, Viral titer

1 Introduction

In autumn 2011, the Friedrich Loeffler Institute in Germany isolated the genetic material of a previously unknown virus linked to an unexplained syndrome involving fever, decreased milk production, and diarrhea in dairy farms in the Netherlands and Northwest Germany [1]. This new virus was named Schmallenberg virus (SBV) after the geographical origin of the first positive samples and was classified within the Orthobunyavirus genus. It was the first time an arbovirus from this group was found to circulate among ruminants in Western Europe. The emergence of SBV was similar to the Bluetongue virus type 8 outbreak in 2006 [2] and highlighted the evolving dynamics of infectious diseases in Europe. Understanding and documenting the emergence and evolution of Schmallenberg virus in this context is crucial.

Hani Boshra (ed.), *Bunyaviruses: Methods and Protocols*, Methods in Molecular Biology, vol. 2893,
https://doi.org/10.1007/978-1-0716-4338-9_17,

Virion

Schmallenberg virus belongs to the Orthobunyavirus genus, within the Bunyaviridae family and the Bunyavirales order. Orthobunyaviruses are characterized by an enveloped virion, glycoprotein heterodimers (Gn and Gc) anchored to the envelope, a nucleocapsid protein (N), and segmented genetic material composed of three RNA molecules [1]. The SBV genome contains three segments of negative-stranded RNA. The large (L) segment encodes RNA-dependent RNA polymerase (RdRp); the medium (M) segment encodes viral glycoproteins Gn and Gc, and a non-structural protein (NSm); and the small (S) segment encodes the nucleocapsid protein (N) and the nonstructural protein NSs. High genetic variability was observed within the 5′-terminal part of the M-segment in virions extracted from the central nervous system of malformed calves and lambs [3]. This region encodes the "head-domain" of the major SBV envelope protein Gc, which is highly immunogenic and crucial for neutralization.

Biology

SBV is an arbovirus transmitted by Culicoides midges during blood feeding on mammals. Species in the Obsoletus complex are the main vectors in Europe [4–6], feeding opportunistically on various hosts including domestic and wild ruminants, birds, mice, and even humans. However, 77% of their blood meals come from cattle [7]. The virus's transmission is seasonal, with Culicoides activity in Europe beginning in early spring and ending in late October. Depending on the gestation period and time of infection, malformed newborns appear several months post-infection.

Unlike the Bluetongue Virus (BTV), SBV relies more on its vector for circulation dynamics than on mammal hosts. This higher efficiency in vector transmission enables SBV to cause a larger number of cases and spread over longer distances compared to BTV-8 [8]. In ruminants, viremia lasts only a few days (4–6 days after experimental infection [9]), and no other transmission routes besides transcutaneous entry have been identified [10, 11].

Disease

In adults, SBV infection causes a low-intensity disease with transient fever and diarrhea, leading to a drop in milk production in dairy herds [1]. SBV is mainly harmful through its placental transmission and subsequent teratogenic effects. The trans-placental passage window is estimated to last from 60 to 150 days of gestation [12]. Following an outbreak, a high percentage of calves (28% of tested animals) are seropositive at birth without malformations [13]. The ability of fetuses to mount a comprehensive immune response after 150 days of gestation likely closes this window [14]. If the virus infects a fetus before 3 months of age, the mortality risk is suspected to be high as it is described for the closely related Akabane virus [15]. After this period, while severe lesions may develop, especially in the central nervous system, the fetus can usually survive to term [16, 17]. Dystocic parturition of severely deformed and contracted newborns is a common outcome in pregnant females infected with SBV.

Macroscopically, arthrogryposis is the dominant lesion, varying in severity from subtle muscle contractures to widespread body deformities. Significant central nervous system damage includes porencephaly and hydranencephaly. The spinal cord often shows a severe reduction in diameter (micromyelia). This reduction in volume, though particularly striking in the spinal cord, is not confined to this area. The entire central nervous system may appear smaller, with cerebellar hypoplasia and hydrocephaly commonly reported as part of this phenomenon.

Microscopically, muscle lesions show atrophy together with fibrous connective tissue and fat infiltration, with some regeneration attempts visible (myotubes). In the central nervous system, the most significant change is a massive decrease in the number of neurons, which is associated with varying degrees of demyelination. This demyelination is particularly notable in the spinal cord, where, in severe cases, myelin is absent along the entire length, with the significant exception of the dorsal tracts. This lesion is emblematic of the consequences of in utero infection.

Our study [16] shows that the severity of muscle alterations correlates with the decrease in neuron density, suggesting that denervation plays a pivotal role in the development of arthrogryposis.

Interest in Studying SBV

SBV infection is limited to ruminants. However, a study demonstrated that the SBV N and L protein duo can transcribe and replicate a wide variety of Orthobunyavirus genomes [18]. The study showed the viability of a reassorted virus with the L and S segments from SBV and the M segment from the Oropouche virus. Since the latter is a human pathogen, this finding underscores the potential public health implications of the reassortment mechanism in Orthobunyaviruses.

The aim of this chapter is to provide a detailed protocol for studying the virus genome from necropsied calves, encompassing diagnosis, lesion characterization, and virus genome sequencing.

2 Materials

1. ***Necropsy.***
   - Victorinox Sticking knife (20 cm, # 5.5603.20).
   - Electric bone saw (Medoc, model STL-350).
   - Scalpel n.24.
   - 5 mL syringe mounted with an 18G needle.
2. ***Sampling.***
   - Tubes.
   - Centrifuge.
3. ***Histopathology.***
4. ***Seroneutralization Test.***
   - Baby Hamster Kidney cells (BHK).
   - BHK culture medium.
     - Glasgow's Modified Eagle's Medium (GMEM) 500 mL.
     - TBP 11 mL.

       Tryptose Phosphate Broth (TBP—Sigma T8782-500 g): For BHK cells 23.6 g in160 mL (Autoclave before using and every month if not used).
     - Penicillin—Streptomycin 2.2 mL.
     - Fetal Bovine Serum (FBS) 50 mL (remove 50 mL of medium before adding the FBS).
   - Dilution medium.
     - Eagle's Minimum Essential Medium (EMEM) 500 mL.
     - Penicillin –Streptomycin 2.2 mL.
   - Infection medium.
     - Eagle's Minimum Essential Medium (EMEM) 500 mL.
     - Penicillin –Streptomycin 2.2 mL.
     - Fetal Bovine Serum (FBS) 5 mL.
   - Crystal violet staining solution.
     - Crystal violet powder (Sigma-Aldrich) 0.5 g.
     - Distilled Water 80 mL.
     - Methanol 20 mL.
   - V-shaped sterile reservoir.
   - V-shaped 96-well plates.
   - Cell culture 96-well plates.
   - Water bath.

- CO_2– incubator.
- Laminar flow hood.
- Electronic multichannel pipette.

5. ***Biomolecular Characterization of the Virus.***
   1. *RNA Extraction.*
      - TRIzol™ Reagent, Invitrogen.
      - Ethanol, molecular biology grade.
      - Isopropanol, molecular biology grade.
      - Liquid nitrogen.
      - Nuclease-free safe-lock microtubes.
      - Mortar and pestel.
      - Precision balance.
      - Microcentrifuge.
      - Heat block.
   2. *Diagnosis Confirmation by qRT-PCR.*
      - Gene Expression Master Mix (Applied Biosystems).
      - AgPath-ID™ One-Step RT-PCR Reagents (Applied Biosystems).
      - RNase, DNase free water.
      - 0.1 × TE (pH 8.0)
      - Nuclease-free 96-well PCR reaction plate.
      - Optically clear plastic seal for PCR plates.
      - Real-time PCR instrument.
   3. *cDNA Synthesis,*
      - SuperScript™ IV First-Strand Synthesis System, Invitrogen.
      - Gene Specific Primers, *see* **Note 12**.
      - Nuclease-free PCR tubes.
      - Thermocycler.
   4. *PCR Amplification, Purification, and Sequencing of the M Fragment of SBV.*
      - DNA polymerase (e.g. GoTaq® G2 Hot Start Taq Polymerase, Promega Corporation).
      - dNTP mix (2 mM each),
      - Molecular Biology grade Agarose.
      - Tris–acetate–EDTA (TAE) buffer (0.5×).
      - Nucleic acid stain (e.g. Midori Green Advance, Nippon Genetics).

- DNA ladder.
- Nuclease-free microtubes.
- Scalpel blade.
- Casting tray.
- Well combs.
- Voltage source.
- Electrophoresis unit.
- UV-light or Blue/Green light source.
- Precision balance.
- Microwavable flask.
- Microwave.
- Thermocycler.

3 Methods

All the figures in this section are reproduced with the kind permission of *Emerging Infectious Diseases journal* [16].

1. ***Necropsy.***

   The necropsy should focus on the three primary systems affected: nervous, muscular, and skeletal. In utero infection typically results in an "arthrogryposis-hydranencephaly" syndrome. While no lesions of the internal organs are typically described, viral RNA may be isolated in some, particularly the lungs and intestine. The necropsy should document the presence and severity of the following lesions:

   - Skeletal deformities: Arthrogryposis, including scoliosis.
   - Muscular degeneration: Atrophy, fibrosis, and fat infiltration (referred to as "end-stage muscles").
   - Nervous lesions:
     - Spinal cord atrophy (micromyelia).
     - Porencephaly and its extreme form, hydranencephaly.
     - Cerebellar hypoplasia.
     - Hydrocephaly.

1. Weigh the calf: this serves as an indirect indication of muscle atrophy and fibrosis.
2. Place the calf in lateral recumbency on the dissection table.
3. Assign a "Whole-Body Deformity Score" as illustrated in Fig. 1 (*see* **Note 1**).

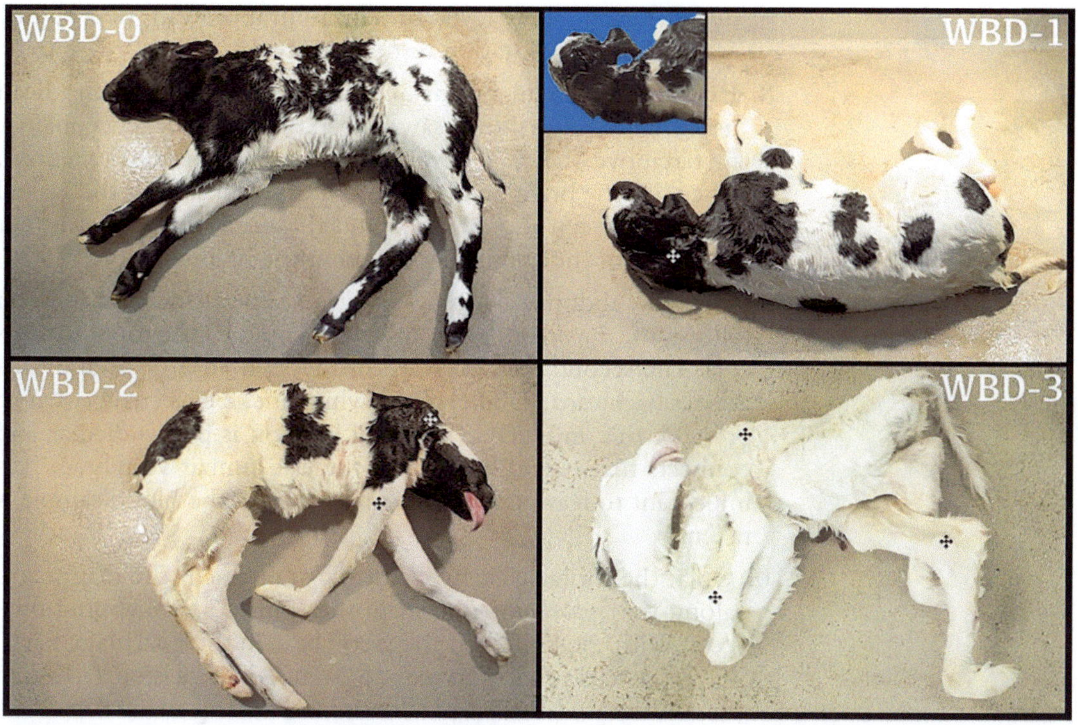

Fig. 1 Definition of whole-body deformity (WBD) scores in SBV cases. Animals with neurological signs and apparently normal body shape were given a WBD score of 0. Deformities can involve three main body parts: the spine, the forelimbs, or the hindlimbs. The WBD was given according to the number of parts involved

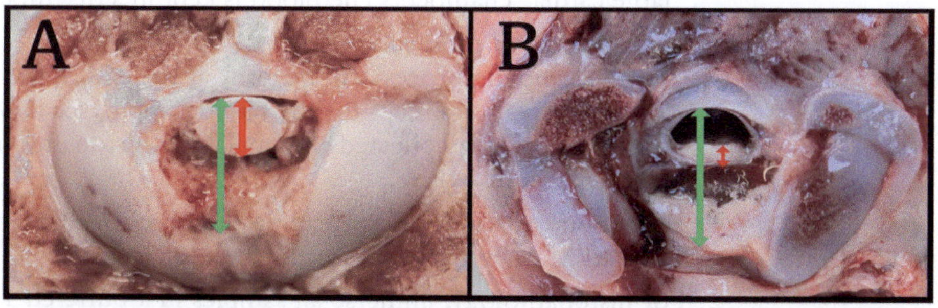

Fig. 2 Occipital ratio modification after in utero infection by the Schmallenberg virus in calf. A control newborn calf (**a**) is compared with a severely affected case (**b**) showing micromyelia. The occipital ratio is the result of the division of the spinal cord diameter (red arrow) by the foramen magnum diameter (green arrow)

4. Separate the head from the body with a Victorinox Sticking knife (*see* **Note 2**).
5. Measure the "Occipital Ratio" as shown in Fig. 2 (*see* **Note 3**).
6. The skeletal deformities are more accurately assessed after the removal of the skin.
7. Cut all the large muscle groups with a knife to analyze their texture and color. Irregular whitish areas may indicate muscle

degeneration. The level of fat infiltration can be roughly assessed using a flotation test (*see* **Note 4**).

8. Examine the brain by making a sagittal section of the head using an electric bone saw. To exteriorize the encephalon, first remove the *falx cerebri* by multiple transversal sections. Then, lift the brain hemisphere by dorsally inserting the scalpel handle between the *cerebrum* and the *dura mater*. Finally, cut the roots of the cranial nerves along the brainstem.
9. Open the abdominal cavity using a scalpel. The first incision follows the *hypochondrium* from the spine to the xiphoid process. The second incision starts at the xiphoid process and progresses backward, parallel to the white line, 2 to 3 cm above it to preserve the umbilicus. The final incision is perpendicular and connects the second incision to the last lumbar vertebrae. It is important to leave the organs in place until the blood sample is taken.
10. Open the thoracic cavity. With the scalpel, cut the diaphragm along the *hypochondrium* from the spine to the xyphoid process. With the knife, cut the sternebrae in their cartilage parts. The way to do that is to make a paramedian incision 1–1.5 cm beside the mid-line of the *sternum*. A strong manual traction on the opened sternum, will break all the ribs insertion on the spine at once. Before removing the internal organs, a blood sample is taken by aspirating with a 5 mL syringe and an 18G needle through the wall of the main thoracic vessels or in the large cardiac chambers. The blood is kept in a dry tube.

2. ***Sampling Technique.***

The sampling process can be done at the end of the macroscopic examination except for the blood sample (see above). All samples should be collected in parallel for histopathology and biomolecular analysis.

1. For blood conservation, centrifuge the tube for 30 minutes at 1800 RPM, then freeze the serum at −20 °C.
2. For histopathology, immerse all samples in a buffered 10% formalin solution (*see* **Note 5**). If the sample is intended for immunohistochemistry, transfer it to 70% alcohol after 24 h in formalin to preserve protein conformation, as prolonged contact with formalin can mask some antibody targets.
3. Samples for viral RNA detection are best taken from the central nervous system: spinal cord, brainstem, and cerebral cortex.
4. Muscle sampling: Given the high correlation between muscle and spinal cord lesions, it is beneficial to sample muscles from regions innervated by the same spinal cord segments. It is important to use new scalpel blades and avoid using forceps when cutting muscle samples to prevent artifactual tearing of

muscle fibers, which is frequently observed under the microscope. It is better to take small blocks of muscle rather than strips to avoid deformation during fixation. The orientation of fibers is not critical in muscle histopathology of SBV cases, as the muscle fiber orientation is already disrupted by the infection.

5. Spinal Cord Sampling:
   - The spinal cord can be sampled from multiple areas along the spine.
   - After evisceration, remove the limbs from the body.
   - Using an electric bone saw, make several transverse sections along the spine to obtain vertebral "slices" approximately 2 cm thick.
   - If necessary, cut the nerve roots with scissors.
   - Gently push small sections of the spinal cord out of the vertebral canal using a finger, ensuring the meninges remain intact around the spinal cord.
   - Place half of each spinal cord section (approximately 1 cm) in formalin for histopathology, and use the other half for biomolecular analysis.
6. Brain sampling (*see* **Note 6**).
   - After sagittal section of the head, and exteriorization of the brain, one half is kept in formalin for histopathology. Because fixation of deep nervous tissue in the central nervous system can take time, after 24 h in formol, cut the areas that will be used for histopathology (4 mm large tranches) and let them fix for another 24 h in formol.
   - The other half is used for biomolecular analysis. Symmetrically with respect to the precise areas which will be selected for histopathology, small pieces of brain are cut out and placed in small tubes.
7. Additional organ sampling in calves.

   Aside from in the central nervous system, viral RNA is rarely detected in other organs of SBV-infected calves. Although skeletal muscles are heavily modified, they are almost always negative. The lungs are the best option for finding viral RNA outside of the central nervous system.

3. ***Histopathology.***

   Characterization of SBV-induced lesions is largely achieved through classical hematoxylin and eosin staining. Two other particularly helpful stains are Masson's Trichrome and Luxol Fast Blue.

   Masson's Trichrome: Stains muscle fibers red and collagen blue (or green), providing a clear view of muscle fiber replacement by collagen.

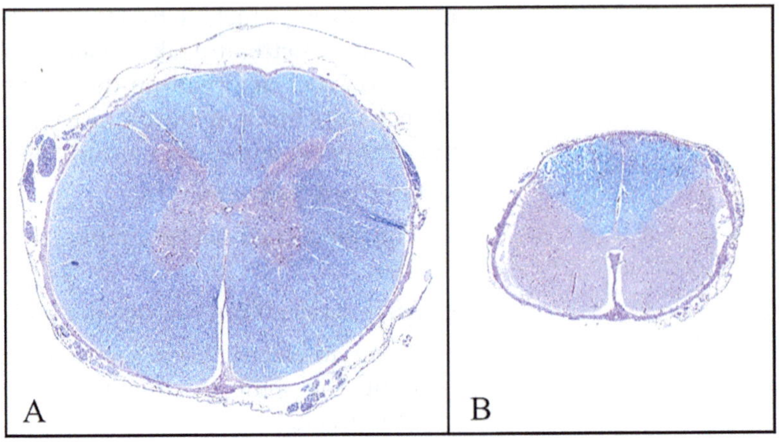

Fig. 3 Micromyelia. Age- and site-matched spinal cord transversal histological section at the level of C4. A, control calf; B, SBV case. Note atrophy/hypoplasia and disappearance of most myelinated cords in ventral and lateral tracts of SBV case. Luxol fast blue staining

Luxol Fast Blue: Stains myelin blue and cells pink, aiding in the diagnosis of demyelination, particularly severe demyelination of the spinal cord. For optimal color density with Luxol Fast Blue, the paraffin section thickness should be increased to 10–15 μm. Figure 3 illustrates typical SBV-associated lesions detectable with Luxol Fast Blue staining.

These staining techniques have been described for several decades in many histotechnology references. Protocols can be found, for instance, in Laboratory Methods in Histotechnology by Edna B. Prophet from the Armed Forces Institute of Pathology (U.S.).

4. ***Seroneutralization Test (SNT).***

Several commercial ELISA kits are available to test the seropositivity of ruminants. Postmortem samples of natural cases may however exhibit an unexpected behavior. It is therefore essential to always validate commercial ELISA results with sero-neutralization tests (SNT) by following the protocol described below.

It is essential to first adjust the infection dose and the concentration of cells. All manipulations are performed under a laminar flow hood, and contaminated waste is disposed of in plastic bags and inactivated with bleach.

1. Determine the necessary viral particle count on day one to achieve 100–200 TCID50 per well (*see* **Note 7**). With an infection volume of 50 μL, the viral concentration after overnight incubation must be 100–200 TCID50/50 μL. Higher concentrations of infectant particles can increase the number of false

negatives, while lower concentrations may increase the false positive rate.

2. Determine the number of cells to plate (*see* **Note 8**).
3. Day 1—cell plating.
   - Calculate the requirement of flat-bottom 96-well cell culture plates. Test each sample in duplicate. If the required information is limited to positive vs. negative, six dilutions are sufficient (ranging from 1/2 to 1/256). One plate allows testing of eight samples. Add a negative and a positive control for each batch of SNT. Prepare an additional plate for back titration.
   - Write the sample identification on the cover. When running multiple plates, it is recommended to label the side of the plate (both cover and plate) to prevent mixing.
   - Prepare the cell solution (*see* **Note 8**). Add 150 μL of solution containing 20,000 cells into each well.
   - Distribute the solution using an electronic pipette and a V-shaped sterile reservoir.
   - Place the plates in the incubator overnight at 37 °C and 5% CO_2 with a humidified atmosphere.
4. Day 1—Virus and serum mixing.
   - Decomplement thawed sera at 56 °C for 30 min. Place 1.5 mL Eppendorf tubes loaded onto foam floating racks in a prewarmed water bath.
   - Fill V-shaped 96-well plates with 60 μL of dilution medium. Each sample requires six wells (each well of the V-plate will be split for duplicate infection on the U-plates). One 96-well plate can accommodate 16 samples. The first well is left empty, and the remaining five wells receive 60 μL of dilution medium.
   - Dispense 120 μL of each sample in the empty wells. It is crucial to pipette the 120 μL into well-mixed samples to ensure homogeneous distribution of antibodies.
   - For the twofold serial dilution, transfer 60 μL from the first well to the next using a multichannel pipette. After transferring the material into the well, it is important to mix by pipetting up and down and to change the tips between each dilution. In the last well of the dilution, discard the final 60 μL to maintain 60 μL in every well of the plate.
   - Add 60 μL of virus solution to each well. Place the viral solution, containing sufficient virus to ensure infection with 100–200 TCID50 (see above), in a sterile V-shaped reservoir for easy use with an electronic multichannel pipette. Load the 6 × 60 μL required for one sample at once and pour them

successively from the 1/256 dilution to the 1/2. Change tips between each sample.

- Retain a 1.5 mL Eppendorf tube of the virus solution for back titration.
- Incubate the plates and the Eppendorf tube overnight at 37 °C.

5. Day 2—Infection.
   - Prepare a bin lined with paper towels to absorb the liquid from the cell culture.
   - Dispose of the culture medium by quickly inverting the culture plates over the bin.
   - Ensure that empty wells are covered promptly to prevent cell desiccation.
   - Transfer 50 μL from the V-shaped plate to each duplicate well using a multichannel pipette.
   - Incubate the infected plates for 2 h at 37 °C.
   - After the incubation period, add 100 μL of prewarmed fresh culture media using an electronic multichannel pipette. Unlike the step for loading the viral solution, the addition of fresh media starts from the 1/2 dilution to reach 1/256. Change tips between samples.
   - Incubate the plates at 37 °C, 5% CO_2, for 72 h until reading.
6. Day —Back titration.
   - Prepare a tenfold serial dilution from 1/1 to 10^{-7}. Infecting eight wells per dilution provides precise results for back titration. On an Eppendorf tube rack, load seven tubes with 900 μL of dilution medium. Transfer 100 μL from the viral solution to the next tube. To ensure consistent results, it is crucial to avoid transferring additional viral particles. Use a new tip for each transfer to prevent contamination. Repeat this process for each tube.
   - After removing the culture media from the last culture plate, transfer 50 μL from each tube into all eight wells of a column using an electronic pipette. Begin with the tube containing the lowest concentration of virus.
   - Incubate the plate at 37 °C with 5% CO_2 for 72 h.
7. Day 5—Reading.
   - Discard the medium from the plates into a bin containing bleach.
   - Load each well with 50 μL of crystal violet solution. After 25 min, remove the crystal violet by flipping the plate over a large bin and vigorously shaking it. Rinse the plate in a basin of plain water.

- Once the plates have dried at room temperature, reading involves macroscopically detecting areas of lysis, which appear as empty spots on the purple cellular layer.
- A sample is considered positive if the serum provides protection from the 1/8 dilution.
- Back titration reading follows the same technique, even if the lysis disappears with dilution, unlike the SNT. The precise calculation of TCID50 is obtained using the Reed and Muench method. However, because back titration serves as a control and does not require precise results, reading can be limited to checking if lysis is present in five to seven wells of the 10^{-2} dilution.

5. ***Biomolecular Characterization of the Virus.***

   Characterizing SBV in utero infection is completed once other important causes of central nervous system defects and arthrogryposis in calves are ruled out:

   - Viral Etiology.

     Differential diagnosis for viral etiology varies depending on the region of the world. In Europe, the main viruses to exclude are Bovine Viral Diarrhea virus (BVDV) and Bluetongue virus (BTV). Recently, another pathogen belonging to the Orbivirus genus, Epizootic Haemorrhagic Disease virus (EHDV), has emerged in Western Europe. It infects cattle and is suspected to cause congenital lesions in calves, although no description of such lesions is yet available.

   - Genetic Origin of Arthrogryposis.

     Congenital arthrogryposis can result from a genetic defect in many cattle breeds, and is often associated with cleft palate defects. It may be worthwhile to investigate whether genetic-induced arthrogryposis has been described in the studied breed. This can be easily achieved using the Online Mendelian Inheritance in Animals (OMIA—omia.org) database from the University of Sydney.

   - Nutritional Cause.

     Deficiencies (selenium, iodine, vitamin E) can induce arthrogryposis of low intensity. This etiology can however be ruled out when central nervous system lesions are present or when the arthrogryposis is generalized and severe.

1. RNA Extraction.
   - Crush frozen tissue with mortar & pestle under liquid nitrogen.
   - Weight 50–100 mg tissue and place in a microtube.
   - Add 1 mL TRIzol™ Reagent.

- Incubate for 5 min to allow complete dissociation of the nucleoproteins complex.
- Add 0.2 mL of chloroform, securely cap the tube, then thoroughly mix by shaking for 30 s.
- Incubate for 2–3 min.
- Centrifuge the sample for 15 min at 12,000 × g at 4 °C.
- Transfer the colorless upper aqueous phase containing the RNA to a new tube by angling the tube at 45° and pipetting the solution out (*see* **Note 9**).
- Add 0.5 mL of isopropanol to the aqueous phase.
- Mix by inverting the tube ten times.
- Incubate for 10 min at 4 °C.
- Centrifuge for 10 min at 12,000 × g at 4 °C.
- Discard the supernatant with a micropipettor being careful not to disturb the white gel-like pellet containing the total RNA.
- add 1 mL of 75% ethanol to the pellet and vortex briefly,
- Centrifuge for 5 min at 7500 × g at 4 °C.
- Discard the supernatant with a micropipettor.
- Air dry the RNA pellet for 10 min.
- Add 50 μL of RNase-free water to the pellet.
- Incubate in a heat block set at 55 °C for 10 min. Gently mix the sample by flicking the tube every 2–3 min.
- Take an aliquot of your RNA sample to measure its concentration (on a Nanodrop) and to check its integrity (on a Bioanalyzer).
- Store the RNA at −80 °C or proceed directly to cDNA synthesis.

2. Diagnosis Confirmation by qRT-PCR.

   Three different qRT-PCRs are performed to detect/exclude the presence of viruses: (i) SBV RNA detection, (ii) BTV exclusion, and (iii) BVDV exclusion. A fourth qRT-PCR amplifies the internal control (β-actin) separately.

   SBV and BVDV are amplified by a one-step PCR, whereas BTV is amplified by a two-steps PCR. Samples should at least be run in duplicate. Sequences of the primers and probes are given in *see* **Note 10** (Table 1).

Table 1
Sequences of the primers and probes necessary for the diagnosis of suspected infected calves after necropsy

Target	Primer/probe name	Sequence (5′ → 3′)[a]	Genome position	Reference
SBV	382-F	TCA GAT TGT CAT GCC CCT TGC	382–402[b]	Bilk [19]
	469-R	TTC GGC CCC AGG TGC AAA TC	450–469[b]	
	P-408	FAM-TTA AGG GAT GCA CCT GGG CCG ATG GT-BHQ1	408–433[b]	
BVDV	106-F	CCA TGC CCT TAG TAG GAC TAG C	106–127[c]	La Rocca [20]
	190-R	GCG TCG AAC CAC TGA CGA CT	190–209[c]	
	P-162	FAM-TGG ATG GCT TAA GCC CTG AGT ACA G-EDQ	162–186[c]	
BTV	F_1-19	GGC AAC YAC CAA ACA TGG A	1–19[d]	Toussaint [21]
	R_76-57	AAA GTY CTC GTG GCA TTW GC	57–76[d]	
	P 49-27	FAM-CYC CAC TGA TRT TGT ATT TTC TCA A-TAMRA	49-27[d]	
ß-actin	F_ 1005-1029	CAG CAC AAT GAA GAT CAA GAT CAT C	1005–1029	Toussaint [21]
	R_1135-1114	CGG ACT CAT CGT ACT CCT GCT TT	1114–1135	
	P_1081-1105	FAM-TCG CTG TCC ACC TTC CAG ATG T-TAMRA	1081–1105	

BHQ1 Black Hole Quencher-1, *FAM* 6-carboxyfluorescein; *EDQ* Eclipse Dark Quencher
[a]Reverse primers complementary to positive-sense target strand
[b]In the SBV genome segment-S (GenBank accession no. HE649914)
[c]In the BVDV NADL genome (GenBank accession no. M31182)
[d]In the BTV genome segment-5

- One-step qRT-PCR for SBV or BVDV amplification.
  - For each qRT-PCR, prepare the specific primer-probe mix by combining:

    20 μL forward primer (100 μM).

    20 μL reverse primer (100 μM).

    3.75 μL probe (100 μM).

    156.25 μL 0,1 × TE (pH 8.0).
  - Combine the following components in a nuclease-free microcentrifuge tube. Scale the reaction up according to the number of samples (including controls, *see* **Note 11**) plus 10% (to account for loss due to pipetting).

    4.5 μL RNase-free water.

    12.5 μL 2 x RT-PCR buffer (AgPath-ID One-Step RT-PCR Reagents kit)..

    1.0 μL 25 × RT-PCR enzyme mix (AgPath-ID One-Step RT-PCR Reagents kit).

2.0 μL specific primer-probe mix (10 μM target-specific primers +1.875 μM target-specific probe) for one reaction.

- Distribute 20 μL of master mix / well of a 96-well PCR plate.
- Add 5 μL of RNA template (or H2O for the negative control).
- Carefully close the plate with an optically clear plastic seal and briefly centrifuge it to collect the content at the bottom of the wells.
- For the SBV qRT-PCR, program your real-time thermocycler to perform the following steps:

  10 min at 45 °C.

  10 min at 95 °C.

  42 cycles of 15 s at 95 °C, 20 s at 56 °C and 30 s at 72 °C.
- For the BVDV qRT-PCR, program your real-time thermocycler to perform the following steps:

  10 min at 50 °C.

  2 min at 95 °C.

  40 cycles of 15 s at 95 °C and 30 s at 60 °C.

• BTV is detected by a pan-BTV two-steps RT-qPCR targeting genome segment-5 of bluetongue viruses. ß-actin is again used as internal control. The first step consists of the cDNA synthesis (see cDNA Synthesis below) and the second step consists of the real-time PCR itself.

- For the BTV real-time PCR, combine the following components in a nuclease-free microcentrifuge tube. Scale the reaction up according to the number of samples plus 10%.

  12.5 μL TaqMan Gene Expression Master Mix (2×).

  0.8 μL primer F_1-19 (12.5 μM).

  0.8 μL primer R_76-57 (12.5 μM).

  0.2 μL probe P_49-27 (6.25 μM).

  6 μL RNase-free water.
- For the ß-actin real-time PCR, combine the following components in a nuclease-free microcentrifuge tube. Scale the reaction up according to the number of samples plus 10%.

  12.5 μL TaqMan Gene Expression Master Mix (2×).

  0,6 μL primer F_1-19 (12,5 μM).

  0.6 μL primer R_76-57 (12.5 μM).

Table 2
Sequences of the gene-specific primers (GSP) used for cDNA synthesis of the entire length of the M fragment of the Schmallenberg virus

GSP name	GSP sequence
SBV.M1.R	ATCCACATGTTACCTCAATGG
SBV.M3.R	CTGTGAATTGTTAAGCCCTTG
SBV.M5.R	CCTTGTATGTGGCGTAAAGC
SBV.M7.R	TTTGATACCTTGTCCCCATTC
SBV.M9.R	GATTGTACCCACGCTTGAAAC
SBV.M11R2	AGTGTTCTACCACATGAAAAAC

0.2 μL probe P_49-27 (6.25 μM).

6.4 μL RNase-free water.

- Distribute 20.5 μL of BTV or ß-actin master mix / well of a 96-well PCR plate.
- Add 4.5 μL of cDNA template (or H2O for the negative control).
- Carefully close the plate with an optically clear plastic seal and briefly centrifuge it to collect the content at the bottom of the wells.
- Program your real-time thermocycler to perform the following steps:

  10 min at 95 °C

  45 cycles of 15 s at 95 °C and 1 min at 58 °C.

3. cDNA Synthesis,
   - Combine the following reagents in a PCR tube:
     - 500 ng—1 μg RNA
     - 1 μL Gene Specific Primer (GSP) (2 μM) (*see* **Note 12**, Table 2)
     - 1 μL dNTP mix (10 mM each).
   - Add RNase-free water to reach a final volume of 13 μL.
   - Mix by pipetting three times up and down.
   - Briefly centrifuge to collect the components at the bottom of the tube.
   - Heat the RNA-primer mix at 65 °C for 5 min.
   - Directly transfer the tube on ice and incubate for at least 1 min.
   - In a microcentrifuge tube, prepare the RT reaction mix by combining:

– 4 μL 5× SSIV Buffer.
– 1 μL 100 mM DTT.
– 1 μL Ribonuclease Inhibitor.
– 1 μL SuperScript™ IV Reverse Transcriptase (200 U/μL).

- Mix and briefly centrifuge the contents.
- Add 7 μl of the RT reaction mix to each PCR tube containing the RNA-primer mix.
- Incubate the combined reaction mixture in a thermocycler as follows:
  – 50 °C for 10 min.
  – 80 °C for 10 min.
- To remove RNA, add 1 μL E. coli RNase H, and incubate 37 °C for 20 min.
- Use your RT reaction immediately for PCR amplification or store it at −20 °C.

4. PCR amplification, purification, and sequencing of the M fragment of SBV.
   - Thaw and vortex Green GoTaq® Flexi Buffer (5X), dNTP (2 mM), $MgCl_2$ (25 mM), forward GSP (10 μM) (*see* **Note 13**), reverse GSP (10 μM) (*see* **Note 13**, Table 3).
   - On ice, combine the following components in a nuclease-free microcentrifuge tube. Scale the reaction up according to the

Table 3
Sequences of the gene-specific primers (GSP) and expected product size for amplification of the entire length of the M fragment of the Schmallenberg virus

Fragment	Forward GSP	Reverse GSP	Size (bp)
M1	AGTAGTGAACTACCACAATCAA	ATCCACATGTTACCTCAATGG	543
M2	CGAATTAACCACTACCGAATC	GCAATGCTTTATAGCCAGAAC	505
M3	TACACTACCACTGAATCACTC	CTGTGAATTGTTAAGCCCTTG	449
M4	CGTGGATTGAAGATAATGGAC	CTTGGAGAGGGCACAACTGT	448
M5	GAACTGTAATGTCATAGATAGG	CCTTGTATGTGGCGTAAAGC	417
M6	CTGGATTAGGTCGAGTCTTG	TCACGTTGCATGATGGTATTC	528
M7	CAGCAGAAAGAACAATAACTAC	TTTGATACCTTGTCCCCATTC	472
M8	CCTTCCAGCAATAATTCCTCA	GGTTATACAAACTTCAACATCC	472
M9	AAGTAATTGGGGCTGTGAAGA	GATTGTACCCACGCTTGAAAC	538
M10	AGTCTTGTCTTCTCTTGGAAC	GTTGCTCGATGTAAGATGTTTC	459
M11	GTAAGTGTGTGCTCAGCTTC	AGTGTTCTACCACATGAAAAAC	474

number of samples (including a negative control, *see* **Note 11**) plus 10% (to account for loss due to pipetting).

- 9 μL H_2O (DNase free).
- 4 μL Green GoTaq® Flexi Buffer.
- 2 μL dNTP.
- 1.3 μL $MgCl_2$.
- 0.8 μL forward GSP.
- 0.8 μL reverse GSP.
- 0.1 μL GoTaq® G2 Hot Start Polymerase.

- Mix and briefly centrifuge.
- For each sample, transfer 18 μL of PCR mix to an annotated nuclease free PCR tube.
- Add 2 μL of cDNA (or H_2O for the negative control) to the PCR tube.
- Mix by pipetting up and down three times.
- Cap tubes securely and centrifuge briefly.
- Transfer to a thermocycler with heated lid and program the following cycling conditions:
  - 10 min at 95 °C.
  - 10 cycles of 30 s at 95 °C, 30 sec at a temperature decreasing by 1 °C/cycle starting from 66 °C (*see* **Note 14**), 40 s at 72 °C,
  - 40 cycles of 30 s at 95 °C, 30 s at 56 °C, and 40 s at 72 °C, and a final extension of 10 min at 72 °C.
- Add a hold step at 4 °C if you are not present at the end of the amplification.
- Proceed with agarose gel electrophoresis or store amplified PCR product at −20 °C.
- Prepare a 2% agarose gel by mixing agarose powder with 0.5× TAE buffer in a microwavable flask. For example, mix 1 g agarose with 50 mL TAE.
- Microwave for 1–3 min until the agarose is completely dissolved. Be very careful while manipulating the flask as eruptive boiling can occur.
- Let agarose solution cool down to about 50 °C (about when you can comfortably keep your hand on the flask).
- Add an appropriate volume of nucleic acid stain e.g. 2.5 μL of Midori Green Advance in 50 mL agarose solution.
- Pour the agarose solution into a gel tray with the well comb in place. Make sure that no bubbles have formed around the

comb. If bubbles are present, you can remove them with a clean paper towel or push them to the sides of the gel.

- Let the gel polymerize for at least 20 min at room temperature.
- Once solidified, place the agarose gel into the electrophoresis unit filled with 0.5× TAE buffer. Make sure the gel is submerged.
- Carefully load the gel by placing a molecular size ladder in the first well and your samples in the following wells.
- Put the lid on the electrophoresis box and plug the wires into the power supply.
- Turn the power supply on and apply a current of 70–100 V for ~45 min (until the tracking front dye is about two-thirds down the gel).
- Turn off the power supply, unplug the wires, and remove the lid from the electrophoresis box.
- Visualize your DNA fragments under UV-light or Blue/Green light source and excise the gel bands with a scalpel blade. If using UV-light, work quickly in order to minimize personal exposure to UV light and to minimize nicking of the DNA. Protect yourself by wearing safety goggles or a face shield, gloves and a lab coat.
- Purify DNA from agarose using silica-membrane-based purification columns e.g. QIAquick Gel Extraction Kit (Qiagen) or Wizard® SV Gel and PCR Clean-Up System (Promega Corporation).
- Send your purified DNA fragments and corresponding GSP to your sequencing facility for Sanger Sequencing.

4 Notes

1. The "Whole-Body Deformity Score" reflects the diversity and severity of arthrogryposis and allows for correlations with histopathological lesions. It ranges from 0 (no lesion) to 3 (the whole body is affected). It is calculated by dividing the body into three parts: the spine (including the head), the forelimbs, and the hindlimbs. One point is given for each part showing arthrogryposis-associated deformities.
2. Hold the head by grasping the ear and make a transverse cut parallel to the ramus of the mandible, approximately 3 cm caudal to the ear. This position allows the blade to go directly behind the occipital condyles and disarticulate the atlanto-occipital joint.

3. The "Occipital Ratio" quantifies one of the main lesion i.e. the spinal cord atrophy. After separating the head, measure the diameters of the foramen magnum and the spinal cord. The ratio is calculated by dividing the spinal cord diameter by the foramen magnum diameter.
4. The flotation test is achieved by dropping small cubes of muscle (1–2 mm edge length) into plain water. Normal muscle samples will sink, while areas with massive fat infiltration will float.
5. 10% formalin is actually 4% formaldehyde, obtained by a tenfold dilution of the stock formaldehyde solution, which is about 37–40%. The solution should be buffered to prevent artefactual precipitates on the tissue.
6. This technique allows visualization of different parts of the brain within the cranial cavity, which is useful in cases of in utero Schmallenberg virus infection. It is often observed that various parts of the encephalon are atrophied and do not fill the cranial cavity space.
7. To validate a SNT, it must be conducted with a constant dose of infectant viral particles, typically 100–200 TCID50 (50% Tissue Culture Infectious Dose) per well. However, due to the necessity of overnight incubation for virus–serum interaction before infection, some viral particles lose their ability to infect cells without any neutralizing antibody. It is therefore essential to determine the necessary viral particle count on day 1 to achieve 100–200 TCID50 per well. With an infection volume of 50 μL, the viral concentration after overnight incubation must be 100–200 TCID50/50 μL. Higher concentrations of infectant particles can increase the number of false negatives, while lower concentrations may increase the false positive rate. In our laboratory's conditions, the required starting concentration of the virus was 600 TCID50/50 μL to achieve the desired 100–200 TCID50/50 μL at infection time. However, this ratio is highly dependent on factors such as the method of thawing the viral solution, the precise duration of incubation, and so on. Therefore, it is crucial to avoid changing these parameters once the adjustment has been made.
8. Because the total duration of the experiment is 96 hours, it is crucial not to overload the wells to prevent cell stress at the end of the experiment. Baby Hamster Kidney (BHK) cells typically have a multiplication rate of approximately threefold every 24 h. In our conditions, we obtained a good cell density by plating 20,000 cells per well on day 1. The total volume of cell solution is calculated as follows: (number of plates × 96 × 150 μL × 1.15)—(volume of cells). The number of cells required is: number of plates × 96 × 20,000 × 1.15. Warm the culture medium to 37 °C. Transfer the cell solution to a sterile

glass bottle equipped with a stirring magnet to avoid cell sedimentation during plating. Complete with culture medium to reach the desired volume.

9. It is important to avoid transferring any of the interphase or organic layer into the pipette when removing the aqueous phase. This tricky step can be simplified by using microtubes pre-dispensed with Phase Lock Gel (5PRIME, Quantabio). Phase Lock Gel migrates to form a tight seal between the colorless upper aqueous phase on one hand and the interphase and lower phenol-chloroform phase on the other hand. This enables complete and easy pipetting of the entire aqueous phase.
10. For all PCRs, it is strongly recommended to add a negative control i.e. a sample that has all reagents in the same final concentrations used in all other samples, but where H_2O is added instead of the cDNA template. The negative sample is used to exclude any cross-contamination or reagent contamination. For the diagnostic qRT-PCRs, it is also a good practice to add a positive control in order to confirm that a negative PCR results truly reflects the non-carrier state of the animal and is not due to failure to perform PCR amplification of the virus. For the two-steps PCR amplification of BTV, we also recommend to add both a negative and a positive reverse transcription control.
11. The freshness of tissue samples collected from the necropsy room and consequently the integrity of the obtained RNA samples is a frequent issue. To overcome this difficulty, one can use gene-specific primers (GSP). GSP are designed to generate cDNA for a specific gene of interest and are particularly useful when RNA concentrations are low. In addition, using several GSP spread over the gene will allow the generation of cDNA fragments covering its entirety even when the RNA is partially degraded. Sequences of GSP for the M fragment of SBV are given in Table 2 below.
12. The entirety of the M fragment of SBV is amplified by 11 overlapping PCR products. These PCR amplicons have small sizes ranging from 417 to 543 bp in order to maximize the chances of amplification even when working with partially degraded samples. Sequences for the 11 primer pairs are given in Table 3 below together with the expected PCR product size.
13. The decreasing annealing temperatures described in this protocol are characteristic of a touchdown PCR. This method decreases off-target priming and hence increases the specificity of PCRs which is particularly important when amplifying weakly represented DNA. The very high hybridization temperatures of the first cycles ensure a strong stringency and

therefore a specific amplification. In subsequent cycles, the annealing temperature is gradually decreased until it is 4–5 °C below the calculated Tm of the primers. By then, the target sequence will have undergone several cycles of geometric amplification and will therefore become the dominant product of the PCR. The remaining cycles are done at this lower temperature to ensure better PCR efficiency.

References

1. Hoffmann B, Scheuch M, Höper D, Jungblut R, Holsteg M, Schirrmeier H, Eschbaumer M, Goller KV, Wernike K, Fischer M et al (2012) Novel orthobunyavirus in Cattle, Europe, 2011. Emerg Infect Dis 18:469–472. https://doi.org/10.3201/eid1803.111905
2. Elbers ARW, Loeffen WLA, Quak S, de Boer-Luijtze E, van der Spek AN, Bouwstra R, Maas R, Spierenburg MAH, de Kluijver EP, van Schaik G et al (2012) Seroprevalence of Schmallenberg virus antibodies among dairy cattle, The Netherlands, winter 2011–2012. Emerg Infect Dis 18:1065–1071. https://doi.org/10.3201/eid1807.120323
3. Coupeau D, Claine F, Wiggers L, Kirschvink N, Muylkens B (2013) In vivo and in vitro identification of a hypervariable region in Schmallenberg virus. J Gen Virol 94:1168–1174. https://doi.org/10.1099/vir.0.051821-0
4. Elbers ARW, Meiswinkel R, van Weezep E, Sloet van Oldruitenborgh-Oosterbaan MM, Kooi EA (2013) Schmallenberg virus in Culicoides spp. biting midges, The Netherlands, 2011. Emerg Infect Dis 19:106–109. https://doi.org/10.3201/eid1901.121054
5. Rasmussen LD, Kristensen B, Kirkeby C, Rasmussen TB, Belsham GJ, Bødker R, Bøtner A (2012) Culicoids as vectors of Schmallenberg virus. Emerg Infect Dis 18:1204–1206. https://doi.org/10.3201/eid1807.120385
6. De Regge N, Madder M, Deblauwe I, Losson B, Fassotte C, Demeulemeester J, Smeets F, Tomme M, Cay AB (2014) Schmallenberg virus circulation in culicoides in Belgium in 2012: field validation of a real time RT-PCR approach to assess virus replication and dissemination in midges. PLoS One 9: e87005. https://doi.org/10.1371/journal.pone.0087005
7. Lassen SB, Nielsen SA, Kristensen M (2012) Identity and diversity of blood meal hosts of biting midges (Diptera: Ceratopogonidae: Culicoides Latreille) in Denmark. Parasit Vectors 5:143. https://doi.org/10.1186/1756-3305-5-143
8. European Food Safety Authority (2014) EFSA assesses current knowledge of Schmallenberg virus. Vet Rec 174:566. https://doi.org/10.1136/vr.g3572
9. Martinelle L, Poskin A, Dal Pozzo F, De Regge N, Cay B, Saegerman C (2015) Experimental infection of sheep at 45 and 60 days of gestation with Schmallenberg virus readily led to placental colonization without causing congenital malformations. PLoS One 10: e0139375. https://doi.org/10.1371/journal.pone.0139375
10. Wernike K, Eschbaumer M, Schirrmeier H, Blohm U, Breithaupt A, Hoffmann B, Beer M (2013) Oral exposure, reinfection and cellular immunity to Schmallenberg virus in cattle. Vet Microbiol 165:155–159. https://doi.org/10.1016/j.vetmic.2013.01.040
11. Martinelle L, Poskin A, Dal Pozzo F, Mostin L, Van Campe W, Cay AB, De Regge N, Saegerman C (2017) Three different routes of inoculation for experimental infection with Schmallenberg virus in sheep. Transbound Emerg Dis 64:305–308. https://doi.org/10.1111/tbed.12356
12. König P, Wernike K, Hechinger S, Tauscher K, Breithaupt A, Beer M (2019) Fetal infection with Schmallenberg virus – an experimental pathogenesis study in pregnant cows. Transbound Emerg Dis 66:454–462. https://doi.org/10.1111/tbed.13045
13. Garigliany M-M, Bayrou C, Kleijnen D, Cassart D, Desmecht D (2012) Schmallenberg virus in domestic cattle, Belgium, 2012. Emerg Infect Dis 18:1512–1514. https://doi.org/10.3201/eid1809.120716
14. Tizard IR (2012) Veterinary immunology. Elsevier Health Sciences
15. Kurogi H, Inaba Y, Takahashi E, Sato K, Satoda K (1977) Congenital abnormalities in newborn calves after inoculation of pregnant cows with Akabane virus. Infect Immun 17:338–343

16. Bayrou C, Garigliany M-M, Sarlet M, Sartelet A, Cassart D, Desmecht D (2014) Natural intrauterine infection with Schmallenberg virus in malformed newborn calves. Emerg Infect Dis 20:1327–1330. https://doi.org/10.3201/eid2008.121890
17. Peperkamp NH, Luttikholt SJ, Dijkman R, Vos JH, Junker K, Greijdanus S, Roumen MP, van Garderen E, Meertens N, van Maanen C et al (2015) Ovine and bovine congenital abnormalities associated with intrauterine infection with Schmallenberg virus. Vet Pathol 52:1057–1066. https://doi.org/10.1177/0300985814560231
18. Tilston-Lunel NL, Shi X, Elliott RM, Acrani GO (2017) The potential for reassortment between Oropouche and Schmallenberg orthobunyaviruses. Viruses 9(8):220. https://doi.org/10.3390/v9080220. PMID: 28800086. PMCID: PMC5580477
19. Bilk S, Schulze C, Fischer M, Beer M, Hlinak A, Hoffmann B (2012) Organ distribution of Schmallenberg virus RNA in malformed newborns. Vet Microbiol 159(1–2):236–238
20. La Rocca SA, Sandvik T (2009) A short target real-time RT-PCR assay for detection of pestiviruses infecting cattle. J Virol Methods 161(1):122–127
21. Toussaint JF, Sailleau C, Breard E, Zientara S, De Clercq K (2007) Bluetongue virus detection by two real-time RT-qPCRs targeting two different genomic segments. J Virol Methods 140(1–2):115–123

Chapter 18

Reverse Genetics System for Crimean-Congo Hemorrhagic Fever Virus

Florine Scholte, Elif Karaaslan, and Éric Bergeron

Abstract

Reverse genetic systems are powerful tools in molecular virology that allow the generation of infectious recombinant virus and the manipulation of viral genomes. Reverse genetic systems enable the incorporation of reporter genes, facilitating many virological assays, including high-throughput screening. Additionally, reverse genetic systems can be used to introduce targeted mutations into the viral genome, allowing investigations of viral genetic elements and protein functions in virus pathogenesis and biology. Here we describe in detail the materials and methods required for the Crimean-Congo hemorrhagic fever virus (CCHFV) reverse genetic system. This system can be used to generate complete infectious recombinant virus, and virus-like replicon particles (VRPs) lacking the M segment but complemented with an exogenous source of glycoprotein precursor (GPC); resulting in single-round replicon particles that can be used to study components of the viral replicative cycle at a lower biosafety level.

Key words CCHFV, Reverse genetics, Transfection, Recombinant virus, Virus-like replicon particle, Nairoviridae, RNA virus

1 Introduction

Reverse genetic systems can be used to engineer recombinant viruses with reporter genes or targeted mutations to study virus biology and pathogenesis both in vitro and in vivo. The development of fast and affordable synthetic biology has greatly facilitated the use of reverse genetic systems and the study of certain viruses for which only genome sequences are available, such as Bombali virus [1]. Reverse genetic systems can be used to add components, like reporter genes or protein tags such as FLAG/V5, to the viral genome to facilitate proteomics studies. Reverse genetic systems can also be used to delete or mutate parts of the viral genome, facilitating the study of viral proteins or domains and their interactions with the host cellular machinery. They can be used to introduce targeted mutations to investigate specific binding sites or catalytic activity, or be used to scramble the whole genome, by,

Hani Boshra (ed.), *Bunyaviruses: Methods and Protocols*, Methods in Molecular Biology, vol. 2893,
https://doi.org/10.1007/978-1-0716-4338-9_18,

for example, deoptimizing codon usage, an approach that has been used to create successful vaccine candidates against several viral diseases. Reverse genetic systems have been applied to studies of pathology, molecular characterization, and vaccine development.

Crimean-Congo hemorrhagic fever virus (CCHFV) is a negative-sense RNA virus with three genome segments, named small (S), medium (M), and large (L), encoding respectively the nucleoprotein (NP), glycoprotein precursor (GPC), and L protein, which functions as the RNA-dependent RNA polymerase. The GPC is a polyprotein that is cotranslationally and posttranslationally cleaved by host proteases to release various structural and nonstructural glycoproteins, including Gn, Gc, and GP38. Gn and Gc, in turn, mediate the entry of viral particles into host cells.

CCHFV belongs to the *Orthonairovirus* genus in the *Nairoviridae* family. The vast majority of orthonairoviruses can be found in ticks which transmit the virus to humans and animals. Several other nairoviruses can cause severe illness in humans, including newly identified Songling virus, Beiji virus and Yezo virus, and older human pathogens such as Dugbe virus and Nairobi sheep disease virus. CCHFV is the most geographically widespread and clinically relevant nairovirus, frequently causing severe disease in humans. Nairoviruses are unique in the bunyavirus order due to the unusual large size of their L proteins, which also contain an N-terminal ovarian tumor (OTU) protease domain [2].

A CCHFV minigenome consisting of a chloramphenicol acetyl transferase reporter gene flanked by noncoding regions of the S segment was first described in 2003, but its transcription and replication strictly relied on CCHFV superinfection of transfected cells, limiting its use to the study of noncoding RNA elements [3]. The generation of transcriptionally active ribonucleoproteins and virus-like particles encoding a reporter gene (minigenomes) using a recombinant source of CCHFV protein (plasmids) was later described, making the study of modifications of the viral protein sequence possible [4–6]. These assays can be used as life cycle modeling systems, examining individual steps of the viral replicative cycle at a lower biocontainment level. In 2015, a reverse genetic system able to generate infectious recombinant CCHFV was described, forming the first such system for nairoviruses [7]. More recently, a similar reverse genetics approach to recover the related and nonpathogenic Hazara virus was reported, suggesting that this method could be applicable to other nairoviruses [8]. Recombinant CCHFV has been used to examine various components of the viral genome, including GPC processing, OTU immune modulation, and NSm functionality in vitro and in vivo [7, 9, 10]. Recombinant CCHFV containing a reporter gene (e.g., ZsGreen) has been used to determine in vivo tropism and is adaptable to high-throughput drug screens [11–13].

For these reverse genetic systems, the plasmids used encode the CCHFV genome segments in positive-sense orientation under the control of a bacteriophage T7 RNA polymerase promoter. Plasmids containing full-length cDNA from the S, M, and L segments were obtained by reverse transcription of CCHFV RNA extracted from virus stocks, using a DNA primer complementary to the 3′ end of the viral genomic RNA, followed by PCR amplification and cloning into the pT7 vector [7]. The pT7 vector was previously described as V (0.0)/B, and contains a 5′ T7 RNA polymerase promoter, a 3′ hepatitis D ribozyme and a T7 RNA polymerase terminator motif, flanking the complete viral segment sequences [14]. When co-transfected with a plasmid expressing the T7 polymerase, negative-sense genomic RNA is produced. The RNA is then encapsidated by CCHFV NP and L proteins expressed by helper plasmids, resulting in the production of all viral proteins and subsequent rounds of production of viral RNA and proteins. Helper plasmids are generated by inserting the open reading frame (ORF) of CCHFV NP and codon-optimized L protein (LCO) into the mammalian expression vector pCAGGS [7].

The CCHFV reverse genetic system can be adapted to generate single-round virus-like replicon particles (VRPs) by omitting the transfection of the pT7-M plasmid and supplying helper GPC *in trans* (Fig. 1). The CCHF VRP system can be used to safely study various steps of the CCHFV replicative cycle at a lower biosafety level. Biosafety studies demonstrated that inoculation of immunocompromised animals like IFNAR$^{-/-}$ mice and suckling mice did not result in clinical signs or disease [15]. Furthermore, immunization of mice with VRPs protected mice from a lethal CCHFV challenge [15–18].

These recently described recombinant infectious viruses and VRPs can be produced in highly transfectable cell lines. The complete system results in recombinant infectious virus that can be used to study the function of various viral proteins and their functions in pathogenesis and biology [7]. The VRP system can be used at biosafety level 2 (BSL-2) and is a convenient tool to study nearly all components of CCHFV's replicative cycle. In principle, VRPs are single-round particles, since they do not contain the M segment. However, particle formation and spread can be studied in cell lines that express the CCHFV GPC *in trans.*

It is important to note that the plasmids used to generate VRPs should be stored physically separated from the plasmids used to generate infectious virus to avoid any risk of contamination and accidental rescue of complete infectious virus.

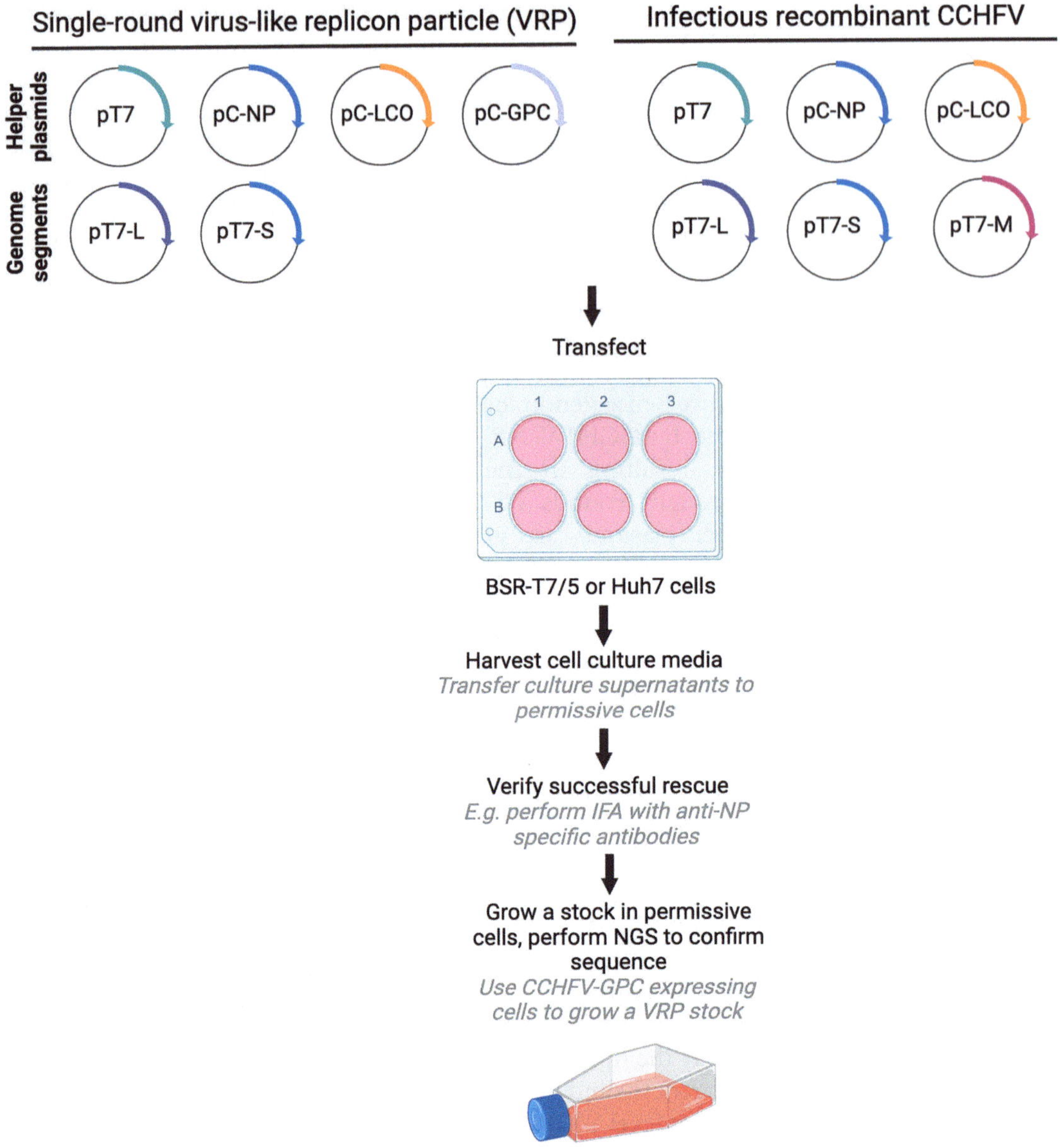

Fig. 1 Schematic overview of CCHFV reverse genetic systems generating full infectious virus or single-round virus-like replicon particles. Highly transfectable cells permissive for CCHFV are transfected with plasmids expressing the CCHFV genome segments under control of a T7 promoter, alongside helper plasmids expressing NP, L and the T7 polymerase. Single-round virus-like replicon particles (VRPs) are generated by replacing the M genome segment with a helper plasmid expressing the GPC. Successful rescue of infectious virus or VRP is verified by transferring supernatant to fresh permissive cells and performing IFA staining for CCHFV NP. Amplification of VRP stocks requires cells expressing CCHFV GPC in trans. Created with BioRender.com

2 Materials

All plasmids should be of high quality, sequence-verified, and endotoxin-free. Only perform transfections resulting in infectious virus in appropriate biosafety containment (i.e., BSL-4 laboratory).

2.1 Cells and Media

1. Use cells that are both highly transfectable and highly permissive to CCHFV. We use either Huh7 or BSR-T7/5 cells (*see* **Note 1**). The BSR-T7/5 cells were a kind gift from K.K. Conzelmann (Ludwig-Maximilians-Universität, Munich, Germany).
2. Cell culture media for Huh7 or BSR-T7 cells: DMEM supplemented with 5% fetal bovine serum, 1 mM sodium pyruvate, 1% penicillin-streptomycin, 1% nonessential amino acids, and 1% Glutamax.
3. For amplification of recombinant infectious CCHFV: a wide range of cell lines is permissive to CCHFV infection, with some cell lines resulting in higher viral titers than others. We routinely use Huh7 or BSR-T7/5 cells to grow and titer virus stocks.
4. For VRP amplification: use cells expressing CCHFV GPC *in trans*. We use Vero cells stably expressing codon-optimized CCHFV GPC. These cells require 30 μg/mL puromycin during maintenance. Alternatively, we have used Huh7 cells transiently transfected with a plasmid encoding codon-optimized CCHFV GPC to generate VRP stocks (*see* **Note 2**).

2.2 Transfection to Generate Infectious Recombinant CCHFV

1. Transfection medium: OptiMEM.
2. TransIT LT1 transfection reagent (Mirus Bio). Use 2.0–2.5 μL per μg of DNA.
3. Helper plasmids encoding T7 polymerase, CCHFV NP, and codon-optimized CCHFV L (LCO) (*see* **Notes 3** and **4**; *see* Table 1).
4. Plasmids encoding the 3 genome segments under control of the T7 polymerase (pT7-L, pT7-M, pT7-S) (*see* **Note 5**; *see* Table 1).

2.3 Transfection to Generate CCHF VRPs (Single Round)

1. Transfection medium: OptiMEM.
2. TransIT LT1 transfection reagent (Mirus Bio). Use 2.0–2.5 μL per μg of DNA.
3. Helper plasmids encoding T7 polymerase, CCHFV NP, and codon-optimized CCHFV L and GPC (*see* Table 1).
4. Plasmids encoding the CCHFV S and L genome segments under control of the T7 polymerase. Do not include any

Table 1
Plasmid quantities required to transfect a single well of a six-well plate (9.5 cm^2)

Plasmid	Infectious CCHFV (ng DNA)	VRP (ng DNA)
pT7-L	1000	1000
pT7-M	2500	–
pT7-S	1000	1000
Helper plasmids		
pT7	1000	1000
pC-NP	660	660
pC-LCO	330	330
pC-GPC	–	1000

plasmid designed to express the complete M segment here to avoid generation of infectious virus (*see* Table 1).

2.4 Verification of Successful Rescue

1. Fixative: 37% formaldehyde solution + 0.1% Triton X-100 (*see* **Note 6**).
2. Rabbit anti-CCHFV NP polyclonal antibody (IBT Bioservices #04-0011; *see* **Note 7**).
3. Goat anti-rabbit IgG (H&L) Alexa Fluor 488.
4. PBS for wash steps.
5. Antibody dilution buffer: PBS containing 0.5% BSA.
6. Inverted fluorescent microscope (e.g., EVOS Cell Imaging System).

3 Methods

3.1 Transfection

1. The day before transfection, prepare BSR-T7/5 or Huh7 cells in six-well plates to reach 70–80% confluence on the day of transfection (*see* **Note 8**).
2. Briefly vortex each plasmid tube and briefly centrifuge the tubes to collect plasmids on the bottom of the tubes.
3. Add 200 μL OptiMEM to a sterile 2.0 mL screwcap tube, and add the appropriate amount of each plasmid: 1 μg pT7-S, 2.5 μg pT7-M, 1 μg pT7-L. Add helper plasmids: 0.66 μg pC-NP, 0.33 μg pC-LCO, and 1 μg pC-T7 (*see* Table 1).
4. For the generation of infectious recombinant virus, move the cells, plasmid mixtures, and LT1 transfection reagent into a BSL-4 laboratory.

5. Bring TransIT-LT1 transfection reagent to room temperature and vortex gently. Add LT1 at a ratio of 2:1 (LT1 to μg plasmids) to the plasmid mixture. Do not exceed 12 μL LT1 per 6-well reaction.
6. Mix thoroughly by flicking the tube or vortexing briefly.
7. Incubate lipid/plasmid mixtures 15–30 min at room temperature.
8. Add transfection mixture dropwise to different areas of the well. Place the plate in a humidified incubator at 37 °C and incubate overnight.
9. The next day, remove the media containing the transfection mixture and replace with fresh culture media (*see* **Note 9**).
10. On days 2, 3, and 4 after transfection, harvest cell culture media and store at −80 °C until later use. Replace media with 2–4 mL fresh culture media per well.

3.2 Verification of Successful Rescue of Recombinant Infectious Virus or VRPs

1. Seed fresh Huh7 cells (or other permissive cells) in a 24-well plate. Seed as many wells as samples harvested, plus one extra well to serve as a negative control (mock) (*see* **Note 10**).
2. The next day, thaw the stored supernatants harvested from the 6-well plates. After removing the supernatant from the 24-well plate wells, add 500 μL of thawed supernatant to each well. Place 24-well plate in the incubator and incubate for 4–5 days.
3. Fix and permeabilize cells by adding 55 μL of 37% formaldehyde containing 0.1% Triton X-100 directly to the well (final concentration 3.7% formaldehyde and 0.01% Triton X-100) and incubate for 20 min at room temperature (*see* **Note 11**).
4. Remove supernatant containing fixative and wash cell monolayer 3× with PBS.
5. Dilute CCHFV NP antibody 1:2500 in PBS/0.5%BSA and add 200 μL to each well. Incubate for 1 h at room temperature or 37 °C.
6. Dilute the secondary antibody (goat anti-rabbit-Alexa488) 1: 2500 in PBS/0.5%BSA. Wash each well 3× with PBS and add 200 μL of the diluted secondary antibody to each well. Incubate for 1 h at room temperature or at 37 °C.
7. Wash each well 3× with PBS and end with leaving 500 μL PBS in each well.
8. Examine NP-positive wells using an inverted fluorescent microscope using the mock well as a negative control (*see* **Note 12**).

3.3 Growing a Stock of Infectious Recombinant Virus or Single-Round VRPs

1. After successful recovery of recombinant virus, grow a p1 stock by adding the remainder of the frozen p0 aliquot(s) to a T75 flask containing fresh permissive cells (e.g., Huh7, BSR-T7/5, or SW13 cells).
2. A stock of VRPs can be grown in cells expressing the CCHFV GPC *in trans*, such as Vero cells stably expressing the Oman GPC.
3. Confirm the genetic sequence of the rescued recombinant virus or VRP stock using (next generation) sequencing.

4 Notes

1. BSR-T7/5 cells are poorly permissive to CCHFV after transfection. Therefore, transfection efficiency needs to be carefully optimized for these cells.
2. High levels of GPC are required for efficient VRP amplification. However, too low or too high levels of GPC can hinder initial VRP generation.
3. When using BSR-T7/5 or another cell line stably expressing the T7 polymerase, plasmid-derived T7 is not an absolute requirement. However, we found that including the T7 helper plasmid enhances rescue efficiency, likely because T7 expression in stable cell lines can vary with passaging and antibiotic selection regimens.
4. In our hands, codon optimization of the helper L protein greatly improved activity and expression [7].
5. The pT7-L plasmid is unstable in *E. coli*, so ensure that new plasmid preps are of good quality (e.g., by running on agarose gel after control digestions). We recommend growing the bacteria at 32 °C instead of 37 °C. After retransformation of the plasmid, pick small colonies rather than large ones.
6. For these methods, we operate under the assumption that no reporter genes are present, and therefore successful rescue of VRP or recombinant virus needs to be confirmed by immunofluorescence staining or by passaging of the stock on a fresh monolayer of SW13 cells and monitoring for the appearance of cytopathic effect (CPE). Cytopathic effect is not reliably observed in Huh7 or BSR-T7/5 cells.
7. Other CCHFV antibodies or antisera can be used, though we recommend antibodies targeting NP for stronger immunofluorescence.
8. Other cell lines could be used but they have to be highly transfectable and highly permissive for CCHFV. In our hands, some cell lines (e.g., BSR-T7/5) are not easily infected

after transfection. Visually inspect cells the day of infection. Cells should not be under- or over-confluent. Inspect Huh7 cells for fatty vacuoles, a sign of stress.

9. In particular, Huh7 cells are fairly sensitive to the prolonged presence of the transfection reagent. Alternatively, the transfection media can also be replaced after 4–6 h if the media cannot be changed the next day.
10. CCHFV NP immunofluorescence is easier to score when the cells are not overconfluent at time of fixation, so seeding a lower concentration of cells is recommended (e.g., 50,000 cells/well).
11. Alternatively, SW13 cells can be used and cytopathic effect can be monitored as an indicator of infectious virus. However, in our hands, CCHFV-induced CPE is not reliably observed; therefore, we prefer to use immunofluorescence to confirm the presence of infectious virus in the absence of a reporter gene. If using CPE as a read-out for successful rescue, include a negative control in which one CCHFV genome segment is excluded (e.g., the pT7-L plasmid).
12. If an inverted microscope is not available, seed cells on glass coverslips to test the p0 supernatants.

Acknowledgement

We thank Tatyana Klimova for assistance with editing the manuscript.

Disclaimer The findings and conclusions in this report are those of the authors and do not necessarily represent the official position of the Centers for Disease Control and Prevention.

Financial Support This work was supported by Centers for Disease Control and Prevention (CDC) Emerging Infectious Disease Research Core Funds and by an appointment to the CDC administered by the Oak Ridge Institute for Science and Education (ORISE) through an interagency agreement between the US Department of Energy (DOE) and the US Department of Agriculture, Agricultural Research Service (to E.K.). ORISE is managed by Oak Ridge Associated Universities under contract with DOE.

References

1. Bodmer BS, Breithaupt A, Heung M, Brunetti JE, Henkel C, Müller-Guhl J et al (2023) In vivo characterization of the novel ebolavirus Bombali virus suggests a low pathogenic potential for humans. Emerg Microbes Infect 12(1):2164216

2. Kuhn JH, Wiley MR, Rodriguez SE, Bào Y, Prieto K, Travassos da Rosa APA et al (2016) Genomic characterization of the genus Nairovirus (family Bunyaviridae). Viruses 8(6):164
3. Flick R, Flick K, Feldmann H, Elgh F (2003) Reverse genetics for Crimean-Congo hemorrhagic fever virus. J Virol 77(10):5997–6006
4. Bergeron E, Albariño CG, Khristova ML, Nichol ST (2010) Crimean-Congo hemorrhagic fever virus-encoded ovarian tumor protease activity is dispensable for virus RNA polymerase function. J Virol 84(1):216–226
5. Zivcec M, Metcalfe MG, Albariño CG, Guerrero LW, Pegan SD, Spiropoulou CF et al (2015) Assessment of inhibitors of pathogenic Crimean-Congo hemorrhagic fever virus strains using virus-like particles. PLoS Negl Trop Dis 9(12):1–23
6. Devignot S, Bergeron E, Nichol S, Mirazimi A, Weber F (2015) A virus-like particle system identifies the endonuclease domain of Crimean-Congo hemorrhagic fever virus. J Virol 89(11):5957–5967
7. Bergeron E, Zivcec M, Chakrabarti AK, Nichol ST, Albariño CG, Spiropoulou CF (2015) Recovery of recombinant Crimean Congo hemorrhagic fever virus reveals a function for non-structural glycoproteins cleavage by Furin. PLoS Pathog 11(5):e1004879
8. Fuller J, Surtees RA, Slack GS, Mankouri J, Hewson R, Barr JN et al (2019) Rescue of infectious recombinant Hazara nairovirus from cDNA reveals the nucleocapsid protein DQVD caspase cleavage motif performs an essential role other than cleavage. J virol 93(15):e00616–e00619
9. Welch SR, Scholte FEM, Spengler JR, Ritter JM, Coleman-McCray JD, Harmon JR et al (2020) The Crimean-Congo hemorrhagic fever virus NSm protein is dispensable for growth in vitro and disease in Ifnar-/- mice. Microorganisms 8(5):775
10. Scholte FEM, Zivcec M, Dzimianski JV, Deaton MK, Spengler JR, Welch SR et al (2017) Crimean-Congo hemorrhagic fever virus suppresses innate immune responses via a ubiquitin and ISG15 specific protease. Cell Rep 20(10):2396–2407
11. Mears MC, Rodriguez SE, Schmitz KS, Padilla A, Biswas S, Cajimat MNB et al (2022) Design and evaluation of neutralizing and fusion inhibitory peptides to Crimean-Congo hemorrhagic fever virus. Antivir Res 207:105401
12. Welch SR, Ritter JM, McElroy AK, Harmon JR, Coleman-McCray JAD, Scholte FEM et al (2019) Fluorescent Crimean-Congo hemorrhagic fever virus illuminates tissue tropism patterns and identifies early mononuclear phagocytic cell targets in IFNAR-/- mice. PLoS Pathog 15(12):e1008183
13. Welch SR, Scholte FEM, Flint M, Chatterjee P, Nichol ST, Bergeron É et al (2017) Identification of 2′-deoxy-2′-fluorocytidine as a potent inhibitor of Crimean-Congo hemorrhagic fever virus replication using a recombinant fluorescent reporter virus. Antivir Res 147: 91–99
14. Albarino CG, Bird BH, Chakrabarti AK, Dodd KA, Erickson BR, Nichol ST (2011) Efficient rescue of recombinant lassa virus reveals the influence of S segment noncoding regions on virus replication and virulence. J Virol 85(8): 4020–4024
15. Spengler JR, Welch SR, Scholte FEM, Coleman-McCray JAD, Harmon JR, Nichol ST et al (2019) Heterologous protection against Crimean-Congo hemorrhagic fever in mice after a single dose of replicon particle vaccine. Antivir Res 170:104573
16. Scholte FEM, Spengler JR, Welch SR, Harmon JR, Coleman-McCray JD, Freitas BT et al (2019) Single-dose replicon particle vaccine provides complete protection against Crimean-Congo hemorrhagic fever virus in mice. Emerg Infect Dis 8(1):575–578
17. Spengler JR, Welch SR, Scholte FEM, Rodriguez SE, Harmon JR, Coleman-McCray JD et al (2021) Viral replicon particles protect IFNAR-/- mice against lethal Crimean-Congo hemorrhagic fever virus challenge three days after vaccination. Antivir Res 191: 105090
18. Scholte FEM, Karaaslan E, O'Neal TJ, Sorvillo TE, Genzer SC, Welch SR et al (2023) Vaccination with the Crimean-Congo hemorrhagic fever virus viral replicon vaccine induces NP-based T-cell activation and antibodies possessing Fc-mediated effector functions. Front Cell Infect Microbiol 13:1233148

Chapter 19

Recombinant Vaccine Production: Production of a Recombinant CCHF MVA Vaccine

Emma Kennedy, Roger Hewson, and Stuart Dowall

Abstract

One of the key interventions against infection is immunization, including an increasing focus on development of vaccines against pathogenic bunyaviruses. Whilst different vaccine development approaches exist, recombinant viral vaccines have a strong safety record, are rapid to produce, are cost-effective, and have been demonstrated to be rolled out in response to outbreaks, including in low- and middle-income countries. One viral vector, modified Vaccinia Ankara (MVA), has been used to develop vaccine candidates against Crimean-Congo Haemorrhagic Fever (CCHF) virus through incorporation of the nucleoprotein (NP) and glycoprotein (GP) regions, with the former candidate having now progressed to being the first vaccine against CCHF virus to enter Phase 1 clinical trials. Herein, we report the method used to generate this MVA-based vaccine construct.

Key words Modified Vaccinia Ankara, Immunogenicity, Vaccine, Recombinant protein

1 Introduction

Bunyaviruses continue to present a threat to human health, causing a significant disease burden and potential to cause large disease outbreaks. Due to mainly being spread by arthropod vectors, their geographical range is continuing to alter as a result of changing climate, resulting in expansion of susceptible populations. One major bunyavirus of concern is Crimean-Congo Haemorrhagic Fever (CCHF) virus, a hazard group 4 tick-borne pathogen which is endemic across a wide geographic range and listed by the World Health Organization in their Research & Development (R&D) Blueprint as a priority pathogen requiring urgent research and development efforts [1]. Whilst various vaccination strategies for CCHF virus exist [2], one based on using the modified Vaccinia Ankara was the first to demonstrate complete protection against disease in a lethal challenge model [3] and the first to enter Phase 1 human clinical trials.

Hani Boshra (ed.), *Bunyaviruses: Methods and Protocols*, Methods in Molecular Biology, vol. 2893,
https://doi.org/10.1007/978-1-0716-4338-9_19,

Vaccinia was successfully used in the eradication campaign of smallpox, but safety complications, particularly in immunocompromised individuals, halted its use as a suitable vaccine vector candidate. Following continual passages of the Vaccinia Ankara strain in Chicken Embryo Fibroblasts (CEF's), the virus became highly attenuated [4, 5], and known as MVA has subsequently been used as a successful smallpox vaccine [6]. Whilst MVA does not fully replicate in mammalian cells lines, it is still able to produce high levels of expression of recombinant proteins. MVA has subsequently been used as a viral vector for vaccine development as an effective method to deliver antigenic target genetic material into a host to elicit an immune response, and to protect against subsequent natural infection with pathogenic viruses. As such, MVA has been demonstrated to be an efficacious vector for several viral antigens including: Lassa virus (LASV) [7], influenza [8, 9], and HIV [10].

During development of vaccine candidates for CCHF virus, MVA was developed to express the two major antigenic encoding regions of the virus: the S segment encoding the nucleoprotein and the M segment encoding the glycoproteins. Whilst both candidates demonstrated strong immune responses specific to the antigenic target, after challenge using a susceptible mouse model the nucleoprotein-targeting vaccine was not protective [11] whereas the glycoprotein-targeting vaccine resulted in 100% protection [3]. Further analysis indicated that protection required both humoral and cellular immunity [12].

To accompany standard methodologies required to produce recombinant MVA vaccine [13], in this chapter we have detailed the specific protocols used for the development of the MVA-based vaccine against CCHF virus. The protocols cover the complete stages of the procedure: (1) Cell culture, (2) Initial transfection of MVA and viral target genome to insert the sequence of choice into MVA, (3) purification of recombinant vaccine, (4) DNA extraction, (5) PCR confirmation of pure vaccine stocks and amplification, (6) purification by sucrose cushion method, and (7) confirmation of vaccine titer using immunostaining to determine a concentration of virus as plaque forming units per mL (pfu/mL).

2 Materials

2.1 Cell Culture

1. Minimum Essential Media Eagle (EMEM) with Earles salts and containing L-glutamine.
2. Penicillin-streptomycin solution containing 10,000 units penicillin and 10 mg streptomycin per mL.
3. Foetal bovine serum of Australian origin (*see* **Note 1**).
4. Phosphate buffered saline (PBS), pH 7.4.

5. 0.25% Trypsin/EDTA solution comprising of 2.5 g porcine trypsin and 0.2 g EDTA per litre in Hanks' Balanced Salt solution containing phenol red.
6. M10 media comprised of EMEM, L-glutamine (1%) or equivalent, penicillin-streptomycin (1%), foetal bovine serum (10%).
7. Chicken embryo fibroblast (CEF) (*see* **Note 2**) or baby hamster kidney fibroblasts (BHK) cells (*see* **Note 3**).
8. 0.4% trypan blue solution.

2.2 Infection/Transfection

1. Tissue culture media components and reagents detailed above (EMEM, L-glutamine or equivalent, penicillin-streptomycin, PBS, FBS, trypsin/EDTA).
2. Lipofectamine™ transfection reagent.
3. Opti-MEM (reduced serum medium).
4. M0 media comprised of EMEM, L-glutamine (1%) or equivalent, penicillin-streptomycin (1%).
5. M2 media comprised of M0 with 2% FBS.
6. M10 media comprised of M0 with 10% FBS.
7. CEF or BHK cells.
8. Modified Vaccinia Ankara virus (MVA) (*see* **Note 4**).
9. Plasmid design containing viral protein genome of choice (*see* **Note 5**).

2.3 rMVA Plaque Purification

1. Tissue culture media components and reagents detailed previously.
2. Minimum Essential Media (MEM), Temin's modification (2×), no phenol red.
3. 2× MEM comprised of MEM ×2, L-glutamine (1%), penicillin/streptomycin (1%), FBS (4%).
4. 2-Hydroxyethylagarose type VII, a low gelling temperature agarose suitable for cell culture.
5. 2% w/v 2-Hyroxyethylagarose comprised of DI water (98%), 2-Hydroxyethylagarose type VII (2%).

2.4 DNA Extraction

1. Genomic DNA purification kit.
2. Molecular biology grade, nuclease-free water.

2.5 PCR Confirmation of Plaques

1. Tris-acetate-EDTA (TAE) buffer containing 40 mM Tris, 40 mM acetate and 1 mM EDTA, pH ~8.3.
2. Molecular biology grade, nuclease-free water.
3. DNA polymerase (KAPA HiFi HotStart Ready mix).
4. UltraPure Agarose.

5. SyberSafe gel stain.
6. DNA loading dye.
7. DNA ladder.
8. Custom oligonucleotides in forward and reverse to detect viral sequence of choice.

2.6 Sucrose Cushion Purification

1. 36% Sucrose solution comprised of 72 g Ultrapure sucrose (≥99.5%) in 200 mL 10 mM Tris-Cl.
2. 100 mM Tris-Cl comprised of Trizma® Preset crystals pH 8.8 in pyrogen free water.
3. 10 mM Tris-Cl.

2.7 Titration and Immunostain

1. Tissue culture media and reagents as previously described.
2. 2% w/v 2-Hydroxyethylagarose.
3. Methanol, anhydrous (99.8%).
4. Primary antibody e.g. *Rabbit anti-VACV* (*see* **Note 6**).
5. Vectastain alkaline phosphatase kit with appropriate specificity to primary antibody (e.g. *Vector labs anti-rabbit)*.
6. Tween® 20, nonionic, aqueous solution, 10% (w/v).
7. Alkaline phosphatase blue/red substrate kit.
8. 100-200 mM Tris-HCl, pH 8.2–8.8.

2.8 Equipment

The following list of equipment is required throughout the methods section.

1. Microbiological safety cabinet class 2 (MSCII).
2. Cup-horn sonicator or equivalent.
3. Vortex mixer.
4. Adjustable temperature water bath.
5. Inverted and Fluorescence microscope.
6. Centrifuge and ultra-centrifuge.
7. Cell culture incubator at 37 °C, 5% CO_2.
8. Thermocycler.
9. Gel electrophoresis power pack.
10. Gel imager.
11. Plate rocker.

2.9 Requisites

1. Sterile Dounce homogenizer.
2. Tissue culture flasks.
3. Tissue culture plates.
4. Serological pipettes.

5. PCR RNA/DNase-free tubes.
6. Bijou.
7. Gel casting mould and comb.
8. Beaker/conical flasks.
9. 50 and 15 mL falcon tubes
10. Oakridge centrifuge tubes—or equivalent.

3 Methods

3.1 Preparation of CEF and BHK-21 Cells

All manipulations with cell cultures and MVA virus are performed within a class II microbiological safety cabinets using aseptic technique.

1. CEFs are a primary cell line and are delivered fresh. On arrival, the CEF cells should, where possible, be processed within 2 h to ensure that viability is not lost (*see* **Note 7**). Transfer the cell suspension to a 50 mL tube and then rinse the universal with 20 mL of M10 media and add to the cell suspension. This is then centrifuged at 500×*g* for 10 min at 20 °C. Remove the supernatant from the pellet and resuspend in 25 mL of M10 media and mix gently by pipetting. Then add a further 25 mL of M10 media to make a final volume of 50 mL. Perform a cell count [14] and add 5 mL cell suspension to each T175 flasks with 20 mL of M10 media. Transfer the flasks to the incubator (37 °C, 5% CO_2) and incubate for 24 h. After 24 h, inspect the cells microscopically. If the cells are 85% or higher confluent, the cells can be transferred to a 31 °C incubator (5% CO_2) and stored until use. If the cells are less than 85% confluent, leave the flasks at 37 °C until this level of confluency has been achieved.
2. BHK cells are used at >85% confluency and prepared using standard cell culture protocols [15, 16].

3.2 Infection/Transfection of Plasmid

1. Seed T25 flasks with CEF cells and incubate at 37 °C with 5% CO_2 until 85–90% confluency is reached.
2. A single T25 flask is sacrificed for cell counting [14].
3. For infection of cells with MVA, calculate the amount of virus to achieve this using the following formula:

   A = cell number calculated in **step 2** (e.g., 3×10^6 cells).

   B = Multiplicity of infection (MOI) required (usually 0.05).

   $C = A \times B$ (pfu required) (e.g.: $3 \times 10^6 \times 0.05 = 1.5 \times 10^5$).

   D = Concentration of stock MVA (e.g.: 4.9×10^6 pfu/mL)

$$\text{Volume of virus required } (\mu L) = (C/D) \times 1000$$

$$\text{e.g.: } = \left(1.5 \times 10^5 / 4.9 \times 10^6\right) \times 1000$$
$$30.6 \ \mu L \text{ of virus/infection}$$

4. Thaw a vial of MVA and vortex for 30 s, then sonicate for 30 s (pulsing 1 s on: 1 s off).
5. Add 2.5 mL M0 media to a 15 mL falcon tube. Gradually dilute, with vortexing, the required volume of MVA (as calculated in **step 3**).
6. Remove all media from the T25 flask for infection.
7. Add the 2.5 mL of diluted virus to the T25 flask, rocking gently to ensure the monolayer is evenly coated.
8. Incubate at 37 °C with 5% CO_2 for 90 min.
9. The flask should be rocked every 15 min to ensure good coverage of the monolayer and increase the efficiency of the infection.
10. Aseptically, add 70 μL of lipofectamine to a sterile bijou containing 280 μL of Opti-MEM.
11. Calculate the amount of DNA required (V1) to achieve a concentration of 0.4 μg of DNA per 25 μL in 350 μL (0.4 μg per 25 μL = 0.4 μg/25 μL = 0.016 μg/μL of DNA) following the formula below:

    C1 = Concentration of DNA (e.g. 0.5 μg/μL).

    C2 = Concentration of DNA required (0.016 μg/μL).

    V1 = Final volume of media required (350 μL for a T25).

    V2 = Volume of DNA in sample required.

    V2 − V1 = Volume of OptimMEM required for dilution

$$V1 = V2 \times (C2/C1)$$

$$\text{e.g. } \quad V1 = 350 \times (0.016/0.5)$$
$$V1 = 11.2 \ \mu L$$

12. Add the required amount of DNA (V1) to the required amount of OptiMEM (V2 − V1) and mix thoroughly by pipetting.
13. Add the DNA mix (**step 12**) to the OptiMEM/Lipofectamine mix (**step 10**) and mix by slowly pipetting the mixture up and down. This should give a final volume of 700 μL.
14. This complex should be incubated at room temperature for 15–45 min, Whilst the complex will remain stable for up to 6 h, it should not be incubated for more than 6 h.

15. Once the T25 flask from **step 8** has been incubated for 90 min, remove the media from the flask.
16. Wash the monolayer with 5 mL of antibiotic free M0 media. It is essential that no antibiotics are present in this step as the presence of antibiotics can cause cellular toxicity due to increased accumulation of plasmids in the cell, causing cell death.
17. In a sterile bijou, add 2.1 mL of OptiMEM to the complex made in **step 13**. This should make a final volume of 2.8 mL. Mix gently by pipetting.
18. Add 2.8 mL of antibiotic free M0 to the flask.
19. Using a gentle swirling motion, slowly add the transfection solution (**step 13**) to the flask and incubate at 37 °C, 5% CO_2 for 5 h.
20. After 5 h, return the flask to the MSCII and remove all the media from the flask.
21. Add 5 mL of M2 media to the flask and incubate at 37 °C, 5% CO_2 for 2–3 days.
22. When ready to harvest, label a cryovial with the following information:

    MVA-Name of insert

    Transfer plasmid details

    I/T harvest

    Date
23. Remove the supernatant from the flask and transfer to a labelled sterile bijou.
24. Transfer 1 mL of the supernatant back to the flask and remove the cells mechanically into the supernatant with a cell scraper, checking microscopically that all cells have been removed.
25. Transfer this cell suspension to a labelled cryovial and snap freeze at below −60 °C. Ensure that the harvest is frozen-thawed-vortexed three times then sonicated for 30 seconds with pulsing (1 s on: 1 s off).
26. The material can now be stored at below −60 °C and is ready to proceed to plaque purification.

3.3 Purifying rMVA Vaccine

1. Ensure the infection/transfection sample and subsequent picks have been frozen and thawed three times with vortexing between each cycle before proceeding with plaque purification. Following the third cycle (i.e., prior to reinfecting cells) sonicate each sample with pulsing for 30 s (1 s on: 1 s off).
2. To prepare for plating out infection/transfection harvest, melt sufficient 2% agarose in a water bath or microwave. Once melted, maintain the agarose at 42 °C prior to use. Pre-warm 2 × MEM to 37 °C.

3. Dilute the infection/transfection harvest to 10^{-2}, 10^{-3}, 10^{-4} in pre-warmed M0.
4. Remove media from a 96-well plate containing cells and add 100 μL per well of each dilution to the required wells (64 wells are required for each dilution).
5. Incubate plates at 37 ± 2 °C 5 ± 1% CO_2 for 90 min.
6. During incubation ensure:
   (a) Molten agarose is cooled to 42 °C.
   (b) 2× MEM is warmed to 37 °C and aliquoted to a falcon tube.
7. Remove media from the 96-well plate containing cells, add equal volume of agarose to 2 × MEM (*see* **Note 8**), and immediately add 100 μL of agarose/2× MEM to each well of the 96-well plate.
8. Cool to room temperature for approximately 10 min to form a solid overlay.
9. Incubate the plates at 37 ± 2 °C with 5 ± 1% CO_2 for 2–3 days until plaques are observed (*see* **Note 9**).
10. Using a fluorescent microscope, mark any wells that contain plaques on a 96-well plate layout.
11. For plaque picking, the preference should be for plaques that are observed in the highest dilution. Pick a minimum of three plaques per passage.
12. Mark the position of the plaque in the well. Using a suitable pen, whilst looking down the microscope mark around the plaque on the underside of the well. Identify the plaque position and using a P1000 pipette, depress the button slightly and carefully lower the pipette tip through the agar layer and press against the cell layer firmly. Steady the pipette tip in place and rotate the pipette to lift the plaque from the cells surface. Release the button slowly to remove the agar and plaque, and keeping the tip steady remove the tip from the well. Wash the pipette tip into a cryovial containing 500 μL M0 media.
13. Observe the well under the microscope to check if the plaque has been removed. If a small portion of the plaque remains—do not repeat pick. If the majority of the plaque remains, repeat the picking process one more time only.
14. Label the cryovial with MVA, the plasmid identity, plaque identity and date.
15. Freeze and thaw plaque picks three times with vortexing between each cycle and sonication after the final freeze/thaw (with pulsing for 30 s 1 s on:1 s 0ff).

16. Plaque picks should be amplified and tested by PCR to confirm presence of recombinant MVA (*see* Subheading 3.5).
17. For subsequent plating out of plaques, dilute the plaques to 10^{-1}, 10^{-2}, 10^{-3} in pre-warmed M0 (24 wells are required for each dilution).
18. Label picks according to the passage number then the pick form which it was passaged and then numerically depending on how many plaques are picked. i.e.: Passage 1 pick one = P1.1, Passage 2 pick 1 from P1.1 = P2.1.1 and passage 2 pick 2 from P1.1 = P2.1.2.
19. Continue plaque purification for up to five passages until QC-PCR confirms pure rMVA.
20. Once a plaque has been confirmed as pure rMVA, it must be amplified to a high titer prior to purification. Infection of CEF cells starting in a small volume (i.e., six-well plate), harvesting and subsequent infection into a larger volume (T25 flask, T75 flask and T175 flask and ultimately 10 × T175 flasks) in stages will ensure a suitable vaccine titer. A final amplification volume should be harvested, pelleted by centrifugation and resuspended in 10 mL of 10 mM Tris-Cl pH 8.8.

3.4 Purifying DNA Extraction

1. Before DNA extraction, inspect infected cells for good GFP/CPE and signs of infection. Remove all media from the wells to waste and add 250 μL of lysis buffer from the DNA purification kit to each well and dislodge the cells. Extract DNA according to the manufacturer's instructions provided with kits.
2. DNA can be stored at −20°c until ready for use.

3.5 PCR for Purity Confirmation of rMVA

This method describes the procedure and details to perform PCR to confirm the purity of rMVA by amplifying the entire recombinant insert. As intended by its design, GFP expression is transient in "GMP compatible" strains. A recombination event can occur after serial passage of the recombinant virus removing the GFP from the viral genome. It is also possible for the region containing the antigen-producing gene to be removed through recombination leaving only GFP. This procedure can be used to determine the presence of wild-type MVA within a batch as well as the state of the recombinant in terms of antigen and GFP presence in the genome.

1. Store lyophilized oligonucleotides at room temperature until reconstituted.
2. Use manufacturer's instructions to determine volume of nuclease-free water required to make 100 μM primer (volume (μL) = weight of oligonucleotides (nmoles) × 10).

3. Vortex-mix reconstituted primers and perform a 1/10 dilution using nuclease-free water to create a 10 μM working stock of primers and aliquot to avoid unnecessary freeze-thaws.
4. Store reconstituted primers and working stock at <−20 °C.
5. Before preparing the reaction, the work area should be fully wiped down with DNase wipes.
6. Dispense the following volumes of each reagent into a clean 50ul PCR tube:

10 μL	KAPA HiFi polymerase
4 μL	Nuclease-free water
1 μL	Each Forward and Reverse Primer (10 μM)
4 μL	DNA template

If testing multiple samples, a master mix of polymerase, water and primers can be made and then dispensed into multiple reaction tubes followed by the addition of template. *N.B If preparing a master mix for multiple reactions, multiply the volumes of reagents by the number of reactions plus two to account for volume lost during pipetting.*

7. Undertake PCR reaction according to the following conditions for 25 cycles of the denaturation, annealing, and extension steps (*see* **Note 10**):

Initial denaturation	95 °C	3 min
Denaturation	98 °C	20 s
Annealing	60 °C	15 s
Extension	72 °C	5 min
Final extension	72 °C	5 min
Hold	4 °C	∞

8. Analyze the PCR product by casting a gel size that is large enough to accommodate all samples (e.g. 150 mL 1×TAE, 1.5 g UltraPure Agarose melted for 2 min in the microwave). A 1% Agarose gel is typically used to resolve bands as the expected size(s) will be in the kilo base range. SybrSafe is added at a 1 in 10,000 dilution (e.g. 15 μL in 150 mL molten agar) once molten agarose has cooled slightly. Gently swirl flask to ensure equal distribution. Allow molten agarose to cool slightly and then pour into appropriate size gel cast with comb in place ensuring no bubbles form in areas in which DNA may run.
9. Allow gel to solidify fully and then place in appropriate size electrophoresis tank filled with 1xTAE buffer.

10. Select appropriate range DNA ladder and add to appropriate well (e.g. high mass DNA ladder into first well).
11. Add sample loading dye to each reaction tube (e.g. 2 μL dye to a 20 μL reaction).
12. Carefully load samples to individual wells ensuring not to spill contents into adjacent wells.
13. Set the power pack to the following conditions and run: 110 V, 400 mA, 45 mins.
14. Analyze gel under UV light.

3.6 Sucrose Cushion Purification

Ensure sample(s) have been freeze-thawed and vortexed three times prior to purification.

1. Precool centrifuges to 4 °C.
2. Sonicate harvested cell-lysate (~15 mL) for 1 min on ice (no pulsing) prior to purification.
3. Transfer to Dounce homogenizer and provide 50 strokes.
4. Transfer sample back to original container and centrifuge for 10 min, 300×*g* at 4 °C.
5. Separate supernatant into three bijous (~5 mL per bijou) making sure not to disturb pellet. Store on ice.
6. Resuspend the pellet in 3 mL 10 mM Tris-Cl, pH 8.8 and mix by thorough pipetting.
7. Centrifuge the tube for 10 min, 300×*g* at 5 °C.
8. Collect the supernatant in a fourth bijou tube and store immediately on ice.
9. Sonicate bijous for 1 min, no pulse, and then cool for 1 min on ice. Repeat for a total of three cycles.
10. Pipette 15 mL 36% sucrose solution into a sterile Oakridge centrifuge tube (one per sample to be purified, include balance tube if required).
11. Layer entire bijou contents on top of 36% sucrose solution using a fine tip pastette.
12. Balance with sterile 10 mM Tris-Cl as appropriate or use a separate Oakridge tube balanced with Tris-Cl if purifying one sample.
13. Centrifuge 80 min, 33,000×*g* (17,000 rpm using ss-34 rotor) at 4 °C.
14. Remove supernatant.
15. Resuspend viral pellet in appropriate buffer e.g. sterile PBS then rinse tube with additional buffer. E.g. Resuspend in 0.25 mL and rinse tube with a further 0.25 mL.

16. Remove a small volume of sample (generally 50 μL) for QC or titration if required and store remaining volume in appropriate size aliquot(s) at ≤−60 °C (*see* **Note 11**).

3.7 Titration and Immunostaining

1. Prepare reagents. CEF cells are seeded in advance (*see* **Note 12**). Melt sufficient 2% agarose in a water bath at approximately 80 °C. Once melted, reduce the temperature of the water bath to 42 °C.
2. Prepare virus dilutions of 10^{-4} to 10^{-9} in pre-warmed M0.
3. Remove media from all wells of the cell assay plate and starting with the most dilute sample, add appropriate virus dilutions in duplicate, 500 μL per well, vortexing tube for a minimum of 5 s prior to addition. Rock plates to ensure cells are coated.
4. Incubate at 37 °C 5% CO_2 for 90 min, rocking periodically (every 15–20 min).
5. After 90 min, remove media from all wells.
6. Add an equal volume of agarose to 2 × MEM (*see* **Note 8**), and immediately add 2 mL of agarose/MEM mix to each well. Allow to cool at room temperature until the overlay is completely solid.
7. Incubate for 2–3 days at 37 °C 5% CO_2.
8. If MVA contains a green-fluorescent protein (GFP) marker, plaques should be enumerated prior to immunostaining by counting under a fluorescence microscope. Plaques should be green, bright, and tightly clustered together. Individual green cells do not count as plaques.
9. For staining of MVA containing the viral antigen, immunostaining is conducted by removing solid overlay from wells to be immunostained.
10. Fix cells by adding 1 mL per well of methanol and incubate at room temperature for 5 min on rocker. Remove methanol and wash cells with 2 mL/well PBS twice.
11. Prepare primary antibody in PBS 3% FBS (e.g., rabbit α-VACV@1:250), invert to mix and add 500 μL/well. Incubate at room temperature on rocker for 30 min.
12. Wash cells twice with 2 mL/well PBS.
13. Prepare secondary antibody in PBS 3% FBS (e.g., biotinylated anti-rabbit or anti-mouse IgG from appropriate Vectastain alkaline phosphatase kit) as outlined in the manufacturer's instructions (*see* **Note 13**). Add 500 μL/well secondary antibody and incubate at room temperature on rocker for 30 min.
14. Wash cells twice with 2 mL/well PBS.
15. Add 500 μL/well enzyme complex (*see* **Note 14**). Incubate at room temperature on rocker for 30 min.

16. Wash cells twice with 2 mL/well PBS.
17. Add 500 μL/well substrate (*see* **Note 15**). Incubate on rocker at room temperature until staining is visible macroscopically (*normally* ~ *5–10 min*). Examine under microscope at **10×** magnification to ensure staining has reached an appropriate level.
18. Remove substrate and wash cells twice with 2 mL/well PBS, leaving final wash on.
19. Enumerate plaques and calculate titer in plaque-forming units (PFU) per mL using the following formula

$$\text{Titere(PFU/mL)} = \frac{\text{Mean number of plaques in well}}{(\text{Volume of diluted virus in well (mL)} \times \text{Dilution factor})}$$

e.g. in the −5 (1/100,000) dilution wells, 48 and 52 (mean = 50) plaques were formed. 0.5 mL per well of diluted virus was added:

$$\frac{50}{(0.5 \times 0.00001)} = 1 \times 10^7 \text{PFU/mL}$$

4 Notes

1. Due to the nature of FBS, differences between batches can occur having an effect on cell culture cultivation [17]. FBS should be obtained from a reputable supplier to ensure absence of contaminants. The source of FBS should also be considered, especially for future use including regulatory approval and international transport where restrictions on animal products are enforced. Replacements are increasingly being used [18] and should be considered.
2. CEF cells should be provided from a reputable source if not produced in-house according to standard protocols [19]. Consideration for future use should be considered, including the use of specific-free eggs and regulatory approval of subsequent vaccine generation.
3. BHK cells should be provided by a reputable supplier (e.g. European Collection of Cell Cultures (ECACC), product 85011433) with full authentication.
4. MVA strains for clinical development should be sourced with appropriate consideration of authentication for clinical use and regulatory approval along with any restrictions for use (e.g. commercial vaccine development).

5. Plasmid design: A DNA cassette for MVACCHFNP was generated by GeneArt (Thermofisher, UK) to contain an orthopoxvirus P11 promotor, Green Fluorescence Protein (GFP) and P7.5 promotor followed by an appropriate Kozak sequence upstream of the NP open reading frame. A stop codon was engineered downstream of NP. The cassette was inserted into an SfiI/SfiI cloning site of plasmid pMK-RK. The plasmid DNA was transformed, amplified in bacteria (*E. coli* K12 DHB10™ T1R), and subsequently purified.
6. For the primary antibody, a polyclonal preparation is preferable due to recognition of multiple epitopes often generating a strong signal, e.g. rabbit anti-Vaccinia virus, purified polyclonal IgG antibody (Bio-Rad, UK).
7. During colder temperatures, CEF cells do not travel as well. As a consequence, we have observed a loss in cell viability preventing good seeding and following cell splitting. We have found that resting the cells at 37 °C upon arrival for 1 h allows the cells to warm prior to processing and increases their use and viability in future assays.
8. It is important to work quickly to prevent the agarose from setting before being added to the plate. It may be necessary to work on one plate at a time.
9. The 96-well plates can be read by eye using a fluorescence microscope. Plaques should fluoresce brightly green and be tightly clustered together. Individual cells or sparse clusters will not successfully passage.
10. HiFi polymerase requires a higher denaturation temperature per cycle than other polymerases and is also inefficient using primers with Tm below 60 °C.
11. Purified viral samples should not be subjected to multiple freeze-thaw cycles.
12. Titration is normally performed using six-well plates and two plates are normally used per titration. CEFs must be healthy and approaching confluence to be used for titration. If comparative titers of different viruses (e.g. wild-type MVA compared to recombinant MVA(s)) are required, viruses must be titrated in the same assay using cells seeded at the same time.
13. For the preparation of the secondary antibody using the Vectastain alkaline phosphatase kit, agitate antibody prior to addition and add 1 drop per 10 mL PBS + 3% FBS, then invert prepared antibody to mix well. Other secondary antibody preparations can be considered.
14. For the preparation of the enzyme complex using the Vectastain alkaline phosphatase kit, agitate reagent A and B bottles, then for every 10 mL of PBS+ 3% FBS add 2 drops of reagent A,

swirl to mix and add 2 drops of reagent B plus 5 μL (0.05%) Tween-20. Invert to mix and incubate at room temperature for minimum of 30 min prior to use. Other enzyme complex preparations can be considered.

15. Immediately before use, prepare substrate (from vector blue or vector red substrate kit), ensuring reagents are thoroughly agitated prior to addition and that solution is mixed well between additions of the different components. Add 2 drops each of reagent 1, 2, and 3 per 5 mL Tris-Cl. Other substrate preparations can be considered.

References

1. Mehand MS, Al-Shorbaji F, Millett P, Murgue B (2018) The WHO R&D Blueprint: 2018 review of emerging infectious diseases requiring urgent research and development efforts. Antivir Res 159:63–67
2. Dowall SD, Carroll MW, Hewson R (2017) Development of vaccines against Crimean-Congo haemorrhagic fever virus. Vaccine 35:6015–6023
3. Buttigieg KR, Dowall SD, Findlay-Wilson S, Miloszewska A, Rayner E, Hewson R, Carroll MW (2014) A novel vaccine against Crimean-Congo haemorrhagic fever protects 100% of animals against lethal challenge in a mouse model. PLoS One 9:e91516
4. Meyer H, Sutter G, Mayr A (1991) Mapping of deletions in the genome of the highly attenuated vaccinia virus MVA and their influence on virulence. J Gen Virol 72:1031–1038
5. Carroll MW, Moss B (1997) Host range and cytopathogenicity of the highly attenuated MVA strain of vaccinia virus: propagation and generation of recombinant viruses in a nonhuman mammalian cell line. Virology 238:198–211
6. von Krempelhuber A, Vollmar J, Pokorny R, Rapp P, Wulff N, Petzold B, Handley A, Mateo L, Siersbol H, Kollaritsch H, Chaplin P (2010) A randomized, double-blind, dose-finding phase II study to evaluate immunogenicity and safety of the third generation smallpox vaccine candidate IMVAMUNE. Vaccine 28:1209–1216
7. Kennedy E, Dowall S, Salguero F, Yeates P, Aram M, Hewson R (2019) A vaccine based on recombinant modified Vaccinia Ankara containing the nucleoprotein from Lassa virus protects against disease progression in a guinea pig model. Vaccine. https://doi.org/10.1016/j.vaccine.2019.07.023
8. Evans TG, Bussey L, Eagling-Vose E, Rutkowski K, Ellis C, Argent C, Griffin P, Kim J, Thackwray S, Shakib S, Doughty J, Gillies J, Wu J, Druce J, Pryor M, Gilbert S (2022) Efficacy and safety of a universal influenza A vaccine (MVA-NP+M1) in adults when given after seasonal quadrivalent influenza vaccine immunisation (FLU009): a phase 2b, randomised, double-blind trial. Lancet Infect Dis 22:857–866
9. Kreijtz JHCM, Goeijenbier M, Moesker FM, van den Dries L, Goeijenbier S, De Gruyter HLM, Lehmann MH, Mutsert G, van de Vijver DAMC, Volz A, Fouchier RAM, van Gorp ECM, Rimmelzwaan GF, Sutter G, Osterhaus ADME (2014) Safety and immunogenicity of a modified-vaccinia-virus-Ankara-based influenza A H5N1 vaccine: a randomised, double-blind phase 1/2a clinical trial. Lancet Infect Dis 14:1196–1207
10. Guimarães-Walker A, Mackie N, McCormack S, Hanke T, Schmidt C, Gilmour J, Barin B, McMichael A, Weber J, Legg K, Babiker A, Hayes P, Gotch F, Smith C, Dally L, Dorrell L, Cebere I, Kay R, Winstone N, Moore S, Goonetilleke N, Fast P (2008) Lessons from IAVI-006, a Phase I clinical trial to evaluate the safety and immunogenicity of the pTHr.HIVA DNA and MVA. HIVA vaccines in a prime-boost strategy to induce HIV-1 specific T-cell responses in healthy volunteers. Vaccine 26:6671–6677
11. Dowall SD, Buttigieg KR, Findlay-Wilson SJ, Rayner E, Pearson G, Miloszewska A, Graham VA, Carroll MW, Hewson R (2016) A Crimean-Congo hemorrhagic fever (CCHF) viral vaccine expressing nucleoprotein is immunogenic but fails to confer protection against lethal disease. Hum Vaccin Immunother 12:519–527
12. Dowall SD, Graham VA, Rayner E, Hunter L, Watson R, Taylor I, Rule A, Carroll MW,

Hewson R (2016) Protective effects of a modified Vaccinia Ankara-based vaccine candidate against Crimean-Congo Haemorrhagic fever virus require both cellular and humoral responses. PLoS One 11:e0156637

13. Pavot V, Sebastian S, Turner AV, Matthews J, Gilbert SC (2017) Generation and production of modified vaccinia virus Ankara (MVA) as a vaccine vector. In: Ferran MC, Skuse GR (eds) Recombinant virus vaccines: methods and protocols. Springer, New York, pp 97–119. https://doi.org/10.1007/978-1-4939-6869-5_6
14. Butler M, Spearman M, Braasch K (2014) Monitoring cell growth, viability, and apoptosis. Methods Mol Biol 1104:169–192
15. Phelan K, May KM (2017) Mammalian cell tissue culture techniques. Curr Protoc Mol Biol 117:A.3f.1–A.3f.23
16. Noel V, Berry MD (2022) Culture of adherent cancer cell lines. Methods Mol Biol 2508:19–29
17. Barosova H, Meldrum K, Karakocak BB, Balog S, Doak SH, Petri-Fink A, Clift MJD, Rothen-Rutishauser B (2021) Inter-laboratory variability of A549 epithelial cells grown under submerged and air-liquid interface conditions. Toxicol In Vitro 75:105178
18. Subbiahanadar Chelladurai K, Selvan Christyraj JD, Rajagopalan K, Yesudhason BV, Venkatachalam S, Mohan M, Chellathurai Vasantha N, Selvan Christyraj JRS (2021) Alternative to FBS in animal cell culture – an overview and future perspective. Heliyon 7: e07686
19. Hernandez R, Brown DT (2010) Growth and maintenance of chick embryo fibroblasts (CEF). Curr Protoc Microbiol Appendix 4:4i

Chapter 20

Efficient Expression of Oropouche Virus Nonstructural Proteins NSs and NSm

Eduardo Jurado-Cobena and Tetsuro Ikegami

Abstract

Oropouche fever, a mosquito- or midge-borne emerging zoonotic disease endemic to South and Central America, manifests as a dengue-like acute febrile illness with occasional occurrences of meningitis or meningoencephalitis. The causative agent, Oropouche virus (OROV), belongs to the genus *Orthobunyavirus* within the family *Peribunyaviridae*. Its tripartite negative-sense RNA genome comprises small (S), medium (M), and large (L) segments, encoding structural N, Gn/Gc, and L proteins, respectively. Additionally, the S- and M-segments encode nonstructural proteins: NSs and NSm, which may act as virulence factors. OROV NSs functions as an interferon antagonist with an unknown mechanism, while the roles of OROV NSm remain elusive. This chapter introduces efficient expression systems for OROV NSm and NSs proteins. Validating the presence of a signal peptide at the N-terminus of NSm protein is essential for its expression. Furthermore, expressing OROV NSs protein independently of an RNA polymerase II promoter is crucial to prevent restricted gene expression, potentially caused by NSs inhibiting cellular RNA polymerase II, as observed in closely related bunyavirus NSs proteins. These protein expression strategies offer insights into the molecular characterization of OROV NSm and NSs proteins, facilitating a deeper understanding of their virulence mechanisms.

Key words Oropouche virus, Rift Valley fever virus, MP-12 strain, Bunyamwera virus, Reverse genetics, NSs protein, NSm protein, Signal peptide

1 Introduction

Oropouche fever (ORO) is an emerging zoonotic disease endemic to South and Central America, transmitted by infected midges (*Culicoides paraensis*) and mosquitoes (*Culex quinquefasciatus*, *Coquillettidia venezuelensis*, and *Aedes serratus*). It presents as a dengue-like acute febrile illness accompanied by severe headache, sometimes leading to meningitis or meningoencephalitis. Since the 1960s, over half a million human infections have been reported.

Although only a few fatalities have been reported, this mostly self-limiting yet incapacitating febrile illness, which has a high morbidity rate, remains a significant public health concern in

Hani Boshra (ed.), *Bunyaviruses: Methods and Protocols*, Methods in Molecular Biology, vol. 2893,
https://doi.org/10.1007/978-1-0716-4338-9_20,

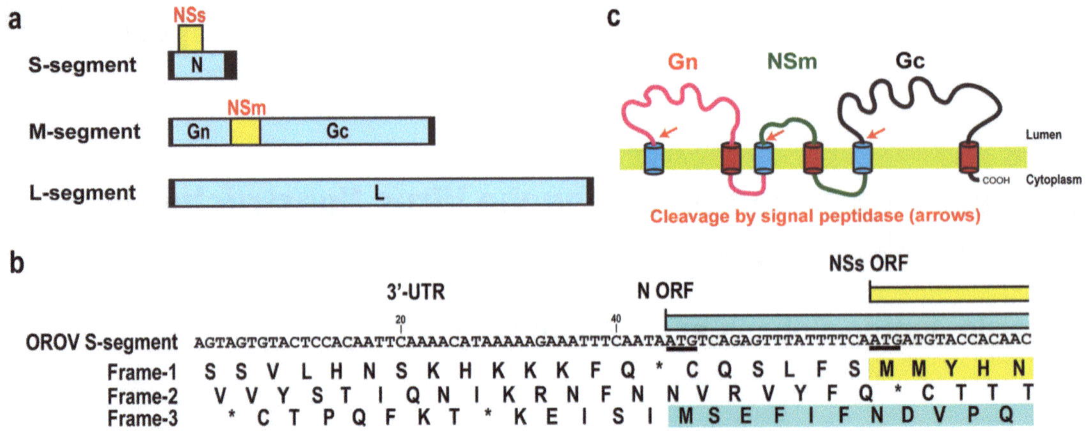

Fig. 1 The coding strategy of Oropouche virus (OROV) nonstructural protein NSs and NSm. (**a**) The genome structure of OROV. The OROV S-segment contains overlapped open reading frames encoding N and NSs genes. The M-segment of OROV encodes a glycoprotein precursor, which undergoes co-translational cleavage into Gn, NSm, and Gc, presumably by signal peptidases. (**b**) The schematic representation illustrates the coding strategy of the NSs gene within the OROV S-segment. The N gene (depicted in blue) and NSs gene (shown in yellow) are encoded in separate open reading frames within overlapped nucleotide sequence. (**c**) The membrane topology of the OROV NSm protein. Cleavage sites of glycoprotein precursor by signal peptidases are indicated by red arrows

endemic countries [1]. The causative agent, Oropouche virus (OROV), belongs to the genus *Orthobunyavirus* in the family *Peribunyaviridae*. Its genome comprises three RNA segments: small (S), middle (M), and large (L). The L-segment encodes the RNA-dependent RNA polymerase (L protein), the M-segment encodes two virion glycoproteins Gn and Gc, along with a non-structural protein (NSm), while the S-segment contains genes for the nucleocapsid (N) protein and a small nonstructural protein (NSs) (Fig. 1a). Isolation of naturally evolved reassortant viruses derived from OROV, with M-segments of unknown origin, such as Iquitos virus (humans, Peru [2]), Jatobal virus (South American coati, Brazil [3]), Perdões virus (marmosets, Brazil [4]), and Madre de Dios virus (moribund monkeys, Venezuela [5], human, Peru [6]), demonstrates the virus's ability to acquire different bunyaviral M-segments from nature. Therefore, assessing the potential virulence phenotypes associated with NSs and NSm proteins of different OROV strains is crucial for public health. Tilston-Lunel NL et al. reported a generation of recombinant infectious clones derived from OROV (rOROV) BeAn19991 Brazilian prototype strain [7]. The rOROV lacking the NSs but not the NSm gene showed reduced viral replication in A549 cells with increased secretion of type-I IFNs. However, the rOROV lacking the NSm gene replicated efficiently in two mosquito cell lines (U4.4 and Aag2) [7]. Thus, OROV NSs proteins can act as antagonists for type-I

IFNs, while the function of OROV NSm proteins remains unknown.

To elucidate the molecular functions of OROV NSs and NSm proteins, our goal here is to develop an experimental approach for efficient expression of these viral proteins in mammalian cells. The OROV S-segment contains overlapping open reading frames (ORFs) encoding N and NSs genes, giving rise to N (231 amino acids, aa) and NSs (92 aa) proteins, respectively (Fig. 1b). The NSs protein of closely related viruses like Bunyamwera virus (BUNV) or La Crosse virus (LACV) has been shown to suppress host RNA polymerase II [8, 9]. Consequently, using a plasmid DNA driven by an RNA polymerase II promoter for BUNV or LACV NSs expression may yield poor NSs protein expression. Therefore, to allow a comparative assessment of OROV NSs with closely related BUNV or LACV NSs, establishing an expression system independent of the RNA polymerase II promoter is crucial. Meanwhile, the OROV M-segment contains a single ORF encoding a glycoprotein precursor, which undergoes co-translational cleavage into Gn, NSm, and Gc proteins (Fig. 1c). The NSm protein (168 aa) comprises three transmembrane domains (TMD1, TMD2, and TMD3), with cleavage likely occurring downstream of TMD1 and TMD3. Since viral M mRNA is translated from the first AUG, confirming the role of TM1 as a signal peptide is essential for designing the exclusive expression of the OROV NSm protein. In this chapter, we describe the experimental materials and methods utilized to efficiently express the OROV NSm and NSs proteins in mammalian cells.

2 Materials

2.1 Expression of Oropouche Virus NSm Protein in Mammalian Cells

1. The DNA fragment encoding the OROV NSm ORF: e.g., custom DNA synthesis.
2. **Plasmid DNA:** pCAGGS-OROV-NSm (wtSP), pCAGGS-OROV-NSm (wtSP-mut), pCAGGS-OROV-NSm (CD40SP), or pCAGGS-OROV-NSm (CD40SP-mut) (*see* **Note 1**).
3. Gibson Assembly Kit: e.g., Gibson Assembly Master Mix, New England Biolabs.
4. Competent *E. coli*.
5. 1% agarose gel and TBE buffer.
6. DNA electrophoresis apparatus.
7. 1% LB agar plates containing 100 μg/mL ampicillin
8. DNA gel purification kit.
9. Liquid LB medium containing 100 μg/mL ampicillin.
10. Plasmid Midi Kit.

11. HEK293 cells: ATCC CRL-1573.
12. DNA transfection reagent.
13. Reduced serum medium.
14. Dulbecco's Modified Eagle Medium (DMEM) high glucose supplemented with 10% FBS, 100 units/mL of penicillin, 100 μg/mL of streptomycin.
15. Protein electrophoresis apparatus.
16. Semi-dry protein transfer system.
17. 10× Tris/Glycine/SDS buffer: 25 mM Tris, 192 mM glycine, 0.1% SDS, pH 8.3.
18. 10× Tris/Glycine Buffer: 25 mM Tris, 192 mM glycine, pH 8.3.
19. TBST: 10 mM Tris-HCl, pH 7.6, 75 mM NaCl, 0.1% Tween 20.
20. TBST-0.5% BSA: 10 mM Tris-HCl, pH 7.6, 75 mM NaCl, 0.1% Tween 20, 0.5% BSA.
21. 12% or 15% polyacrylamide gel for SDS-PAGE.
22. Colored protein size marker.
23. 2× Laemmli Sample Buffer: 65.8 mM Tris-HCl, pH 6.8, 26.3% (w/v) glycerol, 2.1% SDS, 0.01% bromophenol blue
24. Blotting paper.
25. PVDF Transfer Membranes with 0.2 μm pore size.
26. Anti-V5 antibody and anti-GAPDH antibody (*see* **Note 2**).
27. Secondary antibody: the horseradish peroxidase-conjugated antibody, which specifically binds to the primary antibody.
28. Western blot detection substrate.

2.2 Expression of Oropouche Virus NSs Protein in Mammalian Cells

1. **Plasmid DNA:** pProT7-vS(+)-ORONSs-SF, pProT7-vS(+)-BUNNSs-SF, pProT7-vS(+)ΔNSs, pProT7-vM(+)ΔNSm, or pProT7-vL(+), pT7-IRES-vN, pProT7-IRES-vL, or pCAGGS-vG [10] (*see* **Note 3**).
2. Baby Hamster Kidney (BHK) cells stably expressing T7 RNA polymerase: e.g., BHK/T7-9 cells [11] or BSR-T7/5 cells [12].
3. Vero cells: ATCC CCL-81.
4. Minimum Essential Medium (MEM) Alpha supplemented with 10% Fetal Bovine Serum (FBS), 100 units/mL of penicillin, and 100 μg/mL of streptomycin. For BHK/T7-9 cells or BSR-T7-5 cells, MEM Alpha media were additionally supplemented with 600 μg/mL of Hygromycin D or 1 mg/ml geneticin, respectively.

5. Materials for SDS-PAGE and Western blot (*see* Subheading 2.1), including anti-Flag antibody and anti-GAPDH antibody (*see* **Note 4**).

3 Methods

3.1 Expression of Oropouche Virus NSm Protein in Mammalian Cells

The expression pattern of the OROV NSm proteins in mammalian cells is analyzed using the protocol:

1. The DNA fragment encoding the OROV NSm gene is designed as illustrated in Fig. 2a. A comparison of potential signal peptide (SP) sequences for OROV NSm (N-MSFTAAI LTGLILMEFVSPIAG-C) and the mutant (N-MSFTAAILT-GLILMEFVSPRAR-C, with mutations underlined) is conducted to validate SP function. Additionally, the SP sequence of OROV NSm is replaced by that of human CD4 (N-MNRGVPFRHLLLVLQLALLPAATQG-C) or the deletion mutant (N-MNRGVPFRHLLLVLQLALLP-C, with the deletion of "AATQG"). The V5 tag is inserted between the second and third transmembrane domains of the NSm protein (*see* **Note 5**).
2. The plasmid vector, such as pCAGGS, is cut at the multiple cloning site using restriction enzymes. The linearized DNA fragment is separated on a 1% agarose gel using a DNA electrophoresis apparatus and then purified from the gel using a DNA gel purification kit.
3. The synthesized NSm DNA fragments are cloned into a vector plasmid. The linearized plasmid and NSm DNA fragment are combined in the presence of 2× Gibson Assembly Master Mix, and the mixture is incubated at 50 °C for 60 min for the recombination reaction. Competent *E. coli* is transformed with the Gibson Assembly reactions, followed by incubation in SOC medium and spreading onto LB agar plates containing ampicillin. Incubate the LB plates for 16 to 20 h at 37 °C, followed by selection of a few *E. coli* colonies for culture in liquid LB medium containing ampicillin for 20 h at 37 °C.
4. Purifying plasmid DNA from the overnight culture of *E.coli* colony is carried out using a DNA midiprep Kit, followed by validation of the insert sequence. The resulting plasmids are designated as pCAGGS-OROV-NSm (wtSP), pCAGGS-OROV-NSm (wtSP-mut), pCAGGS-OROV-NSm (CD40SP), or pCAGGS-OROV-NSm (CD40SP-mut).
5. HEK293 cells (1×10^5 cells) are transfected with 1 μg plasmid DNA encoding OROV NSm gene. Transfection is achieved by mixing 3 μL of TransIT-293 with reduced serum medium (150 μL), followed by vortex mixing for 5 s. After 5 min of

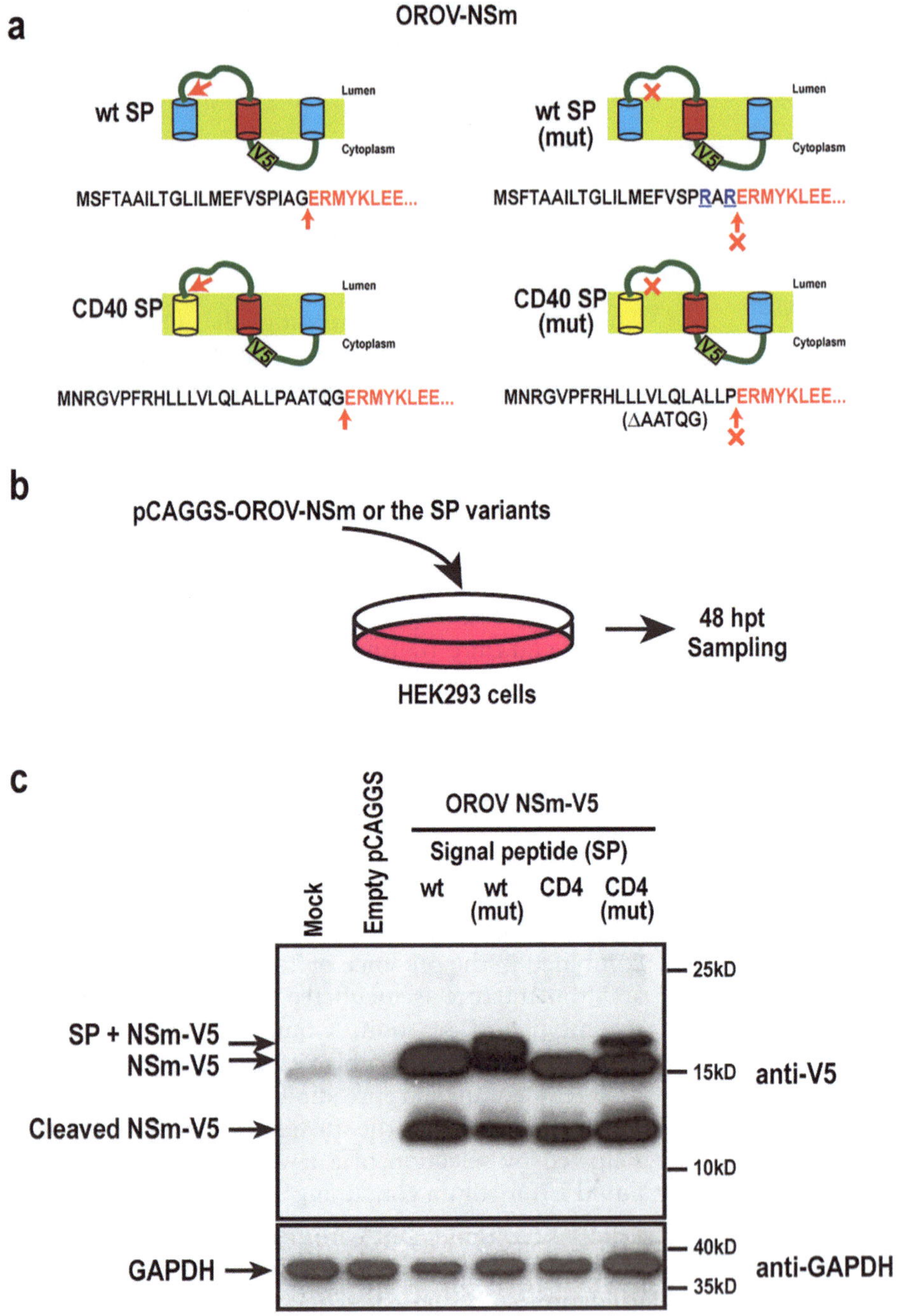

Fig. 2 Expression of Oropouche virus NSm proteins with intact or mutated signal peptides. (**a**) The NSm protein of OROV is combined with intact wildtype (wt) signal peptide (SP), the mutant of wt SP, human CD40 SP, or the mutant of CD40 SP. V5-tag is inserted between the second and third transmembrane domains of NSm protein. Red arrows indicate the cleavage sites. (**b**) Schematics of plasmid transfection experiment. HEK293 cells are mock-transfected or transfected with pCAGGS plasmid encoding OROV NSm gene or the SP mutants. Cell lysates are collected at 48 h post-transfection (hpt). (**c**) Western blot analysis of cell lysates above. Protein lysates in Laemmli buffer containing 5% mercaptoethanol are separated on 15% polyacrylamide by SDS-PAGE. Proteins are blotted onto the PVDF membrane, followed by incubating with an anti-V5 tag antibody or anti-GAPDH antibody (loading control)

incubation at room temperature, this mixture is added dropwise to the plasmid DNA (1 μg) in 1.5 mL Eppendorf tube and incubated at room temperature for 20 min. The transfection mixture is then added to HEK293 cells in DMEM.

6. Cell lysates are collected 48 h after transfection using Laemmli buffer containing 5% mercaptoethanol. Samples are boiled at 100 °C for 10 min.
7. Samples are separated by SDS-PAGE using a 12% or 15% polyacrylamide gel using a protein electrophoresis apparatus in Tris/Glycine/SDS buffer. Colored protein size marker is included.
8. Proteins are transferred onto PVDF membrane using semi-dry protein transfer system with thick blot filter paper soaked in Tris/Glycine buffer.
9. The membrane is blocked in TBST-0.5% BSA for 1 h at room temperature, followed by overnight incubation with the primary antibody (either anti-V5 antibody or anti-GAPDH antibody) at 4 °C. After washing three times with TBST, the membrane is incubated with the secondary antibody for 30 min at room temperature. The membrane is washed three times with TBS.
10. OROV NSm protein tagged with V5 epitope or GAPDH (loading control) can be detected after an incubation with the Western blot detection substrate (Fig. 2b, c) (*see* **Note 6**).

3.2 Expression of Oropouche Virus NSs Protein in Mammalian Cells

The cellular localization of the OROV NSs proteins in mammalian cells is analyzed using the following protocol:

1. The C-terminus of the OROV NSs gene is fused with a tandem Strep-tag and Flag-tag (SF-tag) as illustrated in Fig. 3a. As a control, closely related Bunyamwera virus (BUNV) NSs gene was also fused to the SF-tag. The DNA fragments encoding the OROV NSs or BUNV NSs gene is cloned into the NSs cassette of the pProT7-vS(+) plasmid, which encodes the Rift Valley fever virus (RVFV) MP-12 strain S-segment. Resulting plasmid is designated as pProT7-vS(+)-ORONSs-SF or pProT7-vS(+)-BUNNSs-SF (*see* **Note 7**).
2. For the rescue of recombinant RVFV MP-12 vaccine strain (rMP-12) expressing OROV NSs protein (rMP12ΔNSm-ORONSs-SF), BHK cells stably expressing T7 RNA polymerase are co-transfected with plasmids encoding full-length S-, M-, or L-segments (pProT7-vS(+)-ORONSs-SF 2.2 μg, pProT7-vM(+)ΔNSm 2.2 μg, or pProT7-vL(+) 2.2 μg, respectively), alongside plasmids expressing N, L, or Gn/Gc proteins (pT7-IRES-vN 2.2 μg, pProT7-IRES-vL 1.1 μg, or pCAGGS-vG 1.1 μg, respectively) (Fig. 3b). Similarly, rMP-12 expressing

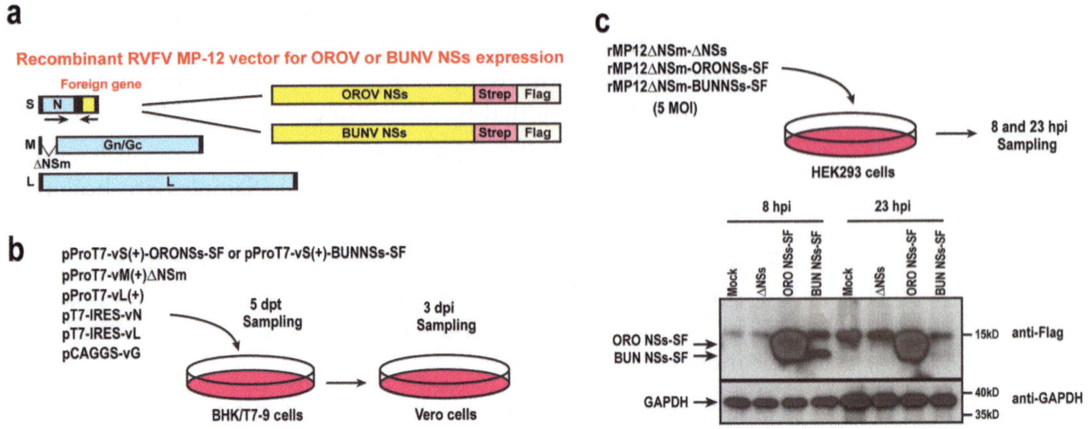

Fig. 3 Expression of Oropouche virus (OROV) NSs proteins using recombinant Rift Valley fever virus MP-12 vaccine strain (rMP-12). (**a**) A schematic representation of the rMP-12 genome is presented, wherein the NSs gene of OROV or Bunyamwera virus (BUNV) is encoded in place of the MP-12 NSs gene. (**b**) BHK/T7-9 cells, constitutively expressing T7 RNA polymerase, are co-transfected with plasmids to express MP-12 S-, M-, or L-segments [pProT7-vS(+)-ORONSs-SF, pProT7-vS(+)-BUNNSs-SF, pProT7-vM(+)ΔNSm, or pProT7-vL(+)], along with those to express MP-12 N, L, or Gn/Gc proteins (pT7-IRES-vN, pProT7-IRES-vL, or pCAGGS-vG). Infectious clones are collected at 5 days post-transfection (dpt) and further amplified in Vero cells, with harvests at 3 days post infection (dpi). (**c**) HEK293 cells undergo mock infection or are infected with rescued rMP-12 viruses expressing OROV NSs (rMP12ΔNSm-ORONSs-SF), BUNV NSs (rMP12ΔNSm-BUNNSs-SF), or no NSs (rMP12ΔNSm-ΔNSs) at an MOI of 5. Cell lysates are collected at 8 or 23 hours post infection (hpi). Western blot analysis is performed using anti-Flag antibody and anti-GAPDH antibody as a loading control. OROV NSs protein accumulates abundantly in infected HEK293 cells, whereas BUNV NSs proteins transiently express in HEK293 cells at 8 hpi

BUNV NSs protein (rMP12ΔNSm-BUNNSs-SF) or rMP-12 lacking the NSs gene (rMP12ΔNSm-ΔNSs) are rescued using pProT7-vS(+)-BUNNSs-SF in place of the pProT7-vS(+)-ORONSs-SF (*see* **Note 8**).

3. Transfection is initiated by combining 33 μL of TransIT-293 with reduced serum medium (385 μL), followed by vortexing for 5 s. After a 5-min incubation at room temperature, this mixture is slowly added to the plasmid DNA (11 μg) and allowed to incubate at room temperature for 20 min. Subsequently, the transfection mixture is added to BHK cells stably expressing T7 RNA polymerase in MEM. The culture supernatants from these cells are replaced 24 h post-transfection, followed by an additional 96-h incubation at 37 °C.
4. Culture supernatant (5 mL) containing rescued virus is collected into 15 mL conical tubes and centrifuged at 250×*g* for 3 min. The resulting supernatant constitutes the recombinant virus stock (passage 0). Further virus amplification in Vero cells at 37 °C for 72 h is recommended for high titer stocks.

5. Infectious virus titers of stock viruses should be determined through plaque assays using Vero cells [13] (*see* **Note 9**).
6. HEK293 cells are infected with rMP12ΔNSm-ORONSs-SF, rMP12ΔNSm-BUNNSs-SF, or rMP12ΔNSm-ΔNSs (a control lacking NSs gene) at 5 multiplicity of infection (MOI). Cell lysates are harvested at 8 and 23 hours post infection (hpi). Expression of OROV NSs proteins is assessed by western blotting using anti-Flag antibody or anti-GAPDH antibody (Fig. 3c) (*see* Subheading 3.1 and **Note 10**).

4 Notes

1. The pCAGGS plasmid, created by Niwa H. et al., allows for the efficient expression of foreign genes at high levels in mammalian cells [14]. The NSm gene of this study is derived from OROV MD023 strain. "SP" represents signal peptide.
2. Alternatively, an anti-OROV NSm antibody can be created for detection via Western blotting.
3. pProT7-vS(+)-ORONSs-SF, pProT7-vS(+)-BUNNSs-SF, and pProT7-vS(+)ΔNSs are used for the rescue of rMP12ΔNSm-ORONSs-SF, rMP12ΔNSm-BUNNSs-SF, or rMP12ΔNSm-ΔNSs, respectively. The OROV NSs gene is derived from OROV MD023 strain.
4. Alternatively, an anti-OROV NSs antibody can be created for detection via Western blotting.
5. Signal peptide cleavage sites can be predicted using software program such as SignalP—6.0 [15].
6. The Western blot image depicts (i) the cleaved form of OROV NSm, (ii) the un-cleaved full-length NSm with the mutation of wildtype SP or CD40 SP, and (iii) additionally cleaved OROV NSm via an uncharacterized cleavage site. The result confirms that wildtype SP (TMD1) serves as the signal peptide for the expression of OROV NSm protein. The biological significance of this uncharacterized cleavage may be further explored by generating recombinant OROV strains encoding the truncated NSm gene [7].
7. The BUNV NSs protein acts as an inhibitor of cellular RNA polymerase II, potentially hindering efficient expression when driven by an RNA polymerase II promoter on plasmid DNA. However, NSs protein expression from the rMP-12 virus operates independently of RNA polymerase II activity.

8. In this protocol, the rMP-12 variants have been modified to exclude the 78kD and NSm genes. The 78kD protein in rMP-12 facilitates viral spread in mosquito vectors, whereas the NSm gene inhibits apoptosis in mammalian cells. By utilizing the rMP-12 strain devoid of the 78kD and NSm genes (rMP-12ΔNSm), a comprehensive assessment of OROV NSs protein functions becomes feasible.
9. Appropriate biosafety measures and laboratory protocols are followed throughout the procedure. The creation and use of recombinant MP-12 virus require approval from the institutional biosafety committee.
10. The difference in NSs protein stability between OROV and BUNV can be further explored through assessments such as indirect immunofluorescence assays to examine cellular localization, interactions with host proteins, or potential formation of inclusion bodies. The transient nature of BUNV NSs protein expression has been reported [16]. Further characterization of OROV NSs protein functions can be evaluated using this expression assays.

Acknowledgments

This study received support from the NIAID Emerging and Tropical Infectious Diseases Training Program Grant, Pre-Doctoral Fellowship T32AI007526-22 (E.J.C.), and a Data Acquisition Grant from the Institute for Human Infections & Immunity and Center for Vector-borne and Zoonotic Diseases (T.I.). It has also received generous research support from the Sealy Institute for Vaccine Sciences at UTMB.

References

1. Azevedo RS, Nunes MR, Chiang JO, Bensabath G, Vasconcelos HB, Pinto AY, Martins LC, Monteiro HA, Rodrigues SG, Vasconcelos PF (2007) Reemergence of Oropouche fever, northern Brazil. Emerg Infect Dis 13(6):912–915. https://doi.org/10.3201/eid1306.061114
2. Aguilar PV, Barrett AD, Saeed MF, Watts DM, Russell K, Guevara C, Ampuero JS, Suarez L, Cespedes M, Montgomery JM, Halsey ES, Kochel TJ (2011) Iquitos virus: a novel reassortant Orthobunyavirus associated with human illness in Peru. PLoS Negl Trop Dis 5(9):e1315. https://doi.org/10.1371/journal.pntd.0001315
3. Saeed MF, Wang H, Suderman M, Beasley DW, Travassos da Rosa A, Li L, Shope RE, Tesh RB, Barrett AD (2001) Jatobal virus is a reassortant containing the small RNA of Oropouche virus. Virus Res 77(1):25–30. https://doi.org/10.1016/s0168-1702(01)00262-3
4. Tilston-Lunel NL, Hughes J, Acrani GO, da Silva DE, Azevedo RS, Rodrigues SG, Vasconcelos PF, Nunes MR, Elliott RM (2015) Genetic analysis of members of the species Oropouche virus and identification of a novel M segment sequence. J Gen Virol 96:1636–1650. https://doi.org/10.1099/vir.0.000108
5. Navarro JC, Giambalvo D, Hernandez R, Auguste AJ, Tesh RB, Weaver SC, Montanez H, Liria J, Lima A, Travassos da Rosa JF, da Silva SP, Vasconcelos JM, Oliveira R, Vianez JL Jr, Nunes MR (2016) Isolation of Madre de Dios virus

(Orthobunyavirus; Bunyaviridae), an Oropouche virus species reassortant, from a monkey in Venezuela. Am J Trop Med Hyg 95(2):328–338. https://doi.org/10.4269/ajtmh.15-0679

6. Ladner JT, Savji N, Lofts L, Travassos da Rosa A, Wiley MR, Gestole MC, Rosen GE, Guzman H, Vasconcelos PFC, Nunes MRT, Kochel TJ, Lipkin WI, Tesh RB, Palacios G (2014) Genomic and phylogenetic characterization of viruses included in the Manzanilla and Oropouche species complexes of the genus Orthobunyavirus, family Bunyaviridae. J Gen Virol 95:1055–1066. https://doi.org/10.1099/vir.0.061309-0
7. Tilston-Lunel NL, Acrani GO, Randall RE, Elliott RM (2015) Generation of recombinant Oropouche viruses lacking the nonstructural protein NSm or NSs. J Virol 90(5):2616–2627. https://doi.org/10.1128/JVI.02849-15
8. Thomas D, Blakqori G, Wagner V, Banholzer M, Kessler N, Elliott RM, Haller O, Weber F (2004) Inhibition of RNA polymerase II phosphorylation by a viral interferon antagonist. J Biol Chem 279(30):31471–31477. https://doi.org/10.1074/jbc.M400938200
9. Blakqori G, Delhaye S, Habjan M, Blair CD, Sanchez-Vargas I, Olson KE, Attarzadeh-Yazdi G, Fragkoudis R, Kohl A, Kalinke U, Weiss S, Michiels T, Staeheli P, Weber F (2007) La Crosse bunyavirus nonstructural protein NSs serves to suppress the type I interferon system of mammalian hosts. J Virol 81(10):4991–4999. https://doi.org/10.1128/JVI.01933-06
10. Ikegami T, Won S, Peters CJ, Makino S (2006) Rescue of infectious rift valley fever virus entirely from cDNA, analysis of virus lacking the NSs gene, and expression of a foreign gene. J Virol 80(6):2933–2940. https://doi.org/10.1128/JVI.80.6.2933-2940.2006
11. Ito N, Takayama-Ito M, Yamada K, Hosokawa J, Sugiyama M, Minamoto N (2003) Improved recovery of rabies virus from cloned cDNA using a vaccinia virus-free reverse genetics system. Microbiol Immunol 47(8):613–617. https://doi.org/10.1111/j.1348-0421.2003.tb03424.x
12. Buchholz UJ, Finke S, Conzelmann KK (1999) Generation of bovine respiratory syncytial virus (BRSV) from cDNA: BRSV NS2 is not essential for virus replication in tissue culture, and the human RSV leader region acts as a functional BRSV genome promoter. J Virol 73(1):251–259. https://doi.org/10.1128/JVI.73.1.251-259.1999
13. Kalveram B, Lihoradova O, Indran SV, Ikegami T (2011) Using reverse genetics to manipulate the NSs gene of the Rift Valley fever virus MP-12 strain to improve vaccine safety and efficacy. J Vis Exp 57:e3400. https://doi.org/10.3791/3400
14. Niwa H, Yamamura K, Miyazaki J (1991) Efficient selection for high-expression transfectants with a novel eukaryotic vector. Gene 108(2):193–199
15. Teufel F, Almagro Armenteros JJ, Johansen AR, Gislason MH, Pihl SI, Tsirigos KD, Winther O, Brunak S, von Heijne G, Nielsen H (2022) SignalP 6.0 predicts all five types of signal peptides using protein language models. Nat Biotechnol 40(7):1023–1025. https://doi.org/10.1038/s41587-021-01156-3
16. van Knippenberg I, Fragkoudis R, Elliott RM (2013) The transient nature of Bunyamwera orthobunyavirus NSs protein expression: effects of increased stability of NSs protein on virus replication. PLoS One 8(5):e64137. https://doi.org/10.1371/journal.pone.0064137

Chapter 21

Systemic Clodronate Liposome Treatment for Depletion of Blood-Contacting Phagocytes and Investigation of Their Role in the Control of Bunyavirus Viremia

Stephanie E. Ander, Bennett J. Davenport, Alyssa B. Evans, and Thomas E. Morrison

Abstract

During viral infection, phagocytic cells participate in numerous immunological processes. A common approach to elucidate the specific contributions of phagocytic cells is through direct comparison of viral pathogenesis of phagocyte-depleted and phagocyte-intact animals. Clodronate liposomes are a versatile and simple means of depleting phagocytes in any animal model. Selection of clodronate liposome route of delivery can yield systemic or local phagocyte depletion as needed. Here we describe the application of clodronate liposome-mediated depletion of blood-contacting phagocytes to investigate their interactions with La Crosse virus (or other bunyaviruses) in a reductionist model of viremia.

Key words Phagocyte depletion, Clodronate liposomes, Blood circulation, La Crosse virus, Viremia, Virus concentration, Plaque assay, TaqMan qPCR

1 Introduction

Phagocytic cells, such as macrophages, contribute to both innate and adaptive immune responses during viral infection. A common approach to studying the role of phagocytes during infection is investigating the effects of phagocyte depletion on host immune responses, viral infection, and pathogenesis. Many strategies exist for targeted depletion of phagocytes depending on genetic tractability of the animal model and interest in general or specific populations of phagocytes. However, a versatile technique that we have employed is clodronate-loaded liposomes.

Clodronate, or dichloromethylene diphosphonate, is a non-nitrogen bisphosphonate used in the treatment of osteolytic bone diseases. Clodronate itself does not easily cross the cell membrane; however, encapsulation of clodronate in liposomes promotes

Hani Boshra (ed.), *Bunyaviruses: Methods and Protocols*, Methods in Molecular Biology, vol. 2893,
https://doi.org/10.1007/978-1-0716-4338-9_21,

both effective delivery of the drug to phagocytes and subsequent phagocyte-depletion [1, 2]. Liposomes are vesicles of amphipathic phospholipids organized in concentric bilayers defining aqueous compartments. The hydrophilic nature of clodronate allows the drug to be encapsulated in these aqueous compartments. Following uptake of the clodronate liposomes by phagocytes, the liposome is degraded by lysosomal phospholipases and the packaged clodronate is released into the cell. Inside the cell, clodronate is metabolized into a cytotoxic ATP analog that leads to phagocyte apoptosis upon sufficient accumulation [3–5]. Thus, this approach selectively depletes cells with high endocytic capacity, such as phagocytes, while other cell types are unaffected [2, 6].

The route of clodronate liposome delivery determines which phagocytic compartments are depleted. Intravascular delivery depletes phagocytes systemically (including those of the liver [7] and spleen [8]), while local injections cause tissue-specific phagocyte depletion (such as in the draining lymph node [9], lung [10], or brain [11]). It should be noted that clodronate liposome-mediated phagocyte depletion is transient, and these compartments can be replenished over time; for example, repopulation of liver Kupffer cells to normal levels following treatment was observed to occur by 14–16 days post depletion in mouse and rat models [8, 12], while macrophage repopulation in murine lymph nodes was observed to require 5 months [9].

Lastly, the experimenter should be aware of a few additional considerations when depleting phagocytes using clodronate liposomes. Firstly, clodronate liposome treatment is not macrophage-specific, and other phagocytic cell populations, such as monocytes and dendritic cells, are also depleted following clodronate liposome treatment [13, 14]. Secondly, proper controls (such as PBS-loaded liposomes) should be incorporated in clodronate liposome-mediated phagocyte depletion experiments; liposomes themselves can trigger innate immune responses and affect macrophage function [15, 16]. Similarly, neutrophil functions (but not numbers) have also been shown to be affected by clodronate liposome treatment [17].

In summary, clodronate liposomes are a simple and effective technique for phagocyte depletion in animal models. Studies in our laboratory have effectively used clodronate liposomes to deplete blood-contacting phagocytes and thereby elucidate a role for these phagocytes in mediating control of arbovirus viremia [7, 18, 19].

2 Materials

2.1 Preparation of Bunyavirus Experimental Stocks

1. Vero cells (ATCC CCL-81).
2. T150 cell culture-treated flasks.

3. Vero cell culture medium: DMEM with high glucose and pyruvate media, supplemented with 10% fetal bovine serum and 100 U/mL penicillin-streptomycin.
4. Virus growth medium: DMEM with high glucose and pyruvate, supplemented with 2% fetal bovine serum and 100 U/mL penicillin-streptomycin.
5. 15 mL polypropylene tubes.
6. Gasketed cryovials rated for storage at −80 °C.

2.2 Concentration of Virus Stock by Ultracentrifugation

1. 20% sucrose: Dissolve 10 g sucrose in 50 mL of phosphate buffered saline (PBS) without Ca^{2+}/Mg^{2+}. Filter-sterilize through a 0.22 μm filter. Store at 4 °C.
2. Polypropylene thin wall ultracentrifuge tubes, 36 mL capacity with max speed capability at or above 28,000×*g*.

2.3 Quantification of Virus by Plaque Assay

1. Vero cells (ATCC CCL-81).
2. Vero cell culture medium: DMEM, high glucose, pyruvate, supplemented with 10% fetal bovine serum and 100 U/mL penicillin-streptomycin.
3. Plaque assay medium (same as virus growth medium in Subheading 2.2): DMEM with high glucose and pyruvate medium, supplemented with 2% fetal bovine serum and 100 U/mL penicillin-streptomycin.
4. 1.5% carboxy methyl cellulose salt (CMC; ThermoFisher) in MEM HEPES, no glutamine medium (Gibco): Portion CMC at 7.5 g in sealed foil packets. Sterilize in autoclave at 121 °C for 20 min and dry cycle for 30 min. Sealed foil packets can be stored at room temperature indefinitely. In sterile hood, add 7.5 g of sterile CMC to 500 mL of MEM. Shake vigorously (300 rpm) overnight at 37 °C until CMC completely dissolves.
5. 10% formaldehyde: Mix 135 mL 32% formaldehyde with 365 mL sterile Milli-Q H_2O.
6. 0.35% crystal violet: Dissolve 1.7 g crystal violet into 75 mL 95% ethanol. Bring volume to 500 mL with sterile Milli-Q H_2O. Solution can be stored at room temperature indefinitely.
7. 24-well tissue culture-treated plates.
8. Disposable polyethylene transfer pipettes.

2.4 Quantification of Viral RNA by Reverse Transcription (RT)-Quantitative PCR (qPCR)

1. TRIzol reagent (Life Technologies).
2. Chloroform.
3. Nuclease-free molecular grade H_2O.
4. 70% ethanol: Dilute 100% (200 proof) ethanol in nuclease-free molecular grade H_2O.

5. PureLink RNA Mini isolation kit (Life Technologies).
6. SuperScript IV first-strand cDNA synthesis kit (Invitrogen).
7. TaqMan Universal Master Mix II with UNG (Applied Biosystems).
8. Primers and probe for TaqMan qPCR to amplify a region within the La Crosse virus segment S nucleocapsid gene. Primers and probes should be resuspended in nuclease-free molecular grade H_2O to a concentration of 100 μM. LACV_S_Fwd, 5′- GGTTAGCCTTCCTCTCTGGCTTA -3′; LACV_S_Rev, 5′- CCTTGCTGCAGTTAGGATCTTCTT -3′; LACV_S_P-robe, 5′-CAGCCTTGGCCTTTGCGGCATT-3′ conjugated with 5′ 6-FAM/ZEN/3′ IBFQ.
9. 20x primer/probe mix: prepare a working stock with Fwd and Rev. primers at 18 μM, and probe at 5 μM using nuclease-free H_2O. This stock can be aliquoted and stored at −20 °C.

2.5 Evaluating the Role of Blood-Contacting Phagocytes in the Clearance of Bunyavirus Virions from the Blood Circulation

1. Control (PBS) and Clodronate liposomes: Liposoma (PBS and clodronate liposomes can be purchased independently if needed). Control and clodronate liposomes can also be generated in the lab following detailed and established protocols [1, 6].
2. 0.5–1.0 cc Insulin syringe 28G1/2.
3. Isoflurane anesthesia.
4. Proparacaine hydrochloride ophthalmic solution, USP 0.5% (dropper bottle).
5. Bunyavirus stock for experimentation.
6. Inoculum diluent: PBS without Ca^{2+}/Mg^{2+} supplemented with 1% FBS.
7. Serum separator tubes (BD Microtainer SST).
8. Gasketed cryovials rated for storage at −80 °C.

3 Methods

All procedures involving live animals should be performed in accordance with the Guide for the Care and Use of Laboratory Animals of the National Institutes of Health and under approved Institutional Animal Use Protocols. All procedures should be carried out with sterile consumables and in an appropriate biosafety cabinet unless otherwise noted.

3.1 Preparation of Bunyavirus Stocks

1. Seed Vero cells in a T150 cell culture-treated flask at a ratio of 1:15 (~1.3×10^6 cells) in 20 mL of Vero cell culture medium. Allow Vero cells to grow until ~85–90% confluent (usually 4 days).

2. Dilute virus in virus growth medium to a concentration of 3.0×10^5 PFU/mL in a 6 mL volume per T150 (*see* **Note 1**).
3. Remove medium via aspiration from T150 and add 6 mL of virus dilution (1.8×10^6 PFU total, MOI $\approx$ 0.1).
4. Incubate the cells plus virus at 37 °C for 1 h. Gently rock flask every ~10 to 15 min during incubation to ensure full coverage and prevent cell drying.
5. Remove virus from flask by aspiration. Wash flask with 6 mL virus growth medium, replenish with 15 mL virus growth medium. Incubate at 37 °C.
6. Harvest virus when CPE ≥85%. Depending on the virus type, this can take 2 to 6 days. Titers typically reach ~1.0×10^6 to 2.0×10^8 PFU per mL, although some viruses have lower titers.
7. Collect virus in 15 mL tube and clarify the supernatant by centrifugation at 5000×*g* for 7 min. If concentration of virus stock is desired, after clarification proceed to Subheading 3.2. If virus stock does not need to be concentrated, aliquot clarified virus into desired volumes in cryovials and store at −80 °C.
8. Avoid freeze-thaw cycles to the virus stocks. It is recommended to use single freeze-thawed virus for animal studies, and up to one additional freeze-thaw cycle for cell culture studies.

3.2 Concentration of Virus Stocks by Ultracentrifugation

1. Concentration of virus by ultracentrifugation should be conducted on the same day that virus stock supernatants are harvested.
2. Following virus supernatant clarification, transfer 15 mL of clarified virus stock to an ultracentrifuge tube. Bring up the total volume to 30 mL using Vero cell culture medium.
3. Using a serological pipette, slowly underlay a 5 mL cushion of 20% sucrose. As the sucrose cushion is added, a noticeable phase-separation between the sucrose and the virus supernatant will develop. Take care not to disturb the interface between the sucrose cushion and virus stock (*see* **Note 2**).
4. Place the ultracentrifuge tube into the swinging bucket, and weigh the mass of the bucket, the ultracentrifuge tube containing the sucrose-underlaid viral supernatant and the bucket cap. Make a counterbalance by filling a second ultracentrifuge tube with water until combined mass (with its own swinging bucket plus the bucket cap) matches.
5. Transfer swinging buckets to the ultracentrifuge rotor, making sure to not disturb the ultracentrifuge tubes. Concentrate the virus by centrifugation at 100,000×*g* for 6 h at 4 °C.

6. After ultracentrifugation, carefully remove swinging buckets. Virus will be pelleted at the bottom of the ultracentrifuge tube.
7. Carefully lift-out the ultracentrifuge tube. With a 25 mL serological pipette, remove and discard the supernatant just above the sucrose cushion. With a smaller pipette (using a P1000 is recommended), remove the remaining liquid, taking care not to disturb the bottom surface of the ultracentrifuge tube.
8. Add 100–1000 μL of PBS to ultracentrifuge tube (exact amount to be used will depend on desired concentration of virus). Place ultracentrifuge tube inside a 50 mL conical, screw on conical cap, and seal with parafilm. Store tube of virus upright overnight at 4 °C to ensure full resuspension of virus.
9. The next day, make aliquots of concentrated virus and store at −80 °C.
10. It can be expected that the virus concentration will increase 10–20-fold from the starting titer.

3.3 Virus Titer Quantification by Plaque Assay

1. Plate Vero cells at ~1.3×10^5 cells per well in a volume of 0.5 mL of Vero cell culture medium per well of a 24-well tissue culture plate. Incubate at 37 °C overnight up to 24 h. Wells should be fully confluent the next day.
2. The next day, preheat the 1.5% CMC in MEM in a 37 °C water bath.
3. Perform desired dilutions of virus in plaque assay medium. For titering virus stocks, typically an initial tenfold dilution (10^{-1}) of 50 μL stock virus +450 μL plaque assay medium is performed followed by serial fivefold dilutions (5^{-1}-5^{-x}) of 120 μL + 480 μL medium. For titering virus samples from tissue or cells, often tenfold serial dilutions of 10^{-1}–10^{-6} are preferred. Depending on the expected titer of the virus stocks or samples, dilutions from 5^{-1}–5^{-12} (entire 24-well plate) or 5^{-1}–5^{-6} or 10^{-1}–10^{-6} are plated (half of a 24-well plate).
4. Dump media from the 24-well plates quickly into a collection vessel (e.g., a stainless steel tray) inside a sterile tissue culture hood.
5. Quickly to avoid cell drying, plate 200 μL of desired dilutions into each well. Plate samples in duplicate. Work from the highest to lowest dilution (*see* **Note 3**).
6. Incubate plates at 37 °C for 1 h.
7. After 1 h, add 0.5 mL of the pre-warmed 1.5% CMC in MEM to each well using a serological pipette (be careful to hold the tip above the wells as to not contaminate the pipette). There is no need to remove the inoculum virus. Incubate the plates at 37 °C.

8. Depending on virus, plates should be fixed and stained at 3–5 days (*see* **Note 4**).
9. Fix cells and inactivate virus by filling each well with 10% formaldehyde. Incubate at room temperature for at least 1 h.
10. Dump out fixative into collection vessel and transfer to a waste container for proper disposal.
11. Wash plates with deionized (DI) water. Turn DI water to a gentle but steady stream. Hold plate vertically so that the well openings are nearly parallel to the stream of water. Run plate under the DI water so that water gently pours onto the edges of the wells and pools into the well bottoms without disrupting the fixed cell layer, then dump out water. Repeat for a total of two washes per plate.
12. With a transfer pipette, add several drops of 0.35% crystal violet to each well until the bottom of the well is fully covered. Incubate at room temperature for 15 min.
13. Wash plates again with DI water twice. Allow plates to air-dry on benchtop by leaning each plate upside down against its corresponding lid to allow water to drip out by gravity as well as evaporation.
14. When plates are dry, count plaques. Viruses may have uniform plaque sizes or mixtures of large and small plaques.
15. Typically, wells with ~3 to 50 plaques are considered reliable plaque counts, depending on plaque size and distribution in the well. Count all dilutions within this range.
16. For each dilution, count duplicate wells and calculate the average plaque number ("number of plaques").
17. Calculate the virus titer as PFU/mL using the equation: Titer = Number of plaques × dilution factor × infection volume. Number of plaques = average of duplicate wells; Dilution factor = inverse of dilution, for example, dilution of 10^{-4} becomes a dilution factor = 10^4; Infection volume = 5 for 200 μL inoculum.
18. If multiple dilutions contained countable plaques, average the titers to get the final virus titer.

3.4 Viral RNA Quantification by RT-qPCR

3.4.1 Generation of Standard Curve

1. Generate a standard curve to quantify viral RNA by PFU equivalents using RT-qPCR. Below we describe the standard curve we use for quantification of La Crosse virus.
2. Prepare virus stock at a concentration of 5×10^7 PFU/mL (*see* **Note 5**) diluted in PBS without Ca^{2+}/Mg^{2+}.
3. Perform a tenfold serial dilution of virus in PBS without Ca^{2+}/Mg^{2+} ranging from 5×10^6 to 5×10^{-0} PFU.

4. Transfer 100 μL of each dilution into 1 mL of TRIzol reagent, yielding a total of eight samples containing the total PFU equivalents of 5×10^6 PFU down to 5×10^{-1} PFU equivalents in each tube.
5. Extract RNA according to PureLink recommended protocol for RNA isolation using TRIzol Reagent.
6. Add 200 μL of chloroform to each tube of sample in TRIzol reagent. Mix by inverting tube 15 times (do not vortex). Incubate at room temperature for 2 min.
7. Centrifuge at 12,000×*g* for 15 min at 4 °C.
8. Transfer upper, aqueous phase to new tube. Add equal volume of 70% ethanol. Mix.
9. Transfer samples to PureLink RNA isolation columns. Follow manufacturer's instructions for binding and washing. Elute RNA from columns using 50 μL of molecular biology-grade, nuclease-free water.
10. Use 10 μL of isolated RNA from each of the virus standard dilutions (which for each standard dilution will represent 1/5 of the starting material, i.e. 10^6 to 10^{-1} PFU equivalents for each standard dilution) for cDNA synthesis using SuperScript IV kit according to manufacturer's instructions using random hexamers (*see* **Note 6**).
11. At the conclusion of cDNA synthesis, dilute each sample to a final volume of 100 μL.
12. Make 13 μL aliquots of the cDNA dilution series and store at −80 °C. This is the single-use standard curve.

3.4.2 RNA Isolation from Serum Samples

1. For each serum sample, add 20 μL serum into 500 μL of TRIzol reagent (*see* **Note 7**).
2. Add 100 μL of chloroform to each tube of serum in TRIzol reagent. Mix by inverting tube 15 times (do not vortex). Incubate at room temperature for 2 min.
3. Centrifuge at 12,000×*g* for 15 min at 4 °C.
4. Transfer upper, aqueous phase to new tube. Add equal volume of 70% ethanol and mix.
5. Transfer samples to PureLink RNA isolation columns. Follow manufacturer's instructions for binding and washing. Elute RNA from columns using 30 μL of molecular biology-grade, nuclease-free water.
6. Use 10 μL of RNA isolated from serum samples for cDNA synthesis using SuperScript IV kit according to manufacturer's instructions using random hexamers.
7. At the conclusion of cDNA synthesis, dilute each sample to a final volume of 100 μL.

3.4.3 Quantification of cDNA by TaqMan qPCR

1. Prepare a qPCR master mix containing 20x primer/probe mix and TaqMan Universal Master Mix II with UNG without cDNA.
2. For a single 20 μL reaction, use the following calculation, and scale up as necessary to accommodate all samples in the qPCR run: 1 μL of 20× primer/probe mix, 10 μL of TaqMan Universal Master mix II with UNG and 4 μL H_2O (the remaining 5 μL will be accommodated by the cDNA sample).
3. Add 15 μL of the master mix prepared above into each well of the qPCR plate. Then add 5 μL of cDNA from the standard or samples to their respective wells containing 15 μL of master mix.
4. Once all the samples have been added to the qPCR plate, seal plate with optical adhesive cover and spin for 5 min at 130×*g* at 4 °C to collect reactions at the bottom of the wells and remove bubbles.
5. Run samples on a qPCR analyzer setup for detection of TaqMan probe using the following conditions: Step 1: 50 °C for 2 min; Step 2: 95 °C for 10 min; Step 3: 95 °C for 15 s; Step 4: 60 °C for 1 min; Repeat step 3–4 for 40 cycles.

3.5 Evaluating the Role of Blood-Contacting Phagocytes in the Clearance of Bunyavirus Virions from the Blood Circulation

1. Prior to the procedure, acquire the weights of each mouse that will be treated with liposomes in order to calculate individual dosages for intravascular delivery (*see* **Note 8**).
2. Aliquot the necessary volume of control and clodronate liposomes needed to conduct the experiment. For depletion of blood-contacting phagocytes, liposomes are delivered in a volume of 10 μL per gram of mouse (the equivalent of 50 μg of clodronate per 1 g of mouse weight). Allow liposomes to come to room temperature and mix by inverting the vial 5–6 times or gentle shaking, do not vortex. Sterilize the rubber stopper lid with 70% ethanol and use a sterile 28G1/2 insulin syringe to puncture the rubber lid and extract the required volume (*see* **Note 9**) and place into a sterile tube.
3. Using a new insulin syringe for each mouse, draw up the necessary volume of control or clodronate liposomes needed to inoculate the mice at 10 μL per g of mouse weight. Prior to drawing up liposomes for injection, flick the aliquot to ensure a homogenous mixture as liposomes can precipitate over time.
4. Anesthetize the mouse via isoflurane (*see* **Note 10**). Once the mouse is fully anesthetized (as determined by the toe pinch method), apply one drop of proparacaine hydrochloride ophthalmic solution to the mouse's eye. Wait 45–60 s for the proparacaine anesthetic to take effect before performing the injection. Deliver the liposome material intravenously via the retro-orbital sinus (*see* **Note 11**). Following the inoculation,

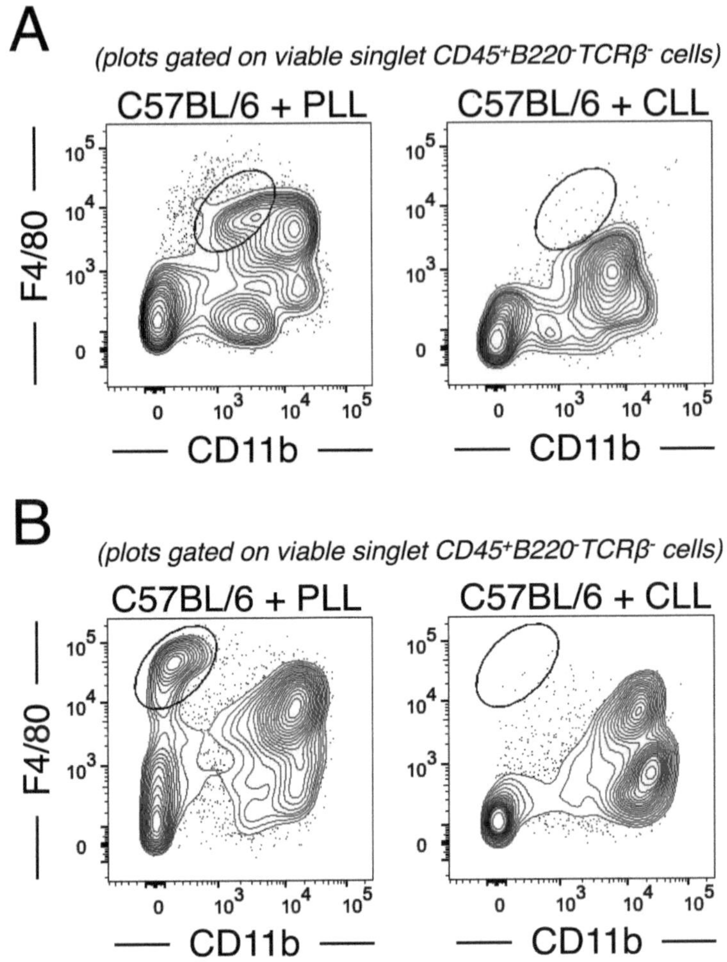

Fig. 1 Naïve 4-week-old C57BL/6 were treated with 10 μL per gram of body weight with control or clodronate-loaded liposomes via intravenous injection (retro-orbital). At 42 h post treatment, liver (**a**) and spleen (**b**) were collected, processed into a single suspension, and stained with surface antigen antibodies for flow cytometric analysis. Tissue resident macrophages were characterized as F4/80^{hi} CD11b^{mid} cells among viable singlet cells that are CD45^{+}B220/TCRβ^{-}

return the mouse to its cage and confirm full recovery from the anesthesia. The mouse should show no ill-effects from the inoculation of control or clodronate-loaded liposomes.

5. Plan to assess the role of blood-contacting phagocytes in bunyavirus clearance from the blood circulation 2 days after liposome treatment, as peak phagocyte depletion is observed at 42–48 h post treatment (Fig. 1). For short-term, transient experiments (less than 2 days; such as described herein), a single treatment of liposomes is sufficient to maintain phagocyte depletion for the duration of the experiment. For longer

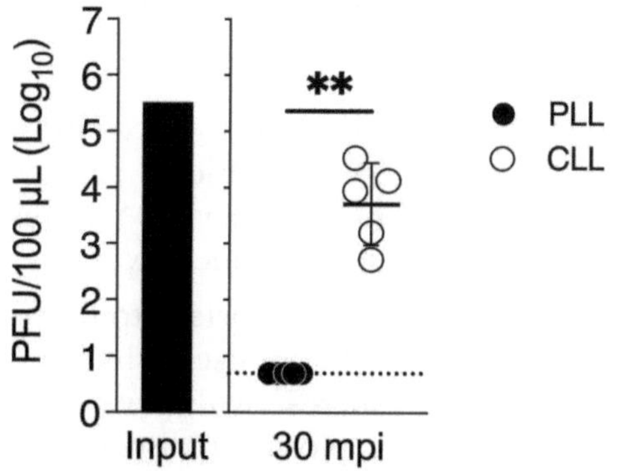

Fig. 2 Naïve 4-week-old C57BL/6 were treated with 10 μL per gram of body weight with PBS-loaded liposomes (PLL) or clodronate-loaded liposomes (CLL) via intravenous injection (retro-orbital). At 2 days post treatment, mice were inoculated with 10^6 PFU of La Crosse virus in a 100 μL volume via intravenous injection (retro-orbital). Thirty minutes post inoculation (mpi), serum was harvested from mice and stored at −80 °C. Virus loads in samples and inoculum (input) were quantified by plaque assay. Data is from a single experiment (5 mice/group); statistics are Kolmogorov-Smirnov test, $^{**}P < 0.01$

experiments (greater than 2 days), it is advisable to continue treatments every 2–3 days to maintain depletion [20, 21].

6. Two days after liposome treatment, prepare virus inoculum. Dilute virus in inoculum diluent to 10^7 PFU/mL; each mouse will receive 100 μL of inoculum. This high dose of virus simulates the viremic state.
7. For serum clearance studies, the virus inoculum is delivered intravenously via retro-orbital injection. This procedure is identical to the procedure detailed in **step 4** but with virus in place of the liposomes. Following the inoculation, return the mouse to its cage and confirm full recovery from the anesthesia. The mouse should show no ill-effects from the inoculation. Store excess inoculum at −80 °C for downstream analysis.
8. Within a 3-h timeframe, euthanize the mouse by isoflurane overdose. Immediately upon euthanasia, open the chest cavity, snip the right atrium, and collect pooling blood from the chest cavity into a serum separator tube.
9. Centrifuge blood at 6000×*g* for 5 min at 4 °C. Transfer serum (located above the wax plug after centrifugation) to a cryogenic vial and store at −80 °C.
10. Quantify virus loads in serum and mouse inoculum by plaque assay or RT-qPCR as described above in Subheadings 3.3 and 3.4 (Fig. 2).

4 Notes

1. To avoid the accumulation of mutations to the viral genome from additional passages through cells, it is recommended to maintain a virus "starter stock" that is used as the inoculum for all subsequent working stocks.
2. It is important to test the pipet aid being used to underlay the 20% sucrose cushion to ensure that the final volume of the 20% sucrose is dispensed at an even flow. Some pipet aids can be explosive in dispensing the final 1 mL volume from a serological pipet.
3. When pipetting samples diluted in plaque assay diluent onto Vero cell monolayers, the same tip for a single virus stock or sample can be used if starting at the highest dilution and working backwards toward the lowest dilution.
4. Depending on the Bunyavirus strain being tested, the incubation time for the plaque assay can vary. We have observed robust plaque formation in Snowshoe Hare virus and Inkoo virus as early as 3 days, La Crosse virus in 4 days, yet Tahyna virus and Jamestown Canyon virus often require 5 days for robust plaque formation to be visualized.
5. If a titer of 5×10^7 PFU/mL cannot be obtained, the RT-qPCR standard generation can start at a lower titer (i.e. 5×10^6 PFU/mL).
6. The cDNA synthesis step of the standard can be scaled up to increase the total volume of each virus dilution to generate more material by using all 50 μL of the RNA from each of the virus standards by increasing the number of cDNA reactions for each virus standard dilution (i.e. ×5 cDNA reactions for each virus standard dilution will ultimately yield 500 μL of total cDNA).
7. Quantification of viral RNA is used as a surrogate for total viral particles. To prevent the quantification of nonencapsulated viral RNA, samples can be treated with RNase prior to addition to TRIzol and subsequent RNA isolation.
8. Ensure that multiple animals that are housed together are identifiable by ear punch or ear tag to keep inoculation volume matched to weight.
9. Use of an insulin syringe for these procedures is preferred as there is little to no dead space in the needle, which reduces waste of liposomes.
10. Multiple methods of isoflurane anesthesia may be used, the drop method is most economical. If available, mechanical isoflurane vaporizers are also very effective.

11. Intravenous inoculation can also be performed by lateral tail vein injection, the same systemic delivery effect will be achieved. We find for smaller or younger mice that the retro-orbital inoculation method is highly effective and efficient and can be mastered more quickly than the tail vein injection method.

Acknowledgements

This work was supported by National Institutes of Health Grant R01 AI148144 to TEM.

References

1. Van Rooijen N (1989) The liposome-mediated macrophage 'suicide' technique. J Immunol Methods 124:1–6. https://doi.org/10.1016/0022-1759(89)90178-6
2. Mönkkönen J, Taskinen M, Auriola SOK, Urtti A (2003) Growth inhibition of macrophage-like and other cell types by liposome-encapsulated, calcium-bound, and free bisphosphonates *in vitro*. J Drug Target 11:279–286. https://doi.org/10.1080/10611860310001636539
3. Frith JC, Mönkkönen J, Blackburn GM, Russell RGG, Rogers MJ (1997) Clodronate and liposome-encapsulated clodronate are metabolized to a toxic ATP analog, adenosine 5′-(β,-γ-dichloromethylene) triphosphate, by mammalian cells in vitro. J Bone Miner Res 12:1358–1367. https://doi.org/10.1359/jbmr.1997.12.9.1358
4. Frith JC, Mönkkönen J, Auriola S, Mönkkönen H, Rogers MJ (2001) The molecular mechanism of action of the antiresorptive and antiinflammatory drug clodronate: evidence for the formation in vivo of a metabolite that inhibits bone resorption and causes osteoclast and macrophage apoptosis. Arthritis Rheum 44:2201–2210. https://doi.org/10.1002/1529-0131(200109)44:9<2201::AID-ART374>3.0.CO;2-E
5. Van Rooijen N, Sanders A, Van Den Berg TK (1996) Apoptosis of macrophages induced by liposome-mediated intracellular delivery of clodronate and propamidine. J Immunol Methods 193:93–99. https://doi.org/10.1016/0022-1759(96)00056-7
6. Rooijen NV, Sanders A (1994) Liposome mediated depletion of macrophages: mechanism of action, preparation of liposomes and applications. J Immunol Methods 174:83–93. https://doi.org/10.1016/0022-1759(94)90012-4
7. Carpentier KS, Sheridan RM, Lucas CJ, Davenport BJ, Li FS, Lucas ED, McCarthy MK, Reynoso GV, May NA, Tamburini BAJ, Hesselberth JR, Hickman HD, Morrison TE (2021) MARCO+ lymphatic endothelial cells sequester arthritogenic alphaviruses to limit viremia and viral dissemination. EMBO J 40: e108966. https://doi.org/10.15252/embj.2021108966
8. Van Rooijen N, Kors N, Ende M, Dijkstra CD (1990) Depletion and repopulation of macrophages in spleen and liver of rat after intravenous treatment with liposome-encapsulated dichloromethylene diphosphonate. Cell Tissue Res 260:215–222. https://doi.org/10.1007/BF00318625
9. Delemarre FG, Kors N, Kraal G, van Rooijen N (1990) Repopulation of macrophages in popliteal lymph nodes of mice after liposome-mediated depletion. J Leukoc Biol 47:251–257. https://doi.org/10.1002/jlb.47.3.251
10. Thepen T, Van Rooijen N, Kraal G (1989) Alveolar macrophage elimination in vivo is associated with an increase in pulmonary immune response in mice. J Exp Med 170: 499–509. https://doi.org/10.1084/jem.170.2.499
11. Han X, Li Q, Lan X, El-Mufti L, Ren H, Wang J (2019) Microglial depletion with clodronate liposomes increases proinflammatory cytokine levels, induces astrocyte activation, and damages blood vessel integrity. Mol Neurobiol 56:6184–6196. https://doi.org/10.1007/s12035-019-1502-9
12. Yamamoto T, Moriyama H, Matsuo H, Kiwada H, Arakawat M (1996) Repopulation of murine Kupffer cells after intravenous

administration of liposome-encapsulated dichloromethylene diphosphonate. Am J Pathol 149:1271

13. Cho WY, Choi HM, Lee SY, Kim MG, Kim H-K, Jo S-K (2010) The role of Tregs and CD11c+ macrophages/dendritic cells in ischemic preconditioning of the kidney. Kidney Int 78:981–992. https://doi.org/10.1038/ki.2010.266
14. Kitamoto K, Machida Y, Uchida J, Izumi Y, Shiota M, Nakao T, Iwao H, Yukimura T, Nakatani T, Miura K (2009) Effects of liposome clodronate on renal leukocyte populations and renal fibrosis in murine obstructive nephropathy. J Pharmacol Sci 111:285–292. https://doi.org/10.1254/jphs.09227FP
15. Szebeni J, Moghimi SM (2009) Liposome triggering of innate immune responses: a perspective on benefits and adverse reactions: biological recognition and interactions of liposomes. J Liposome Res 19:85–90. https://doi.org/10.1080/08982100902792855
16. Li Y, Yao R, Ren M, Yuan K, Du Y, He Y, Kang H, Yuan S, Ju W, Qiao J, Xu K, Zeng L (2022) Liposomes trigger bone marrow niche macrophage "foam" cell formation and affect hematopoiesis in mice. J Lipid Res 63:100273. https://doi.org/10.1016/j.jlr.2022.100273
17. Culemann S, Knab K, Euler M, Wegner A, Garibagaoglu H, Ackermann J, Fischer K, Kienhöfer D, Crainiciuc G, Hahn J, Grüneboom A, Nimmerjahn F, Uderhardt S, Hidalgo A, Schett G, Hoffmann MH, Krönke G (2023) Stunning of neutrophils accounts for the anti-inflammatory effects of clodronate liposomes. J Exp Med 220:e20220525. https://doi.org/10.1084/jem.20220525
18. Carpentier KS, Davenport BJ, Haist KC, McCarthy MK, May NA, Robison A, Ruckert C, Ebel GD, Morrison TE (2019) Discrete viral E2 lysine residues and scavenger receptor MARCO are required for clearance of circulating alphaviruses. elife 8:e49163. https://doi.org/10.7554/eLife.49163
19. Ander SE, Parks MG, Davenport BJ, Li FS, Bosco-Lauth A, Carpentier KS, Sun C, Lucas CJ, Klimstra WB, Ebel GD, Morrison TE (2024) Phagocyte-expressed glycosaminoglycans promote capture of alphaviruses from the blood circulation in a host species-specific manner. PNAS Nexus 3:pgae119
20. Seiler P, Aichele P, Odermatt B, Hengartner H, Zinkernagel RM, Schwendener RA (1997) Crucial role of marginal zone macrophages and marginal zone metallophils in the clearance of lymphocytic choriomeningitis virus infection. Eur J Immunol 27:2626–2633. https://doi.org/10.1002/eji.1830271023
21. Sunderkötter C, Nikolic T, Dillon MJ, Van Rooijen N, Stehling M, Drevets DA, Leenen PJM (2004) Subpopulations of mouse blood monocytes differ in maturation stage and inflammatory response. J Immunol 172:4410–4417. https://doi.org/10.4049/jimmunol.172.7.4410

Correction to: Analysis of *Leishbuviridae* from Trypanosomatids

Danyil Grybchuk, Alexei Yu Kostygov, and Vyacheslav Yurchenko

Correction to:
Chapter 12 in: Hani Boshra (ed.), *Bunyaviruses: Methods and Protocols*, Methods in Molecular Biology, vol. 2893,
https://doi.org/10.1007/978-1-0716-4338-9_12

"The original version of the chapter did not have the acknowledgement section.

It was corrected and the acknowledgement section was added."

The updated version of this chapter can be found at
https://doi.org/10.1007/978-1-0716-4338-9_12

Hani Boshra (ed.), *Bunyaviruses: Methods and Protocols*, Methods in Molecular Biology, vol. 2893,
https://doi.org/10.1007/978-1-0716-4338-9_22,

Index

Hani Boshra (ed.), *Bunyaviruses: Methods and Protocols*, Methods in Molecular Biology, vol. 2893,
https://doi.org/10.1007/978-1-0716-4338-9,

K

L

M

N

O

P

R

S

T

U

V

MIX
Papier aus verantwortungsvollen Quellen
Paper from responsible sources
FSC® C105338
FSC
www.fsc.org

If you have any concerns about our products,
you can contact us on
ProductSafety@springernature.com

In case Publisher is established outside the EU,
the EU authorized representative is:
Springer Nature Customer Service Center GmbH
Europaplatz 3, 69115 Heidelberg, Germany

Printed by Libri Plureos GmbH
in Hamburg, Germany